AN INTRODUCTION TO GENERAL TOPOLOGY

PAUL E. LONG
University of Arkansas

Charles E. Merrill Publishing Company
A Bell & Howell Company
Columbus, Ohio

MERRILL MATHEMATICS SERIES

Erwin Kleinfeld, *Editor*

ISBN: 0-675-09253-1
Library of Congress Catalog Number: 71-138370
AMS Classification Numbers: 5420, 5422, 5423, 5428, 5435, 5440, 5452, 5460

1 2 3 4 5 6 7 8 9 10—79 78 77 76 75 74 73 72 71
PRINTED IN THE UNITED STATES OF AMERICA

Preface

This book is designed for those who are undertaking their first encounter with the subject of topology. The treatment is elementary but rigorous and covers the essential subjects of a beginning course in topology. A background in elementary calculus and a first course in modern algebra are prerequisites for successful study of this book.

While keeping the studies of topological spaces as general as possible, the standard topology for the n-dimensional Euclidean spaces is singled out for special attention throughout. A great deal of emphasis is placed on functions and also the product of topological spaces, as least in the finite case. In this connection, the topological product of an arbitrary family of spaces is defined and used in theorems and exercises where the degree of difficulty of its use is not considered too great. Examples are employed liberally to illustrate definitions and other points. The exercises also frequently ask for examples to be constructed by the reader. There are a minimum of ten exercises at the end of each section ranging from simple to moderate in difficulty. To increase reader participation, the proof of one or two theorems from each section is left as an exercise.

A thorough knowledge of elementary set theory would allow Chapter 1 to be omitted, although reference to this material is used occasionally in later parts of the book. The omission of Sections 4 and 5 of Chapter 7 on filters and nets, respectively, will not affect the remaining development of the book. The same is true of Section 8 of Chapter 8 concerning noncontinuous functions. With these exceptions, it is suggested that this book be read without the omission of any other sections so as to retain the greatest degree of continuity of subject matter.

I would like to express my sincere thanks to Professor Earl E. McGehee, Jr., who read the manuscript and made many valuable contributions to its improvement.

Paul E. Long
Fayetteville, Arkansas

Table of Contents

Basic Notions Concerning Sets

Introduction

It is assumed the reader has studied elementary calculus and an elementary course in modern algebra before attempting this book. Thus, a familiarity with the elementary notions of set theory is presumed so that our own exposition may concentrate on concepts that will be of future importance in our topological studies. As in most cases at this level of mathematical development, we shall not attempt a formal definition of a set nor lay the groundwork for an axiomatic set theory. Instead, we take the intuitive approach that a set is a collection of objects that satisfy some given property. From this, the ideas of set theory which are needed in our studies may be developed. Regardless of the approach, the ideas of set theory are essential to the development of any topology course. In fact, the entire subject matter in this book might be broadly classified as a study of sets. With this in mind, certain basic definitions and notations concerning sets which are used throughout our studies will be reviewed in this chapter. The theorems and exercises presented are intended to bring out the algebraic structure in the algebra of sets and hopefully be of a non-routine and challenging nature to the reader.

The reader is also assumed to have some knowledge of the structure of the real numbers since they have associated with them a very important and useful topology and will be frequently used in examples to point out concepts under discussion. Specifically, we mean such things as elementary facts about the rational and irrational numbers, the fact that the reals under ordinary addition and multiplication form a complete ordered field, the fact that the natural numbers are well ordered, and the Principle of Mathematical Induction. Occasionally we shall make use of the Completion Axiom: Every nonempty set of real numbers which has an upper bound has a least upper bound. On other occasions the Archimedian Property of real numbers will be employed: For any two real numbers a and b with $a \neq 0$, there exists an integer n such that $an > b$.

Perhaps a few words about sentences are appropriate in our introductory remarks. If P and Q are sentences, the compound sentence "P if and only if Q" is a shorthand way of saying both of the sentences "If P, then Q" and "If Q, then P" are true. We shall use the conventional abbreviation "P iff Q" for such a compound sentence. Of course, not every theorem is an "if and only if" type of sentence, but it should be stressed that every definition is of that type. There are times when we can profitably use the contrapositive of a sentence of the form "If P, then Q". The contrapositive is the sentence "If not Q, then not P" and is logically equivalent to the original sentence. That is, a sentence and its contrapositive are either both true or both false.

1. Basic Definitions

For a given set X the objects which make up X will commonly be referred to as *elements* or *points* or *members* of the set. Thus the words "elements", "points" and "members" will be used as synonyms and in view of this a set will sometimes be referred to as a *point set*. Keep in mind that in general, the word "point" need not have geometric significance. As usual, if x is a point or element of the set X, we write

$x \in X$, while, if x is not an element of X, we write $x \notin X$. Along these lines, if $x, y, \ldots \in X$ is written, we mean each of the elements $x, y, \ldots$ belongs to X. Often the elements of a set will simply be listed between braces, while at other times, when the defining property P of a set is of paramount importance, the symbolism $\{x: x$ has property $P\}$ is used. For example, $X = \{x: x$ is a rational number$\}$ is read "X is the set of all x such that x is a rational number." The colon inside the braces is an abbreviation for "such that," and "x is a rational number" is the defining property of the set. In general, sets will be denoted by capital letters and its elements by small letters.

Because of their frequent occurrence throughout the remainder of this book, the following special symbols will be reserved for the sets described:

$$R^1 = \{x: x \text{ is a real number}\}.$$

$$N = \{1, 2, 3, \ldots\} \text{ is the set of all natural numbers.}$$

1.1. Definition: For two sets X and Y, we say that X is a *subset* of Y iff every element of X is also an element of Y. For notation we shall write $X \subset Y$ iff X is a subset of Y. We write $X = Y$ iff $X \subset Y$ and $Y \subset X$ and say that the sets X and Y are *equal*.

In view of Definition 1.1, we see that if $X = Y$, then X and Y are different names for the same set. Continuing along the lines of Definition 1.1, if $X \subset Y$ where $X \neq Y$ and $X \neq \phi$ (ϕ is the empty set), then X is called a *proper* subset of Y. We reserve no special symbol for the fact that X is a proper subset of Y, but rather make a notation in words when such a condition is of importance.

At this point, it might be well to comment briefly on the empty set. In so doing, we are going to hereafter agree that if x and y belong to a given set X and we write $x = y$, then x and y are to be thought of as different names for the same element in the set X. (Notice that this use of the equality sign is different from that between two sets.) Therefore, for a given set X, it is certainly true that $x = x$ for each $x \in X$, so that we may then describe X as $X = \{x: x \in X \text{ and } x = x\}$. Now to describe a subset ϕ_x of X which contains no elements, we make use of the sentence "$x \in X$ and $x \neq x$," which is false for every x. That is, for the set X, we define the empty set ϕ_x to be $\phi_x = \{x: x \in X \text{ and } x \neq x\}$, and this gives us a subset of X which contains no elements because each $x \in X$ satisfies the equality $x = x$. It may now be shown (this is left to the reader as an exercise) that, if

X and Y are any two sets, then $\phi_X = \phi_Y$. Therefore, all empty sets are equal, and we may henceforth speak of *the* empty set. The empty set is denoted by ϕ, and by our discussion, we know ϕ contains no elements and $\phi \subset X$ for every set X.

1.2. Definition: For any two sets X and Y, the symbol $X \backslash Y$ will be used to denote the set of all points in X but not in Y. This set is called the *complement* of Y with respect to X. Thus, $X \backslash Y = \{x : x \in X$ and $x \notin Y\}$.

1.3. Definition: For any two sets X and Y, the *union* of X and Y is the set of all points x such that x belongs to at least one of the two sets and is denoted by $X \cup Y$. In symbols, $X \cup Y = \{x : x \in X$ or $x \in Y\}$.

Remember that the use of the word "or" in mathematics is unlike that of ordinary English. In mathematics, when the sentence "$x \in X$ or $x \in Y$" is written, we mean x belongs to X or to Y or to both X and Y. In other words, x belongs to at least one of the sets X or Y and may belong to both.

1.4. Definition: For any two sets X and Y, the *intersection* of X and Y is the set of all points x such that x belongs to both of the sets X and Y and is denoted by $X \cap Y$. In symbols, $X \cap Y = \{x : x \in X$ and $x \in Y\}$. If $X \cap Y = \phi$, X and Y are said to be *disjoint*.

1.5. Definition: For a given set X, the set of all possible subsets of X is called the *power set* of X. The power set of X is denoted by $\mathscr{P}(X)$.

Thus, each point or element of $\mathscr{P}(X)$ is a subset of X. For example, if $X = \{0, 1\}$, then $\mathscr{P}(X) = \{\phi, \{0\}, \{1\}, \{1, 2\}\}$. Can you give the number of elements in $\mathscr{P}(X)$ if X has three elements? Four elements? The following theorem answers the question for a set having n distinct elements.

1.6. THEOREM: *If X has n distinct elements, then $\mathscr{P}(X)$ has 2^n elements.*

Proof: The task at hand is to count all possible subsets of X. First, we know the empty set ϕ belongs to X. Next, each element $x \in X$ forms a subset $\{x\}$ belonging to $\mathscr{P}(X)$ and there are n of these. Denote the number of these n subsets by $C(n, 1)$, the number of combinations of n objects taken one at a time. Continuing, $C(n, 2)$ names the number of all subsets of X containing exactly two elements. Finally, $C(n, n)$ names the number of subsets with exactly n elements, namely the set X itself. Counting the empty set with $C(n, 0)$, the total number of subsets is represented by $C(n, 0) + C(n, 1) + \ldots + C(n, n)$. Now using the binomial expansion for $(1 + 1)^n$, we have $(1 + 1)^n = C(n, 0) + C(n, 1) + \ldots + C(n, n)$. Consequently, the number of elements in $\mathscr{P}(X)$ is $(1 + 1)^n = 2^n$.

For a given set X, we may think of the union and intersection as closed binary operations on the set $\mathscr{P}(X)$, since either of these operations produces a unique subset from two given subsets of X. In other words, the union or intersection of two elements of $\mathscr{P}(X)$ is another element of $\mathscr{P}(X)$. This would bring about natural questions as to whether these binary operations are commutative, associative, etc. In algebraic terminology we might ask what kind of algebraic structure is induced on $\mathscr{P}(X)$ by the two binary operations union and intersection. We should also note that there are many other types of closed binary operations on $\mathscr{P}(X)$ for which the algebraic structure could be studied. For example, complementation is such an operation, since for any $A, B \in \mathscr{P}(X), A \backslash B \in \mathscr{P}(X)$. (Note that complementation with respect to X may be thought of as a unary operation.) Our next theorem investigates some of the questions about the algebraic structure of $\mathscr{P}(X)$ under the operations of union and intersection. While our discussion at present is limited to elements of $\mathscr{P}(X)$, we might want to think of our sets as being subsets of a *universal set* so that the algebraic structure for our binary operations apply to sets in general, which indeed they do.

1.7. THEOREM: *Let X be a set. For any elements $A, B,$ and C of $\mathscr{P}(X)$ we have*
   (a) *Identities exist:*
       $A \cup \phi = A.$
       $A \cap X = A.$
   (b) *The idempotent properties hold:*
       $A \cup A = A.$
       $A \cap A = A.$
   (c) *The commutative properties hold:*
       $A \cup B = B \cup A.$
       $A \cap B = B \cap A.$

(d) *The associative properties hold:*
   $(A \cup B) \cup C = A \cup (B \cup C)$.
   $(A \cap B) \cap C = A \cap (B \cap C)$.

(e) *The union distributes over the intersection and the intersection distributes over the union:*
   $A \cup (B \cap C) = (A \cup B) \cap (A \cup C)$.
   $A \cap (B \cup C) = (A \cap B) \cup (A \cap C)$.

Proof: Only the proof that union distributes over the intersection will be given. It is to be shown that the set $A \cup (B \cap C)$ is equal to the set $(A \cup B) \cap (A \cup C)$. This means if we show that $A \cup (B \cap C) \subset (A \cup B) \cap (A \cup C)$ and $(A \cup B) \cap (A \cup C) \subset A \cup (B \cap C)$, then by Definition 1.1 our proof will be complete.

If $x \in A \cup (B \cap C)$, there are two possibilities which must be considered: (1) $x \in A$ or (2) $x \in B \cap C$. Possibility (1) implies $x \in A \cup B$ and $x \in A \cup C$, and therefore, $x \in (A \cup B) \cap (A \cup C)$ by the definitions of union and intersection. Possibility (2) implies $x \in B$ and also $x \in C$. It follows that $x \in A \cup B$ and also $x \in A \cup C$. Consequently, $x \in (A \cup B) \cap (A \cup C)$. We have now shown that if $x \in A \cup (B \cap C)$, then each of the two possible cases give $x \in (A \cup B) \cap (A \cup C)$. By Definition 1.1, $A \cup (B \cap C) \subset (A \cup B) \cap (A \cup C)$, which concludes half of the proof.

For the other half, if $x \in (A \cup B) \cap (A \cup C)$, then $x \in A \cup B$, and also $x \in A \cup C$. Thus, $x \in A$ or $x \in B$ and, simultaneously, $x \in A$ or $x \in C$. Considering the possibility $x \in A$, it follows that $x \in A \cup (B \cap C)$. Next, if $x \in B$, then x must also belong to A or C, since $x \in (A \cup B) \cap (A \cup C)$. For $x \in A, x \in A \cup (B \cap C)$, and for $x \in C, x \in B \cap C$ and hence to $A \cup (B \cap C)$. In either subcase, $x \in A \cup (B \cap C)$. The final consideration is for $x \in C$. This implies that $x \in A \cup B$ also and, therefore, $x \in A$ or $x \in B$. For $x \in A$, $x \in A \cup (B \cap C)$, and for $x \in B$, $x \in B \cap C$, and so to $A \cup (B \cap C)$. Again, in either subcase, $x \in A \cup (B \cap C)$. Having covered all possible cases for x, we find that, if $x \in (A \cup B) \cap (A \cup C)$, then $x \in A \cup (B \cap C)$, and so $(A \cup B) \cap (A \cup C) \subset A \cup (B \cap C)$. Now, by the opening paragraph of the proof, we have established the desired equality.

To conclude this section we state two facts about sets which will be of future use and leave their proof to the reader.

1.8. THEOREM: *For any set X and A, $B \in \mathscr{P}(X)$,*
  (a) $X \backslash (X \backslash A) = A$.
  (b) $A \subset B$ iff $X \backslash B \subset X \backslash A$.

Proof: The proof uses only the definitions and is left to the reader.

Exercises:

1. Prove that if A and B belong to $\mathscr{P}(X)$, then
   (a) $A \cup \phi = A$.
   (b) $A \cap \phi = \phi$.
   (c) $A \subset X \backslash B$ iff $A \cap B = \phi$.

2. Consider the elements of $\mathscr{P}(X)$.
   (a) Prove that the relation "is a subset of" is a transitive relation. Is this relation reflexive? Symmetric?
   (b) Is the relation "is disjoint from" reflexive? Symmetric? Transitive?

3. Prove the remaining part of (e) in Theorem 1.7.

4. Prove Theorem 1.8.

5. If $X \subset Y$, prove that $\mathscr{P}(X) \subset \mathscr{P}(Y)$.

6. Give an example of a set X and a binary operation on $\mathscr{P}(X)$ that is
   (a) not commutative.
   (b) not associative.

7. For any two subsets A and B of X, prove
   (a) $A \cap B = B$ iff $B \subset A$.
   (b) $(A \backslash B) \cup B = A$ iff $B \subset A$.

8. (a) Prove that if $A \subset B$, then $A = B \backslash (B \backslash A)$.
   (b) Give an example to show the converse of part (a) need not hold.

9. For any three subsets A, B and C of X, prove
   (a) $A \cap (B \backslash C) = (A \cap B) \backslash (A \cap C)$.
   (b) $(A \backslash C) \cap (B \backslash C) = (A \cap B) \backslash C$.

10. For any two sets A and B, prove that $A \cap B$ and $A \backslash B$ are disjoint and that $A = (A \cap B) \cup (A \backslash B)$. (This gives a way of representing A as a disjoint union.)

11. For any two sets X and Y, prove that $\phi_X = \phi_Y$. (*Hint:* Show $\phi_Y \subset \phi_X$ by using an indirect argument. Similarly, show $\phi_X \subset \phi_Y$.)
12. Prove that for any set X, $X \subset \{X\}$ iff $X = \phi$.

2. The Theorem of DeMorgan and the Cartesian Product

One of the truly important theorems in elementary set theory will be given next. It is known as *DeMorgan's Theorem*, and we will see its use quite often in the future. A Venn diagram drawn and studied by the reader would be helpful in illustrating the validity of the theorem before a proof is attempted.

2.1. THEOREM: (*DeMorgan's Theorem*) *Let X, Y and Z be any sets. Then*

(a) $X\backslash(Y \cup Z) = (X\backslash Y) \cap (X\backslash Z)$.
(b) $X\backslash(Y \cap Z) = (X\backslash Y) \cup (X\backslash Z)$.

Proof: The proof to (b) is given while that of (a) is left to the reader. We first show $X\backslash(Y \cap Z) \subset (X\backslash Y) \cup (X\backslash Z)$. If $x \in X\backslash(Y \cap Z)$, then $x \in X$ but $x \notin Y \cap Z$, which means there are three possibilities for the point x. These are (1) $x \in X$, but x does not belong to either of the sets Y or Z, (2) $x \in X$, x does not belong to Y, but does belong to Z, and (3) $x \in X$, x does belong to Y, but does not belong to Z. A consideration of each of these three possibilities gives $x \in (X\backslash Y) \cup (X\backslash Z)$ in every instance. Thus $X\backslash(Y \cap Z) \subset (X\backslash Y) \cup (X\backslash Z)$.

Now for the reverse inclusion. If $x \in (X\backslash Y) \cup (X\backslash Z)$, then $x \in X$, but $x \notin Y$ or $x \notin Z$. First, if $x \notin Y$, then $x \notin Y \cap Z$ and, therefore, $x \in X\backslash(Y \cap Z)$. Secondly, if $x \notin Z$, then $x \notin Y \cap Z$ and, thus, $x \in X\backslash(Y \cap Z)$. So, in either of the two possible cases we have $x \in X\backslash(Y \cap Z)$. The two subset relationships shown now yield $X\backslash(Y \cap Z) = (X\backslash Y) \cup (X\backslash Z)$, which is the desired conclusion.

Next, we describe still another way of combining two sets to produce a third. This third set is called the Cartesian product of the two given sets and is motivated by the familiar coordinate plane studied in analytic geometry. Recall that each point in the plane is named by a unique *ordered* pair of real numbers, called *coordinates* of the point, and conversely, each *ordered* pair of real numbers names a unique point in the plane. It is clear from experience in this setting that the pair (2, 3) is indeed a different object from (3, 2) and, therefore, the order in which we name the elements in forming the pair is very important. To be a bit more general than forming ordered pairs where the coordinates are real numbers, consider the next definition.

2.2. Definition: Let X and Y be any two sets. The *Cartesian product*, or simply *product*, of X by Y is the set of all possible ordered pairs $\{(x, y) : x \in X \text{ and } y \in Y\}$ and is denoted by $X \times Y$. The elements of $X \times Y$ are subject to the condition that $(x_1, y_1) = (x_2, y_2)$ iff $x_1 = x_2$ and $y_1 = y_2$. For the element $(x, y) \in X \times Y$, it is customary to call x the first coordinate and y the second coordinate.

It might be worthy to note in passing that $X \times Y$ is sometimes read "X cross Y."

2.3. Example: Let $X = \{1, 2, 3\}$ and $Y = \{a, b\}$. Then

$$X \times Y = \{(1, a), (1, b), (2, a), (2, b), (3, a), (3, b)\}.$$
$$Y \times X = \{(a, 1), (a, 2), (a, 3), (b, 1), (b, 2), (b, 3)\}.$$
$$Y \times Y = \{(a, a), (a, b), (b, a), (b, b)\}.$$

Of course, the usual coordinate plane is represented by $R^1 \times R^1$, or when graphing, by the more familiar notation $X \times Y$ where $X = Y = R^1$. In this setting it is easy to sketch the set $A \times B$ where A and B are subsets of the reals. In general, this pictorial representation is not possible, nor is it entirely necessary, except to aid our thinking

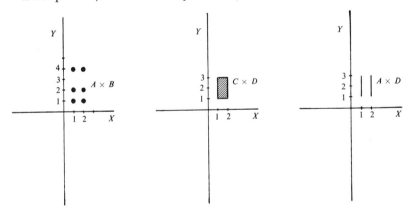

Figure 1

about certain products. Whether we have such an aid or not, we should rely in all cases completely upon our previous definitions and theorems about sets in writing the details of a proof. To give an illustration of sketching the product of two sets, consider the following subsets of R^1: $A = \{1, 2\}$, $B = \{1, 2, 4\}$, $C = \{x : 1 \leq x \leq 2\}$, and $D = \{x : 1 \leq x \leq 3\}$. Figure 1 describes the products of some of these sets as subsets of the plane.

Since it is generally not the case that if $A, B \in \mathscr{P}(X)$, then $A \times B \in \mathscr{P}(X)$, the Cartesian product may not be thought of as a closed binary operation on $\mathscr{P}(X)$. By its definition, we may think of the product as a binary operation on sets within a more general framework and, therefore, ask questions regarding its algebraic properties. Example 2.3, for instance, shows the product is not a commutative operation. In fact, Exercise 7 of this section shows $X \times Y = Y \times X$ iff $X = Y$. Even so, there is a rather natural one-to-one correspondence between $X \times Y$ and $Y \times X$ obtained by pairing $(x, y) \in X \times Y$ with $(y, x) \in Y \times X$, and this fact will enable us to prove later that $X \times Y$ and $Y \times X$ are topologically indistinguishable. So, the fact that the product is not commutative will be of no hinderance to our later studies of topology. Along these same lines, the product is also a non-associative operation with $(X \times Y) \times Z$ being unequal to $X \times (Y \times Z)$ in general. Therefore, to conform with our solid analytic geometry experience and to avoid set theoretic difficulties when forming the product of more that two sets, we take the approach of Definition 2.2 and define the product of three sets as the set of all ordered triples for which $X \times Y \times Z$ is the notation, the product of four sets as the set of all ordered four-tuples for which $X \times Y \times Z \times W$ is the notation, etc., for finite collections of sets. This will be done formally in the next section. It turns out that with three sets, for example, there is a natural one-to-one correspondence between $X \times Y \times Z$, $(X \times Y) \times Z$ and $X \times (Y \times Z)$ by pairing the elements (x, y, z), $((x, y), z)$ and $(x, (y, z))$. As in the case of commutativity, we will later show that these three products are topologically indistinguishable so that again no cumbersome problems arise from the non-associativity of the product in our studies of topology.

The next theorem gives more information about the algebraic nature of our binary operations on sets.

2.4. THEOREM:　　*Let X, Y and Z be any three sets. Then*
(a) The product distributes over the union:
$$X \times (Y \cup Z) = (X \times Y) \cup (X \times Z).$$

(b) *The product distributes over the intersection:*
$$X \times (Y \cap Z) = (X \times Y) \cap (X \times Z).$$
(c) *The product distributes over the complement of Z with respect to Y:*
$$X \times (Y \backslash Z) = (X \times Y) \backslash (X \times Z).$$

Proof: The proof of (b) will be given while those for (a) and (c) will be left as exercises. First, we show $X \times (Y \cap Z) \subset (X \times Y) \cap (X \times Z)$. If $a \in X \times (Y \cap Z)$, then a must be an ordered pair (x, p) where $x \in X$ and $p \in Y \cap Z$ according to Definition 2.2. Since $p \in Y$, it follows that $(x, p) \in X \times Y$. Also, $p \in Z$, from which it follows that $(x, p) \in X \times Z$. Thus, $a = (x, p) \in (X \times Y) \cap (X \times Z)$. Therefore, $X \times (Y \cap Z) \subset (X \times Y) \cap (X \times Z)$.

To show the reverse inclusion, suppose $a \in (X \times Y) \cap (X \times Z)$. Then $a \in X \times Y$, and hence, a must be an ordered pair (x, y) where $x \in X$ and $y \in Y$. Also, $a = (x, y)$ belongs to $X \times Z$ which means $y \in Z$ by Definition 2.2. Therefore, $y \in Y \cap Z$, so that $a = (x, y)$ is an element of $X \times (Y \cap Z)$. Consequently, $(X \times Y) \cap (X \times Z) \subset X \times (Y \cap Z)$, and this, with the previous inclusion, concludes the proof of (b).

Finally, we state and prove a fundamental fact concerning the product of two sets.

2.5. THEOREM: *For any two sets X and Y, $X \times Y = \phi$ iff $X = \phi$ or $Y = \phi$.*

Proof: Consider first the hypothesis that $X \times Y = \phi$. If we assume $X \neq \phi$ and $Y \neq \phi$, there would exist at least one point $x \in X$ and at least one point $y \in Y$. It then follows that $(x, y) \in X \times Y$ so that $X \times Y \neq \phi$, which contradicts our hypothesis. We conclude our assumption is false and $X = \phi$ or $Y = \phi$ must hold.

Conversely, take the hypothesis that $X = \phi$ or $Y = \phi$. The assumption $X \times Y \neq \phi$ implies there is at least one point $(x, y) \in X \times Y$, which in turn means $x \in X$ and $y \in Y$. Therefore, $X \neq \phi$ and $Y \neq \phi$, which contradicts the hypothesis. The conclusion $X \times Y = \phi$ follows.

Exercises:

1. (a) Prove part (a) of Theorem 2.1 in a manner similar to that of part (b) of that theorem.
   (b) Prove part (a) of Theorem 2.1 by using the results of part (b)

as applied to the sets $X \backslash Y$ and $X \backslash Z$. (*Hint:* Intersect these sets, then complement.)

2. Prove parts (a) and (c) of Theorem 2.4.

3. (a) If X has n distinct elements and Y has m distinct elements, how many elements does $X \times Y$ have?
   (b) The product $X \times X$ of real numbers has sixteen elements among which are found $(1, 3)$ and $(5, 7)$. Find the remaining elements and determine the set X.

4. If $A = \{x : 1 \leq x \leq 2\}$ and $B = \{x : 0 \leq x \leq 1\} \cup \{x : 2 \leq x \leq 3\}$ are subsets of R^1, sketch the graph of $A \times B$ and $B \times A$ as subsets of $R^1 \times R^1$.

5. Prove, or give a counterexample to disprove, each of the following statements:
   (a) If $D \subset X \times Y$, then there are subsets $A \subset X$ and $B \subset Y$ such that $D = A \times B$.
   (b) If $X \subset Z$, then for any set $Y, X \times Y \subset Z \times Y$.

6. (a) Prove that if $A \times B \neq \phi$, then $A \times B \subset X \times Y$ iff $A \subset X$ and $B \subset Y$.
   (b) Give an example to show the condition $A \times B \neq \phi$ in part (a) may not be omitted from the hypothesis.

7. Prove that $X \times Y = Y \times X$ iff $X = Y$.

8. Prove that for any four sets X, Y, Z and W,
   (a) $(X \times Y) \cap (Z \times W) = (X \cap Z) \times (Y \cap W)$.
   (b) $(X \times Y) \cup (Z \times W) \subset (X \cup Z) \times (Y \cup W)$.
   (c) Give an example to show the reverse inclusion in part (b) need not hold.

9. If $A \subset X$ and $B \subset Y$, prove that $(X \times Y) \backslash (A \times B) = ((X \backslash A) \times Y) \cup (X \times (Y \backslash B))$.

10. An ordered pair $(x, y) \in X \times Y$ may be defined as a set in the following manner: $(x, y) = \{\{x\}, \{x, y\}\}$. Using this definition, prove that $(x, y) = (z, w)$ iff $x = z$ and $y = w$.

3. Indexed Sets

We now turn to more general formations of unions, intersections and products than those with just two sets. To this end, let Δ be a given *nonempty* set. Suppose, furthermore, that with each element $\alpha \in \Delta$ there is identified a particular set denoted by X_α. Then the collection of sets $\{X_\alpha : \alpha \in \Delta\}$ is called an *indexed family* of sets and the set Δ is called an *indexing set*. Examples 3.2 and 3.3 will offer some concrete insight into this idea and help point out that an indexing set may be any nonempty set whatever, finite or infinite. For notation, we will generally use capital letters from the Greek alphabet to denote indexing sets and corresponding lower case letters to denote elements of such a set. In the frequent case when the indexing set is the set N of natural numbers, the notation $\{X_n : n \in N\}$ will be used. With this background in mind, our next step will be to generalize the definition of union, intersection, and product. It is easy to see that these new definitions are identical to their previous counterparts whenever $\Delta = \{1, 2\}$.

3.1. Definition: Let $\{X_\alpha : \alpha \in \Delta\}$ be an indexed family of sets indexed by the set Δ. Then the *union* of this family is the set $\{x : x \in X_\alpha$ for at least one $\alpha \in \Delta\}$ and is denoted by $\bigcup_{\alpha \in \Delta} X_\alpha$. The *intersection* of the family is the set $\{x : x \in X_\alpha$ for every $\alpha \in \Delta\}$ and is denoted by $\bigcap_{\alpha \in \Delta} X_\alpha$. If $\Delta = \{1, 2, \ldots, n\}$ we shall sometimes write $\bigcup_{k=1}^{n} X_k$ and $\bigcap_{k=1}^{n} X_k$ for the union and intersection, respectively, of the collection $\{X_k : k \in \Delta = \{1, 2, \ldots, n\}\}$.

3.2. Example: Let N be the natural numbers and identify with each $n \in N$ the subset $X_n = \{(x, y) : 0 \le x \le n, 0 \le y \le n\}$ of the plane.

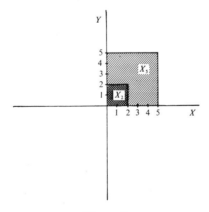

Figure 2

Figure 2 shows the sets X_n for $n = 2$ and $n = 5$. Using Definition 3.1, we have $\bigcap_{n \in N} X_n = X_1$ while $\bigcup_{n \in N} X_n$ represents the entire first quadrant in the coordinate plane.

3.3. Example: Let Δ be the set of all real numbers and $X_\alpha = \{(\alpha, y): 0 \leq y \leq 1\}$ be a subset of the plane for each $\alpha \in \Delta$. Figure 3 shows a few of the sets X_α. Here, $\bigcup_{\alpha \in \Delta} X_\alpha = \{(x, y): x$ is a real number, and $0 \leq y \leq 1\}$, while $\bigcap_{\alpha \in \Delta} X_\alpha = \phi$.

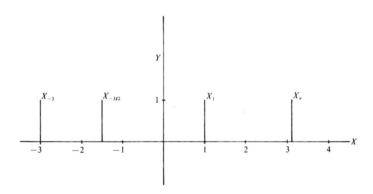

Figure 3

The concept of the product of sets can also be generalized by the use of indexed sets. For our present purposes, the indexing set will be restricted to finite subsets $\Delta = \{1, 2, \ldots, n\}$ of N. Later on, an even more general definition will be given where the indexing set may be any set whatever.

3.4. Definition: Let $\{X_k: k \in \Delta\}$ be an indexed family of sets indexed by the set $\Delta = \{1, 2, \ldots, n\}$. Then the *product* of $\{X_k: k \in \Delta\}$ is the set of all ordered n-tuples $\{(x_1, x_2, \ldots, x_n): x_1 \in X_1, x_2 \in X_2, \ldots, x_n \in X_n\}$. The elements of the product are subject to the condition that $(x_1, x_2, \ldots, x_n) = (y_1, y_2, \ldots, y_n)$ iff $x_1 = y_1, x_2 = y_2, \ldots, x_n = y_n$. The product will be denoted by $X_1 \times X_2 \times \ldots \times X_n$, or $\prod_{k=1}^{n} X_k$. For each $1 \leq k \leq n$, x_k is called the kth coordinate of the point $(x_1, x_2, \ldots, x_n)$, and the set X_k is called the kth factor of the product.

3.5. Example: The set $R^1 \times R^1 \times R^1$ is the usual 3-dimensional coordinate space so familiar in solid analytic geometry. A point in this set will appear as (x_1, x_2, x_3) where each x_k, $k = 1, 2, 3$, is a real number.

3.6. Example: Let $X_1 = \{x : 0 \leq x \leq 1\}$, $X_2 = \{x : 0 \leq x \leq 1\}$ and $X_3 = \{x : 0 \leq x \leq 2\}$ be subsets of the real numbers. Then $X_1 \times X_2 \times X_3$ is a subset of $R^1 \times R^1 \times R^1$ having a geometric description as a rectangular parallelepiped with a square base and height of two units. If X is a circle in $R^1 \times R^1$ and Y a line segment, then $X \times Y$ is a hollow cylinder in $(R^1 \times R^1) \times R^1$.

Throughout our studies the set R^1 of all real numbers and finite products of the real numbers play an important role. For each natural number n, the set $R^1 \times R^1 \times \ldots \times R^1$ (n factors) will henceforth be denoted by R^n. That is, points in $R^n = R^1 \times R^1 \times \ldots \times R^1$ (n factors) have the form $(x_1, x_2, \ldots, x_n)$ where each x_k, $k = 1, 2, \ldots, n$ is a real number and, conversely, each n-tuple $(x_1, x_2, \ldots, x_n)$ of real numbers represents a point in R^n. As we shall see later, there is a rather natural topology for each of the sets R^n and one of our major topological goals will be to investigate them thoroughly.

Our next result is not only a useful theorem concerning products of sets indexed by a finite indexing set, but also serves to illustrate the line of reasoning involved in some of the exercises concerning such products.

3.7. THEOREM: *Let $\{X_k : k \in \Delta\}$ and $\{Y_k : k \in \Delta\}$ be two families of sets each indexed by $\Delta = \{1, 2, \ldots, n\}$. Then*

(a) $(\prod_{k=1}^{n} X_k) \cap (\prod_{k=1}^{n} Y_k) = \prod_{k=1}^{n} (X_k \cap Y_k)$ *and*

(b) $(\prod_{k=1}^{n} X_k) \cup (\prod_{k=1}^{n} Y_k) \subset \prod_{k=1}^{n} (X_k \cup Y_k)$.

Proof: Only the proof of (a) will be given. Let us notice first that, if any X_k or Y_k, $k = 1, 2, \ldots, n$, is the empty set, Exercise 5 of this section gives both sides of (a) as the empty set so that equality does indeed hold. Otherwise, suppose $x \in (\prod_{k=1}^{n} X_k) \cap (\prod_{k=1}^{n} Y_k)$. Then $x = (x_1, x_2, \ldots, x_n)$ and, since x belongs to both $\prod_{k=1}^{n} X_k$ and $\prod_{k=1}^{n} Y_k$, it follows that $x_1 \in X_1$, $x_2 \in X_2, \ldots x_n \in X_n$ and also $x_1 \in Y_1$, $x_2 \in$

$Y_2, \ldots, x_n \in Y_n$. Therefore, $x_1 \in X_1 \cap Y_1$, $x_2 \in X_2 \cap Y_2, \ldots, x_n \in X_n \cap Y_n$ so that $x = (x_1, x_2, \ldots, x_n) \in \prod_{k=1}^{n} (X_k \cap Y_k)$, which shows set inclusion one way. Now let $x \in \prod_{k=1}^{n} (X_k \cap Y_k)$. Then $x = (x_1, x_2, \ldots, x_n)$ where $x_k \in X_k \cap Y_k$ for $k = 1, 2, \ldots, n$ and, consequently, $x_k \in X_k$ and $x_k \in Y_k$ for $k = 1, 2, \ldots, n$. By definition of the product, $x = (x_1, x_2, \ldots, x_n) \in (\prod_{k=1}^{n} X_k) \cap (\prod_{k=1}^{n} Y_k)$ showing the reverse set inclusion holds. The conclusion of part (a) follows.

In view of Definition 3.1, it seems plausible that several of the earlier results concerning unions and intersections could be extended to the more general setting of indexed families of sets. The next theorem shows this to be the case for DeMorgan's Theorem, and some of the exercises will show other generalizations.

3.8. THEOREM: (General form of DeMorgan's Theorem.) Let X be a set and $\{A_\alpha : \alpha \in \Delta\}$ an indexed family of sets. Then
   (a) $X \backslash \bigcup_{\alpha \in \Delta} A_\alpha = \bigcap_{\alpha \in \Delta} (X \backslash A_\alpha)$.
   (b) $X \backslash \bigcap_{\alpha \in \Delta} A_\alpha = \bigcup_{\alpha \in \Delta} (X \backslash A_\alpha)$.

Proof: The proof of (a) is given while (b) is left for the reader. It will be shown first that $X \backslash \bigcup_{\alpha \in \Delta} A_\alpha \subset \bigcap_{\alpha \in \Delta} (X \backslash A_\alpha)$. If $x \in X \backslash \bigcup_{\alpha \in \Delta} A_\alpha$, then $x \in X$ but $x \notin \bigcup_{\alpha \in \Delta} A_\alpha$. By the definition of union in 3.1, $x \notin A_\alpha$ for every $\alpha \in \Delta$. Thus, $x \in X \backslash A_\alpha$ for every $\alpha \in \Delta$, which implies $x \in \bigcap_{\alpha \in \Delta} (X \backslash A_\alpha)$ by the definition of intersection in 3.1. Therefore, $X \backslash \bigcup_{\alpha \in \Delta} A_\alpha \subset \bigcap_{\alpha \in \Delta} (X \backslash A_\alpha)$.

For the reverse inclusion, suppose $x \in \bigcap_{\alpha \in \Delta} (X \backslash A_\alpha)$. Then $x \in X \backslash A_\alpha$ for every $\alpha \in \Delta$, which implies $x \notin A_\alpha$ for every $\alpha \in \Delta$. The definition of union in 3.1 now gives $x \notin \bigcup_{\alpha \in \Delta} A_\alpha$ and so $x \in X \backslash \bigcup_{\alpha \in \Delta} A_\alpha$. Consequently, $\bigcap_{\alpha \in \Delta} (X \backslash A_\alpha) \subset X \backslash \bigcup_{\alpha \in \Delta} A_\alpha$. The two established set inclusions prove part (a) of the theorem.

Exercises:

1. Prove part (b) of Theorem 3.8.
2. Let $\{A_\alpha : \alpha \in \Delta\}$ an indexed family of sets indexed by Δ. Prove
   (a) $\bigcap_{\alpha \in \Delta} A_\alpha \subset A_\alpha$ for every $\alpha \in \Delta$.

(b) For any set B, $B \cup \bigcup_{\alpha \in \Delta} A_\alpha = \bigcup_{\alpha \in \Delta} (B \cup A_\alpha)$.

3. Let $A_n = \{x : -1/n < x < 1/n\}$, $n \in N$, be an indexed family of subsets of R^1.
   (a) Find $\bigcap_{n \in N} A_n$ and prove that your result is correct.
   (b) Find $\bigcup_{n \in N} A_n$ and prove that your result is correct.

4. Let $\{A_\alpha : \alpha \in \Delta\}$ be an indexed family of sets indexed by Δ and $\{B_\beta : \beta \in \Omega\}$ an indexed family of sets indexed by Ω. Prove
   (a) $(\bigcup_{\alpha \in \Delta} A_\alpha) \cap (\bigcup_{\beta \in \Omega} B_\beta) = \bigcup_{\substack{\alpha \in \Delta \\ \beta \in \Omega}} (A_\alpha \cap B_\beta)$.
   (b) $(\bigcap_{\alpha \in \Delta} A_\alpha) \cup (\bigcap_{\beta \in \Omega} B_\beta) = \bigcap_{\substack{\alpha \in \Delta \\ \beta \in \Omega}} (A_\alpha \cup B_\beta)$.

5. Prove that $\prod_{k=1}^{n} A_k = \phi$ iff one of the factor sets X_j, $1 \le j \le n$, is empty.

6. (a) Prove part (b) of Theorem 3.7.
   (b) Give an example to show the reverse inclusion of part (b) in Theorem 3.7 need not hold.

7. Let X be a set and $\{A_\alpha : \alpha \in \Delta\}$ be an indexed family of subsets of X indexed by Δ.
   (a) Prove that $\bigcap_{\alpha \in \Delta} \mathscr{P}(A_\alpha) = \mathscr{P}(\bigcap_{\alpha \in \Delta} A_\alpha)$. (This shows $\bigcap_{\alpha \in \Delta}$ and $\mathscr{P}$ commute.)
   (b) Give an example to show that $\bigcup_{\alpha \in \Delta}$ and $\mathscr{P}$ do not commute. Is there a subset inclusion one way? If so, state and prove.

8. Let $\{A_k : k \in \Delta\}$ and $\{B_k : k \in \Delta\}$ be two families of sets each indexed by $\Delta = \{1, 2, \ldots, n\}$.
   (a) Prove that if $A_k \subset B_k$ for each $k \in \Delta$, then $\prod_{k=1}^{n} A_k \subset \prod_{k=1}^{n} B_k$.
   (b) If $A_k \ne \phi$ for every $k \in \Delta$ and $\prod_{k=1}^{n} A_k \subset \prod_{k=1}^{n} B_k$, then $A_k \subset B_k$ for every $k \in \Delta$.
   (c) Why does part (b) fail if for some $k \in \Delta$, $A_k = \phi$?

9. Let $\{A_\alpha : \alpha \in \Delta\}$ be an indexed family of sets.
   (a) Prove that $B \subset \bigcap_{\alpha \in \Delta} A_\alpha$ iff $B \subset A_\alpha$ for every $\alpha \in \Delta$.
   (b) Prove that $\bigcup_{\alpha \in \Delta} A_\alpha \subset B$ iff $A_\alpha \subset B$ for every $\alpha \in \Delta$.

10. Let $\{X_k : k \in \Delta\}$ be an indexed family of sets indexed by $\Delta = \{1, 2, \ldots, n\}$ and for each $k \in \Delta$, let $A_k \subset X_k$. For each $k \in \Delta$,

define $B_k = X_1 \times X_2 \times \ldots \times X_{k-1} \times A_k \times X_{k+1} \times \ldots \times X_n$.
Prove

(a) $\displaystyle\prod_{k=1}^{n} A_k = \bigcap_{k=1}^{n} B_k$.

(b) $(\displaystyle\prod_{k=1}^{n} X_k)\backslash B_k = X_1 \times X_2 \times \ldots \times X_{k-1} \times (X_k\backslash A_k) \times X_{k+1} \times \ldots \times X_n$ for $k = 1, 2, \ldots, n$.

4. Relations

Generally speaking, the product of sets plays an especially important role throughout the remainder of our studies. In this section the product of two sets is used to study the very fundamental mathematical concept of a relation. The concept of a relation from one set to another has probably already been studied in algebra or some previous course, but we shall begin in a very elementary way so as to pave the way for the study of functions in the next chapter.

4.1. Definition: Let X and Y be two sets. A *relation* R from X to Y is a subset of $X \times Y$. If $(x, y) \in R$, then we say that x is R related to y and express this as xRy. If $(x, y) \notin R$, then x is not R related to y and $x\cancel{R}y$ is used to denote this fact.

4.2. Example: Let N represent the natural numbers and $Y=\{a, b, c\}$. Then

$R_1 = \{(3, a), (7, c)\} \subset N \times Y$ is a relation from N to Y.

$R_2 = \{(1, 1), (1, 2), (1, 3), \ldots\} \subset N \times N$ is a relation from N to N.

$R_3 = \{(a, 5), (b, 6), (c, 7)\} \subset Y \times N$ is a relation from Y to N.

$R_4 = \phi \subset Y \times Y$ is a relation from Y to Y.

For a given element $(x, y) \in X \times Y$, either $(x, y) \in R$ or $(x, y) \notin R$ and, therefore, the description of R can take on any form as long as it actually describes R as a subset of $X \times Y$. This may be as a set or by a sentence implying a set of ordered pairs in $X \times Y$. For example, if $X = Y = R^1$, then the sentence "xRy iff $x < y$" implies the relation under discussion is $R = \{(x, y) : x < y\} \subset R^2$. The relation is illustrated in Figure 4.

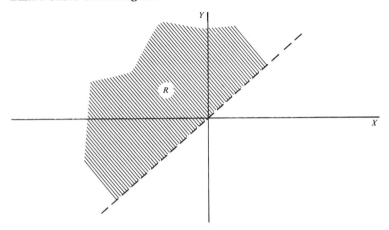

Figure 4

Many of the relations we shall study are from R^1 to R^1 and are, therefore, subsets of the plane R^2. In this setting it is often convenient to shade in the parts of the plane which represent the relation and, in so doing, we usually say we are sketching the graph of the relation. From our viewpoint, then, a relation and its graph are one and the same.

A way to combine two given relations to produce a third relation is now given.

4.3. Definition: Let R_1 be a relation from X to Y and R_2 a relation from Z to W. Then the *composition relation* $R_2 R_1$ is the relation from X to W given by $R_2 R_1 = \{(x, w) : \text{there is an element } y \in Y \cap Z \text{ for which } (x, y) \in R_1 \text{ and } (y, w) \in R_2\}$.

4.4. Example: Let $R_1 = \{(1, 3), (1, 4), (2, 2)\}$ and $R_2 = \{(3, 5), (2, 1), (2, 8), (5, 7), (6, 6)\}$ be relations from N to N. We first form $R_2 R_1$. Considering $(1, 3) \in R_1$, we search R_2 for elements whose first coordinate is 3. In this case there is one, namely $(3, 5)$, and hence $(1, 5) \in R_2 R_1$. Continuing this process for each element of R_1, we have $R_2 R_1 = \{(1, 5), (2, 1), (2, 8)\}$. For the two given relations we may also form the composition $R_1 R_2$, which is $R_1 R_2 = \{(2, 3), (2, 4)\}$. Also, $R_1 R_1 = \phi$ and $R_2 R_2 = \{(3, 7), (6, 6)\}$.

Example 4.4 shows that the composition of relations is not a commutative operation in general. This example also sets the pattern for forming the composition R_2R_1. In so doing, we first select an element $(x, y) \in R_1$ and then search R_2 for all elements whose first coordinate is y, i.e., find $\{(y, w) : (y, w) \in R_2\}$. Then the set $\{(x, w) : (y, w) \in R_2\}$ will belong to R_2R_1. Continuing this process until R_1 is exhausted yields the relation R_2R_1. Notice that if $R_2R_1 \neq \phi$, then $Y \cap Z \neq \phi$.

4.5. Example: Let $X = Y = W = R^1$. Let the relation R_1 from X to Y be described as $R_1 = \{(x, y) : x < y\}$ and R_2 from Y to W be described as $R_2 = \{(y, w) : w = 2y\}$. To see how to describe R_2R_1, let $(1, y) \in R_1$, for instance. Then $y > 1$ and there is an element $(y, 2y) \in R_2$ which implies $(1, 2y) \in R_2R_1$ for all real numbers $y > 1$. In general, for a fixed x, where $(x, y) \in R_1$ and $y > x$, there is an element $(y, 2y) \in R_2$ and hence $(x, 2y) \in R_2R_1$ for every $y > x$. Thus, R_2R_1 consists of all points in the plane above the line $y = 2x$. The relations involved are sketched in Figure 5.

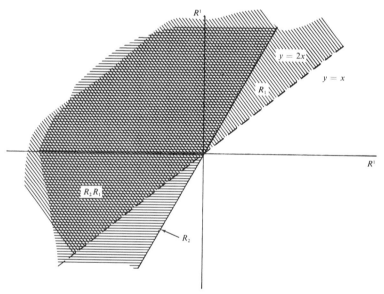

Figure 5

4.6. Definition: Let R be a relation from X to Y. The *domain* of R is the subset of X consisting of $\{x \in X : \text{there is some } y \in Y \text{ so that } (x, y) \in R\}$, and the *range* of R is the subset of Y consisting of $\{y \in Y : \text{there is some } x \in X \text{ so that } (x, y) \in R\}$.

In example 4.4, the domain of R_1 is $\{1, 2\}$ and the range of R_1 is $\{2, 3, 4\}$. For R_2, the domain is $\{2, 3, 5, 6\}$ and the range is $\{1, 5, 6, 7, 8\}$. In examble 4.5, the domain and range of both R_1 and R_2 is the set of all real numbers. Furthermore, the domain and range of $R_2 R_1$ is the set of all real numbers. Generally speaking, in order for the composition relation defined in Definition 4.3 to be nonempty, there must be at least one element in the range of R_1 which is also in the domain of R_2.

4.7. Definition: If R is a relation from X to Y, then the *inverse relation* of R is a relation from Y to X defined by the set $\{(y, x): (x, y) \in R\}$ and is denoted by R^{-1}.

Definition 4.7 tells us that whenever the element (x, y) is found in a relation R from X to Y, then the element (y, x) is always found in R^{-1}. This means R^{-1} is found from R by taking each element of R, interchanging the first and second coordinates, and then listing this new element in R^{-1}. In the particular case of elements in R^2, we know that (x, y) and (y, x) are symmetric with respect to the diagonal line $y = x$. It then follows that for any relation $R \subset R^2$, R^{-1} may be found by reflecting R about the line $y = x$. Consideration of Definition 4.6 along with 4.7 leads us to the fact that the range of R is the domain of R^{-1} and the domain of R is the range of R^{-1}. The next two examples will illustrate the points of our present discussion.

4.8. Example: Let $R = \{(3, 1), (2, 2), (2, 3)\}$ be a relation from N to N. Then $R^{-1} = \{(1, 3), (2, 2), (3, 2)\}$. In Figure 6, the points of R

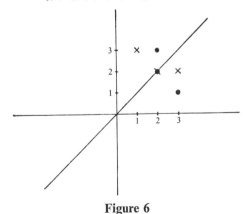

Figure 6

are represented by black dots, while those of R^{-1} are represented by a cross mark.

4.9. Example: Let $R = \{(x, y) : y = x^2\}$ be a relation from R^1 to R^1. Then $R^{-1} = \{(y, x) : y = x^2\} = \{(x, y) : x = y^2\}$. To see that the first of these three equalities hold, recall that according to Definition 4.7, the defining relationship for R, namely $y = x^2$, must still hold in R^{-1}, and to find R^{-1}, we merely interchange the first and second coordinates of points in R. To see that the last two sets are indeed equal, observe that in each we are describing points in the plane R^2 having the property that the first coordinate of the point is the square of the second coordinate of the point. The representation $R^{-1} = \{(x, y) : x = y^2\}$ has the advantage of listing the elements with their x-coordinate first so that graphing is a straightforward matter. Figure 7 shows the relationship between R and R^{-1}.

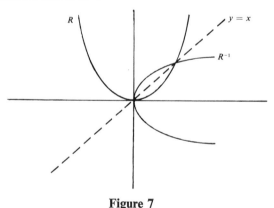

Figure 7

Suppose now that we have given a relation R from X to Y. For any subset $A \subset X$, the relation R affords a rather natural way of selecting a certain subset of Y as described in the next definition. This subset is dependent upon both R and the subset $A \subset X$.

4.10. Definition: Let R be a relation from X to Y. For each $x \in X$, the *image* of x under R is defined to be $R(x) = \{y \in Y : (x, y) \in R\}$. If $A \subset X$, the image of A under R is $R(A) = \bigcup_{x \in A} R(x) = \{y \in Y : x \in A \text{ and } (x, y) \in R\}$. If $B \subset Y$, the *inverse image* of B under R is the image of B under the relation R^{-1} and is denoted by $R^{-1}(B)$. In other words, $R^{-1}(B) = \{x \in X : R(x) \in B\} = \{x \in X : y \in B \text{ and }$

$(x, y) \in R\}$. The inverse image of $B \subset Y$ under R is often called the *preimage* of B under R.

4.11. Example: Let $R = \{(3, 6), (3, 7), (2, 1), (3, 1), (5, 6)\}$ be a relation from N to N. Then $R^{-1} = \{(6, 3), (7, 3), (1, 2), (1, 3), (6, 5)\}$. If $A = \{2, 3\} \subset X$, then $R(A) = \{1, 6, 7\} \subset Y$. If $A = \{2, 6, 10\}$, then $R(A) = \{1\}$. If $B = \{2, 3, 7\} \subset Y$, then $R^{-1}(B) = \{3\} \subset X$ and if $B = \{8, 10, 12, \ldots\}$, then $R^{-1}(B) = \phi$.

Our definitions tell us that for any relation R from X to Y, $R(X)$ is the range of R and $R^{-1}(Y)$ is the domain of R. For relations R which are subsets of R^2, some geometric observations about the image of a set under R are in order. If we reflect on Definition 4.10 for a moment, it is not difficult to see that $R(A)$ would be found by considering the subset of R for which elements of A are first coordinates and then projecting this subset horizontally onto Y. The resulting subset of Y would then be $R(A)$. To be more specific, consider the relation R from $X = R^1$ to $Y = R^1$ given by $R = \{(x, y) : x < y\}$ and let $A = \{x : 1 \leq x \leq 2\} \subset X$. Figure 8 describes the sets involved.

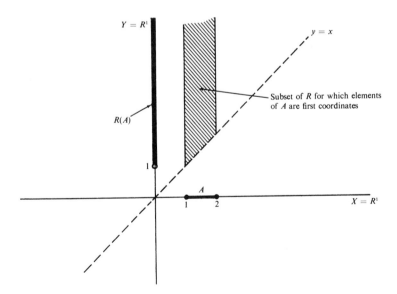

Figure 8

In forming the image of A under a relation R, it is often useful to think of "applying R to A" or of "R acting on A" to obtain the resulting elements of Y as described in Definition 4.10. In particular, if the relation under consideration is a composition $R_2 R_1$ where R_1 is from X to Y and R_2 is from Y to Z and $A \subset X$ is given, then $(R_2 R_1)(A) \subset Z$ is just the set $R_2(R_1(A))$. That is, we first apply R_1 to A to get the image $R_1(A)$ in Y, and then to this set apply R_2 to obtain the subset $R_2(R_1(A))$ of Z. A formal proof to this fact is given next.

4.12. THEOREM: *Let R_1 be a relation from X to Y and R_2 a relation from Y to Z. If $A \subset X$, then $R_2(R_1(A)) = (R_2 R_1)(A)$.*

Proof: Let us first consider $R_2(R_1(A))$ as applying the relation R_2 to the set $R_1(A) \subset Y$. Then according to Definition 4.10, $R_2(R_1(A)) = \{z \in Z : \text{there exists an element } y \in R_1(A) \text{ for which } (y, z) \in R_2\}$. This same definition tells us that $y \in R_1(A)$ iff there is an element $x \in A$ such that $(x, y) \in R_1$. Therefore, we may write $R_2(R_1(A)) = \{z \in Z : \text{there exists a } y \in Y \text{ and } x \in X \text{ such that } (x, y) \in R_1 \text{ and } (y, z) \in R_2\}$. But this last expression for $R_2(R_1(A))$ is precisely $\{z \in Z : \text{there exists an } x \in X \text{ for which } (x, z) \in R_2 R_1\}$, and is consequently another name for $(R_2 R_1)(A)$ by definition of the composition relation.

To conclude this section we give a theorem about relations which has applications to our study of functions in Chapter 2. Our present theorem is stated for any relation R, which means the same results hold (and no more) for the relation R^{-1}. Looking to Chapter 2, however, the particular restriction placed on a relation R to make it a function will enable us to prove a more general result about R^{-1}.

4.13. THEOREM: *Let R be a relation from X to Y and let A and B be subsets of X. Then*
    (a) $R(A \cup B) = R(A) \cup R(B)$ and
    (b) $R(A \cap B) \subset R(A) \cap R(B)$.

Proof: Only the proof of part (a) will be given here. To this end, we first show $R(A \cup B) \subset R(A) \cup R(B)$. Let $y \in R(A \cup B)$. Then there exists an $x \in A \cup B$ such that $(x, y) \in R$, according to Definition 4.10, from which it follows that $x \in A$ or $x \in B$. If $x \in A$, then $y \in R(A)$ so that $y \in R(A) \cup R(B)$. If $x \in B$, then $y \in R(B)$ and again $y \in R(A) \cup R(B)$. This means $R(A \cup B) \subset R(A) \cup R(B)$.

For the reverse inclusion, let $y \in R(A) \cup R(B)$. If $y \in R(A)$, there exists an $x \in A$ such that $(x, y) \in R$, and this implies $x \in A \cup B$ so that $y \in R(A \cup B)$. The same result follows if $y \in R(B)$ and this completes the verification of the reverse inclusion. Part (a) of our theorem is now proved.

Exercises:

1. Consider the two relations $R_1 = \{(x, y) : y = 2x + 1\}$ and $R_2 = \{(x, y) : y < 3\}$, each from R^1 to R^1.
   (a) Express R_1^{-1} and R_2^{-1} as sets and sketch both in the plane.
   (b) What is the relation $R_2 R_1$?
   (c) What is the relation $R_2 R_1^{-1}$?
   (d) Describe the sets $R_1(A)$, where $A = \{x : 1 \leq x \leq 3\}$, and $R_2^{-1}(B)$ where $B = \{y : 0 \leq y \leq 5\}$.

2. Let X represent the set of all integers, and $Y = W = R^1$. Define R_1 from X to Y as $R_1 = \{(x, y) : x = 3\}$ and R_2 from Y to W as $R_2 = \{(y, w) : -4 \leq y \leq 4, 0 \leq w \leq 2\}$.
   (a) What is the domain and range of R_1? Of R_2?
   (b) Find R_1^{-1} and R_2^{-1}?
   (c) What is $R_2 R_1$?
   (d) If $A = \{x : 0 < x \leq 10\}$, what is $R_2(A)$.
   (e) If $B = \{y : 0 < y < 50\}$, what is $R_1^{-1}(B)$?

3. (a) Give an example of a relation R where $R = R^{-1}$.
   (b) Give an example of a relation R from R^1 to R^1 such that for each $x \in R^1$, both $R(x)$ and $R^{-1}(x)$ are single points.
   (c) Let R be a relation from X to Y. If $x \in X$ and x is not in the domain of R, what is $R(x)$?

4. (a) Prove part (b) of Theorem 4.13.
   (b) Give an example to show equality in part (b) of Theorem 4.13 need not hold.

5. If X is any set, define the relation $I_X = \{(x, x) : x \in X\}$ from X to X as the *identity relation* on X. Now let R be any relation from X to Y. Prove or disprove:
   (a) $R^{-1}R = I_X$.
   (b) $RR^{-1} = I_Y$.
   (c) $(R^{-1})^{-1} = R$.

6. (a) Since the empty set is a relation, what is its inverse?
   (b) Since $X \times Y$ is a relation from X to Y, what is its inverse?

7. Prove that the composition relation has the associative property.

8. Consider two relations, R_1 from X to Y and R_2 from Y to Z,

having the property that for each $x \in X$ and $y \in Y$, $R_1(x)$ and $R_2(y)$ are single points. Prove that for each $x \in X$, $(R_1 R_2)(x)$ is a single point.

9. Let R be a relation from X to Y. Prove or disprove:
   (a) For each subset $A \subset X$, $R(X \backslash A) = Y \backslash R(A)$.
   (b) If $A \subset B \subset X$, then $R(A) \subset R(B)$.

10. Suppose we define a relation R from X to Y to be *bijective* iff for each $x \in X$, $R(x)$ is a single point in Y, and for each $y \in Y$, $R^{-1}(y)$ is a single point in X. If the relation R_1 from X to Y and the relation R_2 from Y to Z are both bijective, prove that $R_2 R_1$ is bijective.

5. Equivalence Relations

There are two special types of relations which occur frequently throughout our studies here and, indeed, throughout all of mathematics. These are the *equivalence relation* and a relation called a *function* which will be studied in the next chapter. In each case, we are applying certain restrictions to a general relation to obtain a more mathematically useful relation. In this sense, we are obtaining a relation for which more facts are known and which, hopefully, will help unify or shed more light on parts of existing mathematics. For an equivalence relation, the restrictions are listed in Definition 5.1. The usefulness of equivalence relations to our studies will be apparent in succeeding chapters.

5.1. Definition: A relation R from X to X is an *equivalence relation on X* iff all three of the following properties hold:
   (a) $(x, x) \in R$ for every $x \in X$. This is known as the *reflexive property* for R.
   (b) If $(x, y) \in R$, then $(y, x) \in R$. This known as the *symmetric property* for R.
   (c) If $(x, y) \in R$ and $(y, z) \in R$, then $(x, z) \in R$. This is known as the *transitive property* for R.

We might restate the three conditions by saying R is reflexive iff xRx for every $x \in X$, R is symmetric iff whenever xRy, then yRx, and R is transitive iff whenever xRy and yRz, then xRz.

5.2. Example: Let $R = \{(x, y) : x \leq y\}$ be a relation from R^1 to R^1. Then for each $x \in R^1$, $(x, x) \in R$ because $x \leq x$ for all $x \in R^1$ and, therefore, R is reflexive. If $(x, y) \in R$, it does not always follow that $(y, x) \in R$ and we conclude that R is not symmetric. For example, $(1, 3) \in R$ since $1 \leq 3$, but $(3, 1) \notin R$ because $3 \not\leq 1$. If $(x, y) \in R$ and $(y, z) \in R$, then $x \leq y$ and $y \leq z$ which implies $x \leq z$ so that $(x, z) \in R$. This means R is transitive. According to Definition 5.1, our relation R is not an equivalence relation.

An important feature of an equivalence relation is that while the relation itself is a subset of $X \times X$, there is implied by the relation a natural grouping of elements in X into subsets called equivalence classes. These equivalence classes are defined next.

5.3. Definition: Let R be an equivalence relation on X. For each element $x \in X$, the set $R(x) = \{y \in X : yRx\}$ is called the *equivalence class* in X determined by x with respect to R.

In other words, the equivalence class determined by x with respect to R is simply the image of x under R. It is important to observe that if $X \neq \phi$, restriction 5.1(a) assures us $R(x) \neq \phi$ for each $x \in X$.

5.4. Example: Let $X = \{2, 4, 6, 7\}$ and let R be a relation from X to X given by $R = \{(x, y) : x, y \in X \text{ and } x + y \text{ is an even integer}\}$. Then R is an equivalence relation, as may be easily verified, and $R(2) = \{2, 4, 6\} = R(4) = R(6)$ and $R(7) = \{7\}$.

When talking about equivalence relations, the term *partition* of a set will be a helpful addition to our mathematical vocabulary. Its definition will be given next.

5.5. Definition: Let X be a nonempty set. The indexed family of subsets $\{A_\alpha : \alpha \in \Delta\}$ of X form a *partition* of X iff
  (a) $A_\alpha \neq \phi$ for each $\alpha \in \Delta$,
  (b) $\bigcup_{\alpha \in \Delta} A_\alpha = X$ and
  (c) For $\alpha, \beta \in \Delta$, either $A_\alpha = A_\beta$ or $A_\alpha \cap A_\beta = \phi$.

In Example 5.4, $R(2)$ and $R(7)$ form a partition of the set X. An important consequence of the definition of equivalence classes is that equivalence classes form a partition of the set X. This is the essence of the next theorem.

5.6. THEOREM: *Let R be an equivalence relation on X. Then for any two elements x and y of X, either*

(a) $R(x) = R(y)$ *or*

(b) $R(x) \cap R(y) = \phi$.

Proof: If $R(x) \cap R(y) = \phi$, the theorem is proved. If $R(x) \cap R(y) \neq \phi$, we shall show $R(x) = R(y)$. To this end, let $z \in R(x) \cap R(y)$ and consider any $a \in R(x)$. Then zRx which implies xRz by the symmetric property of R. Now $a \in R(x)$ gives aRx and the transitive property of R, applied to aRx and xRz, gives aRz. But also, $z \in R(y)$ so that zRy. Therefore, the transitive property of R, applied to aRz and zRy, gives aRy, which implies $a \in R(y)$. Thus, $R(x) \subset R(y)$. In a similar manner, it may be shown that $R(y) \subset R(x)$, so that the desired conclusion $R(x) = R(y)$ is reached.

In conclusion, suppose the indexed family $\{A_\alpha : \alpha \in \Delta\}$ forms a partition of X. Then we could define a relation R from X to X as follows: For each pair of elements $x, y \in X$, $(x, y) \in R$ iff both x and y belong to the same subset A_α. Since each $x \in X$ belongs to A_α for some $\alpha \in \Delta$ by Definition 5.5(b), it follows that $(x, x) \in R$ and hence R is reflexive. Also, if $(x, y) \in R$, then x and y belong to the same subset A_α, which means y and x belong to the same A_α and, therefore, $(y, x) \in R$. The transitive property for R is equally easy to see so that R as defined is an equivalence relation on X. Because the equivalence classes in X are precisely the sets A_α of the original partition, we say the partition "defines" the equivalence relation just described. The information at hand, coupled with Theorem 5.6, leads us to our final theorem.

5.7. THEOREM: *The equivalence classes of an equivalence relation on a set X form a partition of X, and conversely, any partition of X defines an equivalence relation on X for which the equivalence classes are the elements of the partition.*

Exercises:

1. Which of the reflexive, symmetric, and transitive properties do each of the following relations have?
   (a) Let the set under consideration be all integers Z. For $m, n \in Z$, define mRn iff $m - n$ is a multiple of 3.

(b) Let the set under consideration be R^2. For (a, b) and (c, d) in R^2, define $(a, b)R(c, d)$ iff $a = -c$ and $b = -d$.

(c) Let the set under consideration be R^1. For $x, y \in R^1$, define xRy iff $x - y$ is an integer.

(d) Let the set under consideration be all 2 by 2 matrices with elements from R^1. For a, b in this set, define aRb iff $a^{-1} = b$.

2. Describe the equivalence classes for those relations in Exercise 1 which are equivalence relations.

3. Give examples of relations in which one but not the other two conditions in Definition 5.1 hold.

4. Give examples of relations in which two but not the remaining condition in Definition 5.1 holds.

5. Let R be an equivalence relation on X. Prove that $R(x) \cap R(y) = \phi$ iff $(x, y) \notin R$.

6. Let R be an equivalence relation on X. Prove R^{-1} is also an equivalence relation on X.

7. Let R be a reflexive relation from X to X. Prove that R is an equivalence relation on X iff $RR = R$ and $R = R^{-1}$.

8. For the set $N \times N$, define $(a, b)R(c, d)$ iff $a + d = b + c$. Prove that R is an equivalence relation on $N \times N$.

9. If R_1 and R_2 are each equivalence relations on a set X, prove or disprove that $R_2 R_1$ is an equivalence relation on X.

10. Let R be a relation on X which is both reflexive and transitive. Prove that $R \cap R^{-1}$ is an equivalence relation on X.

2

Functions

1. Defining a Function

In this chapter we are going to study the concept of a function. Without a doubt, this is one of the most important single concepts that runs through all of mathematics. It is encountered at every level of mathematics (whether specifically recognized or not), and will be of utmost importance to the remainder of our studies in this book. As we shall see in a moment, a function is just a special kind of relation so that in a sense, we are continuing our study of relations in this chapter. Since special relations are under consideration and they are so fundamental in mathematics, it would seem natural to use special symbols to denote them. Following convention, we will generally use

f, g, h, F, G, etc., to denote relations that are also functions. One final word before defining a function: Functions are sometimes called *mappings*, or simply *maps*, in the mathematical literature and for that reason, we shall henceforth consider the words function, mapping and map as synonyms, although function will be preferred.

1.1. Definition: A relation f from X to Y is called a *function* iff the following two conditions hold:
    (a) The domain of f is X.
    (b) For each $x \in X$, $f(x)$ is a single element in Y.

In other words, a function f is a subset of $X \times Y$ such that for each $x \in X$ there is a $y \in Y$ for which $(x, y) \in f$, and, furthermore, if $(x, y) \in f$, then x cannot appear as the first coordinate of any other element of f. Aside from requiring the domain of f to be X, another way of saying $f(x)$ is a single point is to say that if $(x, y_1) \in f$ and $(x, y_2) \in f$, then it must follow that $y_1 = y_2$. Still differently, a function corresponds or pairs with each $x \in X$ precisely one $y \in Y$. We should notice, however, that the definition allows different elements of X to be paired with the same element of Y. What is specifically prohibited by the definition is a single element of X being paired with two or more elements of Y.

1.2. Example: Let $X = \{0, 2, 7\}$ and $Y = \{1, 3, 5, 7\}$. Among the several functions that may be defined from X to Y is $f = \{(0, 1), (2, 3), (7, 7)\}$, which is equally well described by writing $f(0) = 1$, $f(2) = 3$, and $f(7) = 7$. Another function from X to Y is $g = \{(0, 1), (2, 1), (7, 1)\}$. The relation $R = \{(0, 1), (0, 3), (2, 7), (7, 1)\}$ from X to Y is not a function since $R(0) = \{1, 3\}$ is not a single point in Y.

The relation in Figure 9(a) is not a function, because for each $x \in X$, $R(x)$ consists of infinitely many points. The relation in (b) is a function, while the one in (c) is not, due to the fact that for some points in the domain the images are not single points.

Definition 1.1 defines a function by what is ordinarily called its graph. That is, from our present point of view, a function and its graph are indistinguishable. Suppose we now discuss for a moment a way of describing a function which does not explicitly mention ordered pairs. This outlook will greatly help our thinking with many problems concerning functions where our interest lies in sets and their images rather than in ordered pairs. As we have already observed, if

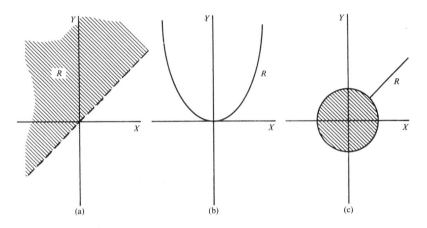

Figure 9

a function f from X to Y is given, then for each $x \in X$ we know the single element $f(x) \in Y$ that is its image. We might say then that to each $x \in X$ there corresponds a single element $f(x)$ in Y with the function showing us exactly how this correspondence takes place. In particular, if $(x, y) \in f$, then $f(x) = y$ and the element x corresponds to the element y. On the other hand, suppose we are given two sets X and Y and in some manner told a way of corresponding each $x \in X$ with exactly one element $y \in Y$. Then for a given $x \in X$, we could think of the single element $y \in Y$ to which it corresponds as being the image $f(x)$ of x under a function f. Now knowing the image $f(x)$ for each $x \in X$, the set $\{(x, f(x)) : x \in X\} \subset X \times Y$ fulfills the conditions of Definition 1.1, and we therefore have a function f from X to Y. (Figure 10 illustrates the two viewpoints.)

The consequence of our discussion is that we may think of a function f from X to Y as being a given correspondence between points of X and Y having the property that each $x \in X$ corresponds to a single $y \in Y$. In doing this, we are focusing attention on what happens to points in the sets X and Y under the function, rather than the behavior of the ordered pairs in f. In view of this, we shall often describe a function with the symbolism $y = f(x)$, or just simply, $f(x)$. For example, $y = x^2$, $f(x) = x^2$ and $f = \{(x, y) : y = x^2\}$ are all notations for the same function from R^1 to R^1. Regardless of how we tend to think about describing a function, the conventional notation $f : X \rightarrow Y$ will be used to mean f is a given function from X to Y.

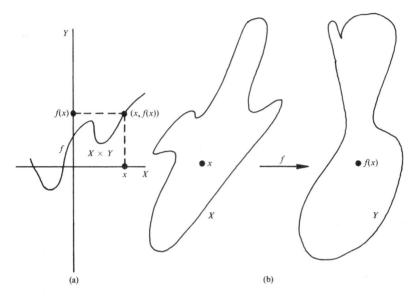

(a)

(b)

Figure 10

If $f: X \to Y$ and $g: X \to Y$ are given functions and we write $f = g$, then we are simply saying the two subsets f and g of $X \times Y$ are equal. However, in view of the observations we have just made about functions, it follows that $f = g$ iff $f(x) = g(x)$ for every $x \in X$. This latter image characterization of the equality of functions will often be more convenient to use than showing set equality in the product set $X \times Y$.

A given function $f: X \to Y$ induces another function from $\mathscr{P}(X)$ to $\mathscr{P}(Y)$ in a rather natural way as described earlier for relations in general. In our present notational setting, whenever $A \subset X$, the image of A under f, denoted by $f(A)$, is the set of all $y \in Y$ such that there is some $x \in A$ for which $(x, y) \in f$. Notationally, $f(A) = \bigcup_{x \in A} f(x)$. Thus, for each $A \in \mathscr{P}(X)$, $f(A)$ is a single element of $\mathscr{P}(Y)$, so that we may think of the original f as inducing a function from $\mathscr{P}(X)$ to $\mathscr{P}(Y)$. Even though the domain of the original function is X and we are inducing a function whose domain is $\mathscr{P}(X)$, it is still customary to use f to denote the induced function.

Our next theorem is similar to Theorem 4.13 of Chapter 1, except that we are now using a function rather than a general relation and considering an indexed family of subsets of X rather that just two subsets. However, the importance of the theorem in our work with functions warrants going through the proof in the present setting.

1.3. THEOREM: *If $f: X \to Y$ is any given function and $\{A_\alpha : \alpha \in \Delta\}$ is a family of subsets of X indexed by Δ, then*

$\quad$ (a) $f(\bigcup_{\alpha \in \Delta} A_\alpha) = \bigcup_{\alpha \in \Delta} f(A_\alpha)$ *and*

$\quad$ (b) $f(\bigcap_{\alpha \in \Delta} A_\alpha) \subset \bigcap_{\alpha \in \Delta} f(A_\alpha)$.

Proof. The proof for (b) will be given while that of (a) is left as an exercise. Suppose y is an element of $f(\bigcap_{\alpha \in \Delta} A_\alpha)$. Then there is some $x \in \bigcap_{\alpha \in \Delta} A_\alpha$ such that $f(x) = y$ by definition of image of $\bigcap_{\alpha \in \Delta} A_\alpha$ under f. Thus, $x \in A_\alpha$ for every $\alpha \in \Delta$, which implies $f(x) \in f(A_\alpha)$ for every $\alpha \in \Delta$. This means that $f(x) = y \in \bigcap_{\alpha \in \Delta} f(A_\alpha)$ and gives the desired conclusion.

Just as the name function denotes a special kind of relation, there are certain names applied to special kinds of functions. Some of these are now defined.

1.4. Definition: Let $f: X \to Y$ be a function.
$\quad$ (a) If the image of every $x \in X$ under f is the same element $y_0 \in Y$, then f is called a *constant function*. That is, f is constant iff $f(X) = \{y_0\}$.
$\quad$ (b) If $X \subset Y$ and $f(x) = x$ for every $x \in X$, then f is called an *inclusion function*.
$\quad$ (c) If $X = Y$ and $f(x) = x$ for every $x \in X$, then f is called an *indentity function*. The identity function from X to X will henceforth be denoted by 1_X. That is, $1_X = \{(x, x) : x \in X\}$.
$\quad$ (d) If $f(X) = Y$, then f is called a *surjective* function. In other words, $f: X \to Y$ is surjective iff each $y \in Y$ is the image of at least one element in X under f.
$\quad$ (e) If given any $x_1, x_2 \in X$ where $x_1 \neq x_2$, it follows that $f(x_1) \neq f(x_2)$, then f is called an *injective* function. That is, f is injective iff distinct elements in X have distinct images in Y.
$\quad$ (f) If f is both injective and surjective, then f is called *bijective*. If X and Y are sets and there exists a bijective function $f: X \to Y$, we sometimes say X and Y are put into a one-to-one correspondence by f.

The function f in Example 1.2 is injective but not surjective because $f(X) = \{1, 3, 7\} \neq Y$. The function g in the same example is not injective because there are distinct elements in X which have the same image in Y. For instance, $g(0) = g(2) = 1$. This same function is not surjective because the range of g is $\{1\} \neq Y$. Among familiar functions from the reals to the reals, the one given by $f(x) = x^2$ is neither injective nor surjective. This function would be injective, however, if we require the domain to be the non-negative reals only. The function described by $f(x) = 2x + 1$ from R^1 to R^1 is bijective. The injective condition holds because if $x_1 \neq x_2$, then $f(x_1) = 2x_1 + 1 \neq 2x_2 + 1 = f(x_2)$. The surjective condition holds because given *any* $y \in Y$, there is an $x \in X$ such that $f(x) = y$, namely $x = (y - 1)/2$, which is a well-defined number in X. Therefore, $f(X) = Y$.

In practice, a convenient way of proving the injective property of a function is to show the contrapositive of Definition 1.4(e). For instance, if the function of the previous paragraph given by $f(x) = 2x + 1$ is under consideration and we assume $f(x_1) = f(x_2)$, then $2x_1 + 1 = 2x_2 + 1$, which implies $x_1 = x_2$. This means no two distinct elements of R^1 can have the same image and, therefore, f is injective.

We are now in a position to state a simple condition which will make the equality hold in Theorem 1.3(b).

1.5. THEOREM: *Let $f : X \to Y$ be an injective function and $\{A_\alpha : \alpha \in \Delta\}$ an indexed family of subsets of X. Then $f(\bigcap_{\alpha \in \Delta} A_\alpha) = \bigcap_{\alpha \in \Delta} f(A_\alpha)$.*

Proof: The proof is left for the reader.

Another special type of function that is very useful is the projection function. Projection functions exist anytime a product of sets is formed.

1.6. Definition: Let $X_1 \times X_2 \times \ldots \times X_n$ be the product of the family $\{X_k : k = 1, 2, \ldots, n\}$. Then for each $k, 1 \leq k \leq n$, there is a *projection function* $p_k : (X_1 \times X_2 \times \ldots \times X_n) \to X_k$ defined as $p_k(x_1, x_2, \ldots, x_n) = x_k$.

If only two sets X and Y are involved, we shall often write p_X to indicate the projection function from $X \times Y$ to X given by $p_X(x, y) =$

x. Also $p_Y : X \times Y \to Y$ is given by $p_Y(x, y) = y$. For some motivation behind the name *projection function*, consider the familiar coordinate plane R^2 where $X = Y = R^1$. Then the projection function p_X may be described as a vertical projection of points in the plane onto the X axis. Similarly, p_Y is a horizontal projection of points in the plane onto the Y axis.

Exercises:

1. List all of the possible functions from $X = \{a, b, c\}$ to $Y = \{0, 1\}$.

2. Prove part (a) of Theorem 1.3.

3. Give an example of a function $f : X \to Y$ and two subsets A and B of X which show that
   (a) $f(A \cap B) \neq f(A) \cap f(B)$.
   (b) $f(A \backslash B) \neq f(A) \backslash f(B)$.

4. For a given $f : X \to Y$, prove that the induced function from $\mathscr{P}(X)$ to $\mathscr{P}(Y)$ has the following properties:
   (a) $f(\phi) = \phi$.
   (b) It need not be injective by use of an example.

5. Give examples of functions from the reals to the non-negative reals that are
   (a) Injective.
   (b) Injective but not surjective.
   (c) Surjective but not injective.

6. Let X be the set of all two by two matrices whose elements are real numbers. Define $f : X \to R^1$ as $f(A) = \det(A)$ for all $A \in X$. (Det (A) is the determinant of A.) Prove or disprove that f is injective. Prove or disprove f is surjective.

7. (a) Let $X = \{3, 5, 7\}$ and $Y = \{a, b\}$. Describe the projection functions p_X and p_Y.

   (b) Prove that if $\prod_{k=1}^{n} X_k \neq \phi$, then every projection function p_k :
   $$\prod_{k=1}^{n} X_k \to X_k \text{ is surjective.}$$
   (c) Give an example of a projection function that is injective and of one that is not injective.

8. Prove Theorem 1.5.

9. Prove that a relation R from X to X is both a function and an equivalence relation iff $R(x) = x$ for every $x \in X$.

10. Prove that the function $f : X \to Y$ is injective iff for every subset $A \subset X$, $f(X \backslash A) \subset Y \backslash f(A)$.

11. (a) Consider the sets ϕ and Y. Prove there is only one function from ϕ to Y.
    (b) Consider the set ϕ and $X \neq \phi$. Prove there exists no function from X to ϕ.
    (c) Is the function ϕ from ϕ to Y injective? Surjective?

2. Function Inverses and Inverse Images

As described in Chapter 1, every relation R from X to Y induces a relation R^{-1} from Y to X called the inverse relation. Since a function $f : X \to Y$ is just a special kind of relation, we may form the inverse relation f^{-1} from Y to X and study its properties. In functional notation, $f^{-1} = \{(y, x) : (x, y) \in f\}$. One of the first questions to arise is whether f^{-1} is a function or not. Since examples are easy to find showing f^{-1} need not be a function from Y to X, our attention turns to finding out if there are conditions under which f^{-1} is always a function. Before attempting to formulate any theorem along these lines, suppose we first study some simple examples on the subject.

2.1. Example: Let $X = \{1, 3, 5, 7\}$ and $Y = \{2, 4, 6\}$ and define $f : X \to Y$ as $f = \{(1, 2), (3, 2), (5, 4), (7, 6)\}$. Then $f^{-1} = \{(2, 1), (2, 3), (4, 5), (6, 7)\}$ and is not a function. Examine several other functions from X to Y and consider their inverse relations. Will *any* function $f : X \to Y$ be such that f^{-1} is a function from Y to X? Why?

Now consider $Z = \{2, 4, 6, 8\}$ and define $g : X \to Z$ as $g = \{(1, 8), (5, 6), (3, 4), (1, 2)\}$. Then $g^{-1} = \{(8, 1), (6, 5), (4, 3), (2, 1)\}$ is a function from Z to X and we may therefore use the notation $g^{-1} : Z \to X$ to denote it.

Finally, let $W = \{2, 4, 6, 8, 10\}$ and define $h : X \to W$ as $h = \{(1, 2), (3, 4), (5, 6), (7, 8)\}$. This gives $h^{-1} = \{(2, 1), (4, 3), (6, 5), (8, 7)\}$ which is not a function from W to X since the domain of h^{-1} is not W. However, h^{-1} is a function from the range of h, $h(X) =$

$W\backslash\{10\}$, to X. Can any function from X to W be such that its inverse is a function from W to X? Why?

These examples point out the two difficulties which are keeping f^{-1} from being a function from Y to X when $f: X \to Y$ is given. The first is that the domain of f^{-1} need not be all of Y, and the second is that f need not be injective. Consideration of these facts leads us to the following result.

2.2. THEOREM: *If $f: X \to Y$ is any injective function, then f^{-1} is a function from $f(X)$ to X and $f^{-1}(y) = x$ iff $y = f(x)$ for every $x \in X$ and $y \in f(X)$.*

Proof: Certainly the domain of f^{-1} is precisely the range $f(X)$ of f. Now if y is any point in $f(X)$, then $\phi \neq f^{-1}(y) = \{x \in X : (x, y) \in f\}$ is a set consisting of a single point. The reason is that if $x_1, x_2 \in f^{-1}(y)$, then (x_1, y) and (x_2, y) both belong to f, which implies $x_1 = x_2$ by the injective property of f. Therefore, $f^{-1}(y)$ is the single point $x \in X$ for which $(x, y) \in f$ or $f(x) = y$. This completes the proof that f^{-1} is a function from $f(X)$ to X. Since $(y, x) \in f^{-1}$ iff $(x, y) \in f$, it follows that $f^{-1}(y) = x$ iff $y = f(x)$ for all $y \in f(X)$ and $x \in X$.

2.3. Corollary: If $f: X \to Y$ is bijective, then f^{-1} is a function from Y to X.

Proof: The bijective hypothesis implies $f(X) = Y$ so that the conclusion follows from Theorem 2.2.

The results of Theorem 2.2 allow us to make the following definition.

2.4. Definition: Let $f: X \to Y$ be any injective function. Then the function $f^{-1}: f(X) \to X$ is called the *function inverse* of f.

Even though $f: X \to Y$ may not be injective, it may still lead to the consideration of function inverses which are closely related to the original function. As an example, we might think of the inverse trigonometric functions, although there are numerous others. What we have in mind is to select a subset $A \subset X$ for which the function $g: A \to Y$ defined by $g = \{(x, f(x)) : x \in A\}$ is injective and then form the function inverse $g^{-1} : g(A) \to A$ of g. From the definition of g, the evident connection between g and f is that $g(x) = f(x)$ for every $x \in A$ so that the function inverse g^{-1} agrees with f^{-1} if the domain of f^{-1} is restricted to the subset $f(A) = g(A)$ of Y. This shows just how closely the functions under consideration are related. In defining g we are in a sense restricting f to the subset A and ignoring the rest of the domain. Since the idea of restriction of a function will be of future importance, it is now set out in a formal definition.

2.5. Definition: Let $f : X \to Y$ be any function and $A \subset X$. The *restriction* of f to A is the function $g : A \to Y$ defined by $g(x) = f(x)$ for every $x \in A$. It is customary to write $g = f|A$ to denote the restriction of f to A.

2.6. Example: Let $X = Y = R^1$ and suppose $f : X \to Y$ is given by $f(x) = x^2$. Then f is not injective, but if we take $A = \{x \in Y : x \leq 0\}$, we see $f|A$ is injective and the function inverse $(f|A)^{-1}$ of $f|A$ is given by $(f|A)^{-1}(y) = -\sqrt{y}$ for each $y \in (f|A)(A) = \{y \in Y : y \geq 0\}$. If $B = \{x \in X : 0 \leq x \leq 2\}$, then $f|B$ is injective and $(f|B)^{-1}(y) = \sqrt{y}$ for each $y \in (f|B)(B) = \{y \in Y : 0 \leq y \leq 4\}$.

2.7. Example: Let $X = Y = R^1$ where $f : X \to Y$ is given by $f(x) = \sin x$. This function is not injective, but it is easy to find subsets of X for which f restricted to that subset is injective. In the study of trigonometry it is customary to select the subset $A = \{x \in X : -\pi/2 \leq x \leq \pi/2\}$ for which $f|A$ is injective and call the function inverse of $f|A$ the Arcsin function. That is, $(f|A)^{-1}(y) = \text{Arcsin}(y)$ for each $y \in (f|A)(A)$. Since $(f|A)(A) = f(X) = \{y \in Y : -1 \leq y \leq 1\}$, we then have the fundamental relation $\text{Arcsin } y = x$ iff $\sin x = y$ for all $-\pi/2 \leq x \leq \pi/2$ and $-1 \leq y \leq 1$.

For any function $f : X \to Y$ and $B \subset Y$, recall that $f^{-1}(B) = \{x \in X : f(x) \in B\}$ according to Definition 4.10 of Chapter 1. This means the relation f^{-1} from Y to X may *always* be interpreted as a function from the set $\mathscr{P}(Y)$ to the set $\mathscr{P}(X)$ and in this sense we say $f^{-1} : \mathscr{P}(Y) \to \mathscr{P}(X)$ is induced by $f : X \to Y$. Surprisingly, the properties of this function are much nicer than those of the induced function from $\mathscr{P}(X)$ to $\mathscr{P}(Y)$ given in Theorem 1.3. The next theorem gives the properties of $f^{-1} : \mathscr{P}(Y) \to \mathscr{P}(X)$.

2.8. THEOREM: *Let $f: X \to Y$ be a function and $\{B_\alpha : \alpha \in \Delta\}$ be an indexed family of subsets of Y. Then the induced function $f^{-1}: \mathscr{P}(Y) \to \mathscr{P}(X)$ has the following properties :*

(a) $f^{-1}(\bigcup_{\alpha \in \Delta} B_\alpha) = \bigcup_{\alpha \in \Delta} f^{-1}(B_\alpha)$.

(b) $f^{-1}(\bigcap_{\alpha \in \Delta} B_\alpha) = \bigcap_{\alpha \in \Delta} f^{-1}(B_\alpha)$.

(c) $f^{-1}(B_1 \backslash B_2) = f^{-1}(B_1) \backslash f^{-1}(B_2)$ for any two subsets B_1 and B_2 of Y.

Proof: Only the proof of (b) will be given while those of (a) and (c) appear as exercises. First, the inclusion $f^{-1}(\bigcap_{\alpha \in \Delta} B_\alpha) \subset \bigcap_{\alpha \in \Delta} f^{-1}(B_\alpha)$ is shown. If $x \in f^{-1}(\bigcap_{\alpha \in \Delta} B_\alpha)$, then $f(x) \in \bigcap_{\alpha \in \Delta} B_\alpha$. Thus, $f(x) \in B_\alpha$ for every $\alpha \in \Delta$. Then by the definition of f^{-1}, we have $x \in f^{-1}(B_\alpha)$ for every $\alpha \in \Delta$ so that $x \in \bigcap_{\alpha \in \Delta} f^{-1}(B_\alpha)$. For the reverse inclusion, let $x \in \bigcap_{\alpha \in \Delta} f^{-1}(B_\alpha)$. Then $x \in f^{-1}(B_\alpha)$ for every $\alpha \in \Delta$. This means that $f(x) \in B_\alpha$ for every $\alpha \in \Delta$ and hence $f(x) \in \bigcap_{\alpha \in \Delta} B_\alpha$. The definition of f^{-1} then states that $x \in f^{-1}(\bigcap_{\alpha \in \Delta} B_\alpha)$.

In conclusion, we show some ways in which the relation f^{-1} from Y to X may be helpful in giving information about the function $f: X \to Y$.

2.9. THEOREM: *The function $f: X \to Y$ is surjective iff for each $y \in Y$, $f^{-1}(y) \neq \phi$.*

Proof: If f is surjective, then $f(X) = Y$ and $y \in Y$ implies there exists an $x \in X$ such that $(x, y) \in f$. Therefore $x \in f^{-1}(y) \neq \phi$. For the converse, if $f^{-1}(y) \neq \phi$ for each $y \in Y$, then there exists an $x \in X$ such that $(x, y) \in f$ for each $y \in Y$ and, consequently, $f(X) = Y$.

2.10. THEOREM: *The function $f: X \to Y$ is injective iff for each $y \in Y$, $f^{-1}(y)$ is a single point in X or $f^{-1}(y) = \phi$.*

Proof: The proof is left as an exercise.

2.11. THEOREM: *The function $f: X \to Y$ is bijective iff for each $y \in Y$, $f^{-1}(y)$ is a single point in X.*

Proof: This proof is also left as an exercise.

Exercises:

1. Let $X = Y = R^1$ and consider the following functions:
   (1) $f: X \rightarrow Y$ given by $f(x) = 2x^2 - 1$ for each $x \in X$.
   (2) $f: N \rightarrow N$ given by $f(n) = 2n + 3$ for each $n \in N$.
   (3) $f: X \rightarrow Y$ given by $f(x) = e^x$ for each $x \in X$.
   (4) $f: X \rightarrow Y$ given by $f(x) = |x|$ for each $x \in X$.
   (5) $f: X \rightarrow$ given by $f(x) = \begin{cases} 1 & \text{if } x \geq 0 \\ -1 & \text{if } x < 0 \end{cases}$.

   (a) Describe f^{-1} for those functions f having f^{-1} as a function inverse.
   (b) For those functions f not having f^{-1} as a function inverse, restrict f to some subset of the domain so that f will be injective and find the function inverse to your restricted function.
   (c) For the functions (1), (3) and (5), let $B = \{y \in Y : -1 \leq y \leq 1\}$. Find $f^{-1}(B)$ in each case.

2. Consider the collection X of all polynomials having real coefficients. Does the derivative function $d/dx : X \rightarrow X$ have a function inverse? Discuss indefinite integration as an inverse operation to differentiation on the set X.

3. Let $f: R^2 \rightarrow R^2$ be given by $f(x, y) = (x + y, y - 2x)$ for each $(x, y) \in R^2$. Prove f is bijective and find the function inverse to f.

4. (a) If $f: X \rightarrow Y$ is a function and B_1 and B_2 are subsets of Y where $B_1 \subset B_2$, prove $f^{-1}(B_1) \subset f^{-1}(B_2)$.
   (b) If $f: X \rightarrow Y$ is a function and $a, b \in Y$ where $a \neq b$, prove $f^{-1}(a) \cap f^{-1}(b) = \phi$.

5. (a) Prove that for any $f: X \rightarrow Y$, the function $f^{-1}: \mathscr{P}(Y) \rightarrow \mathscr{P}(X)$ is always injective.
   (b) Prove the function inverse of a function is always injective.

6. Prove Theorem 2.8.

7. Prove Theorem 2.10.

8. Prove Theorem 2.11.

9. Prove that the function $f: X \rightarrow Y$ is surjective iff for each subset $A \subset X$, $Y \backslash f(A) \subset f(X \backslash A)$.

10. Let $f : X \rightarrow Y$ be a function and define $G : X \rightarrow X \times Y$ as $G(x) = (x, f(x))$ for each $x \in X$.

(a) Prove G is injective.

(b) Since G is injective by (a), $G^{-1} : G(X) = f \rightarrow X$ is the function inverse of G. Now prove $p_X|f = G^{-1}$.

11. If $f : X \rightarrow Y$ is a function, $A \subset X$ and $B \subset Y$, prove that $f(A \cap f^{-1}(B)) = f(A) \cap B$.

3. The Composition of Functions

Since we have already discussed the composition of relations in Chapter 1, the concept itself for functions will present nothing new. Even so, there are several observations of interesting facts and relationships which we will make that are pertinent to the study of functions. To review in our present function terminology, if $f : X \rightarrow Y$ and $g : Z \rightarrow W$ are given functions, then the composition relation gf is given by $gf = \{(x, w) : \text{there is some } y \in Y \cap Z \text{ for which } (x, y) \in f \text{ and } (y, w) \in g\}$. Theorem 4-12 of Chapter 1 tells us that if A is any subset of X, then $(gf)(A) = g(f(A))$ and in particular $(gf)(x) = g(f(x))$ for every $x \in X$. The composition gf will, in general, be a relation from X to W since there is no assurance the domain of gf will be all of X. However, if the range of f is a subset of the domain of g, gf will be a function from X to W, and we write $gf : X \rightarrow W$ to denote this fact. The reason for this is that the domain of gf will then be X and for each $x \in X$, $(gf)(x) = g(f(x))$ is a single point in W. This in turn leads us to the fact that the relation gf is always a function from the domain of gf to the set W.

3.1. Example: Let $f : R^1 \rightarrow R^1$ be given by $f(x) = x^2$ for each $x \in R^1$. Let $Z = \{x \in R^1 : x \geq 1\}$ and $g : Z \rightarrow R^1$ be defined by $g(x) = \sqrt{x - 1}$ for each $x \in Z$. Then the range of f is $\{x \in R^1 : x \geq 0\}$ and is not a subset of the domain of g so that gf is a relation, but not a function, from R^1 to R^1. The relation gf is, however, a function from the domain $\{x \in R^1 : x \geq 1\} \cup \{x \in R^1 : x \leq -1\}$ of gf to R^1 and is given by $(gf)(x) = g(f(x)) = g(x^2) = \sqrt{x^2 - 1}$ for each element x in the domain. If we consider the relation fg from R^1 to R^1, we see that the range of g, namely $\{x \in R^1 : x \geq 0\}$, is a subset of the domain R^1 of f so that fg is a function from the domain Z of g to

R^1 and is given by $(fg)(x) = f(g(x)) = f(\sqrt{x-1}) = x - 1$ for each $x \in Z$.

When dealing with the composition of two functions in the future, we will restrict ourselves to working only with compositions of the form gf where $f: X \to Y$ and $g: Y \to W$ so that $gf: X \to W$ will always be a function from X to W. This means that the composition of two functions will always be a function so that worries about having to restrict the domain of the relation gf to make it a function will be avoided. With this in mind, let us begin consideration of some facts about the composition of two functions.

3.2. THEOREM: *Let $f: X \to Y$ and $g: Y \to X$ be such that $gf = 1_X$. Then f is injective and g is surjective.*

Proof: Before beginning the proof we might remind ourselves that $gf = 1_X$ iff $(gf)(x) = 1_X(x) = x$ for each $x \in X$. To show that f is injective, suppose $f(x_1) = f(x_2)$ for some $x_1, x_2 \in X$. Then $x_1 = 1_X(x_1) = (gf)(x_1) = g(f(x_1)) = g(f(x_2)) = (gf)(x_2) = 1_X(x_2) = x_2$. Consequently, f is injective. Since $x = 1_X(x) = g(f(x))$ for each $x \in X$, it follows that each element $x \in X$ is the image of an element $f(x)$ in Y under the function g. Therefore $g(Y) = X$ and g is surjective.

There is an interesting observation we may make in the context of function compositions about Theorem 2.2. If $f: X \to Y$ is bijective, then $f^{-1}: Y \to X$ is a function and $f^{-1}(y) = x$ iff $f(x) = y$. This means $x = f^{-1}(y) = f^{-1}(f(x)) = (f^{-1}f)(x)$ for each $x \in X$ and $y = f(x) = f(f^{-1}(y)) = (ff^{-1})(y)$ for each $y \in Y$ or that $1_X = f^{-1}f$ and $1_Y = ff^{-1}$. In fact, these last conditions could be used as the defining property of the function inverse because we may prove the converse holds. It is given next in theorem form.

3.3. THEOREM: *Let $f: X \to Y$ be a function. If $g: Y \to X$ is a function such that $1_X = gf$ and $1_Y = fg$, then f is bijective and $g = f^{-1}$.*

Proof: The hypothesis $gf = 1_X$ implies f is injective by Theorem 3.2. Now, interchanging the roles of g and f and of X and Y and using the fact that $fg = 1_Y$, the same theorem gives f surjective. These results imply f is bijective so that one of our desired objectives is reached. From this conclusion we now know $f^{-1}: Y \to X$ is a function and $f^{-1}(y) = x$ iff $f(x) = y$. We still need to show $g = f^{-1}$. To do this, let $x \in X$ and $f(x) = y$. Then we have $x = 1_X(x) = (gf)(x) = g(f(x)) = g(y)$. Also, if $y \in Y$ and $g(y) = x$, then $y = 1_Y(y) = (fg)(y) = f(g(y)) = f(x)$. As a consequence of these results, we may write $g(y) = x$ iff $f(x) = y$. But this means $g(y) = f^{-1}(y)$ for each $y \in Y$ giving $g = f^{-1}$ for our final conclusion.

To summarize our thoughts on the function inverse we state the following theorem.

3.4. THEOREM: *Let $f: X \to Y$ and $g: Y \to X$ be functions. Then g is the function inverse of f iff $gf = 1_X$ and $fg = 1_Y$.*

Proof: Theorem 3.3 and the discussion immediately preceeding it.

Perhaps it is worthy at this point to mention a rather fundamental relationship between functions and projection functions. If we are given $f: X \to Y$, the function $G: X \to X \times Y$ given by $G(x) = (x, f(x))$ may then always be formed. From its definition, the range of G is the set $f \subset X \times Y$ and for each $x \in X$, $(p_Y G)(x) = p_Y(G(x)) = p_Y(x, f(x)) = f(x)$ where p_Y is the projection function from $X \times Y$ to Y. Therefore, $p_Y G = f$.

From Example 3.1 it is clear that the composition of functions need not be commutative. We may, however, prove the composition is always an associative operation.

3.5. THEOREM: *Let $f: X \to Y$, $g: Y \to Z$ and $h: Z \to W$ be given functions. Then $h(gf) = (hg)f$.*

Proof: The proof is left as an exercise.

For functions $f: X \to Y$ and $g: Y \to Z$, f^{-1} is a function from $\mathscr{P}(Y)$ to $\mathscr{P}(X)$ and g^{-1} is a function from $\mathscr{P}(Z)$ to $\mathscr{P}(Y)$ so that $(gf)^{-1}$ is a function from $\mathscr{P}(Z)$ to $\mathscr{P}(X)$. The next theorem shows $(gf)^{-1}$ is the same as the composition function $f^{-1}g^{-1}$.

3.6. THEOREM: *Let $f: X \to Y$ and $g: Y \to Z$ be given functions. Then for each subset $B \subset Z$, $(f^{-1}g^{-1})(B) = (gf)^{-1}(B)$.*

Proof: First, it will be shown that $(gf)^{-1}(B) \subset (f^{-1}g^{-1})(B)$. To this end, suppose $x \in (gf)^{-1}(B)$. Then by definition of $(gf)^{-1}$, it follows that $(gf)(x) \in B$. But $(gf)(x) = g(f(x)) \in B$ and hence $f(x) \in g^{-1}(B)$, again by definition of the inverse image. Applied once more, we obtain $x \in f^{-1}(g^{-1}(B))$ or that $x \in (f^{-1}g^{-1})(B)$, completing half of the proof.

Now let $x \in (f^{-1}g^{-1})(B)$. Then $x \in f^{-1}(g^{-1}(B))$, which means $f(x) \in g^{-1}(B)$, and this in turn implies $g(f(x)) \in B$ or that $(gf)(x) \in B$. But $(gf)(x) \in B$ means that $x \in (gf)^{-1}(B)$. Therefore, $(f^{-1}g^{-1})(B) \subset (gf)^{-1}(B)$. The definition of set equality now gives $(f^{-1}g^{-1})(B) = (gf)^{-1}(B)$.

Suppose we now consider any set X upon which has been defined an equivalence relation R. We make the following definition.

3.7. Definition: Let X be a set and R an equivalence relation on X. Then the *quotient set* of X relative to R is the set of all equivalence classes in X determined by R. The quotient set is denoted by X/R.

Our definition means that each equivalence class $R(x)$ is considered as a single point in X/R. Here, as in the case of factor groups in the study of algebra, there is a *natural map* or *projection function* from X onto X/R given by $p(x) = R(x)$ for each $x \in X$. The projection function is always surjective, but not necessarily injective. The context will always make it clear whether we are talking about the present projection function or the ones from the product into one of the factor sets. The concept of projection function may lead to involvement with the composition of functions as illustrated in the following example.

3.8. Example: Let $X = \{2, 4, 6, 7\}$ and $Y = \{a, b\}$ and consider the function $f: X \to Y$ given by $f(2) = f(4) = f(6) = a$ and $f(7) = b$. Let $R = \{(x, y) : x + y \text{ is an even integer}\}$ be an equivalence relation from X to X. Then $X/R = \{\{2, 4, 6\}, \{7\}\}$ and the projection function p from X onto X/R is given by $p(2) = p(4) = p(6) = R(2) = \{2, 4, 6\}$ and $p(7) = \{7\}$. In this particular case, we may define a function $h: X/R \to Y$ by $h(\{2, 4, 6\}) = a$ and $h(\{7\}) = b$ so as to make $hp = f$.

This fact is easily verified and a diagram of the situation is shown in Figure 11.

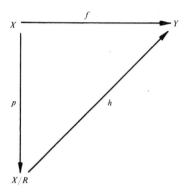

Figure 11

The fact that $hp = f$ in the above example is usually noted by saying the diagram in Figure 11 is *commutative*. Generally speaking, in any diagram of this nature, if for any two sets X and Y in the diagram and for any two paths given by arrows representing functions beginning at X and ending at Y, the resulting function compositions are equal, we say the diagram is commutative.

Exercises:

1. Prove Theorem 3.5.

2. Give an example of a function $f: X \to Y$ to show
   (a) that $(f^{-1} f)(A) \neq A$ for some set $A \subset X$.
   (b) that $(ff^{-1})(B) \neq B$ for some set $B \subset Y$.

3. Let $f: X \to Y$ be any function and $A \subset X$.
   (a) Prove that $A \subset (f^{-1} f)(A)$.
   (b) Prove that f is injective iff $(f^{-1} f)(A) = A$ for every $A \subset X$.

4. Let $f: X \to Y$ be a function and $B \subset Y$.
   (a) Prove that $(ff^{-1})(B) \subset B$.
   (b) Prove that f is surjective iff $(ff^{-1})(B) = B$ for every $B \subset Y$.

5. If $f: X \to Y$ and $g: Y \to Z$ are two functions, then prove
   (a) f is injective if gf is injective.

(b) g is surjective if gf is surjective.

(c) gf is injective if both g and f are injective.

(d) gf is surjective if both g and f are surjective.

6. Let $f: X \to Y$ be a function and define a relation on X as follows: $x_1 R x_2$ iff $f(x_1) = f(x_2)$.

(a) Prove the relation R is an equivalence relation on X.

(b) Let p be the projection function from X onto X/R and $h: X/R \to Y$ be defined as $h(R(x)) = f(x)$. Show $hp = f$.

(c) If f is surjective, prove h is surjective.

7. Let $p_1: R^1 \times R^1 \to R^1$ be the projection function onto the first coordinate. Describe geometrically the equivalence classes given by the relation of Exercise 6 as applied to p_1, the set $(R^1 \times R^1)/R$, the projection function p and the function h.

8. Let the following diagram be commutative. Prove that g is surjective and f is injective.

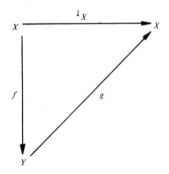

9. Let $A \subset X$ and $g: A \to Y$ a function. Define $f: X \to Y$ as the *extension* of g iff $f(x) = g(x)$ for every $x \in A$. If $i: A \to X$ is the inclusion function, prove that $g = f|A$ iff $g = fi$.

10. Let R^1 be an equivalence relation on X and R_2 an equivalence relation on Y. Then $f: X \to Y$ is called *relation-preserving* iff for $x, z \in X$ and $x R_1 z$, then $f(x) R_2 f(z)$.

(a) Give an example of a relation-preserving function.

(b) If $f: X \to Y$ is relation-preserving, prove there is one and only one function g such that the following diagram commutes (that is, $p_Y f = g p_X$):

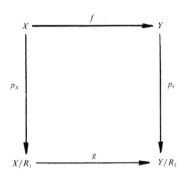

4. Finite and Infinite Sets

We can all give examples of sets with an infinite number of elements and have at least an intuitive idea of some properties that distinguish sets with an infinite number of elements from those with only a finite number of elements. In most cases, our intuition is a reliable enough guide on this matter, even though mathematical rigor is lacking. Even so, it is desirable to have a definition to help us decide not only which sets are finite and which are infinite, but to use in proving some useful properties of both types of sets. The properties we have in mind are again ones our intuition would probably tell us are true. Our proposed definition involves the use of functions.

4.1. Definition: A set X is *infinite* iff there exists an injective function $f: X \to X$ such that $f(X)$ is a proper subset of X. A set is *finite* iff it is not infinite.

From this definition it follows that every set is either finite or infinite, but not both. Even though this is the case, the description of a given set may make it extremely difficult to actually exhibit the injective function called for or to show that none exists. Since the empty set and a set consisting of a single element have no proper subsets, both must be finite sets.

4.2. Example: The set $X = \{0, 1, 2\}$ is finite. The reason is that any injective $f: X \to X$ must have $f(0) \neq f(1)$, $f(0) \neq f(2)$ and $f(1) \neq f(2)$ so that $\{f(0), f(1), f(2)\} = X$. Therefore, it is impossible to find an injective function $f: X \to X$ where $f(X)$ is a proper subset of X.

4.3. Example: The set N is infinite. To prove this, we must exhibit an injective function $f: N \to N$ where $f(N)$ is a proper subset of N. There are many ways of doing this. One such way is to define $f(n) = 2n$ for each $n \in N$, while another is to take $f(n) = n + 1$ for each $n \in N$.

We spoke a moment ago about our intuition leading us to certain properties of finite and infinite sets. Any list of such properties would be almost certain to contain those given in the next two theorems. We prove these properties by use of Definition 4.1.

4.4. THEOREM: *If $X \subset Y$ and Y is finite, then X is finite.*

Proof: If $X = \phi$, then X is finite by the remark following Definition 4.1. On the other hand, if $X \neq \phi$, let us assume X is infinite. Then there exists an injective function $f: X \to X$ such that $f(X) \neq \phi$ and $f(X) \neq X$. Now define a function $g: Y \to Y$ as follows:

$$g(x) = f(x) \text{ if } x \in X$$
$$g(x) = x \text{ if } x \in Y \backslash X.$$

The function g is injective and, since $X \subset Y$, $\phi \neq f(X) = g(X) \subset g(Y)$, from which it follows that $g(Y) \neq \phi$. Furthermore, $g(Y) = g(X) \cup g(Y \backslash X) = f(X) \cup Y \backslash X$, which means $g(Y) \neq Y$ because $f(X) \neq X$. We conclude Y is an infinite set according to Definition 4.1. But this contradicts the hypothesis that Y is finite. Therefore, the assumption that X is infinite is false and our conclusion follows.

4.5. THEOREM: *If X is an infinite set and $a \in X$, then $X \backslash \{a\}$ is infinite.*

Proof: By definition of an infinite set, there exists an injection $f: X \to X$ such that $f(X)$ is a proper subset of X. There are now two possible cases for the point a: (1) $a \in X \backslash f(X)$ or (2) $a \in f(X)$. In either case, we must exhibit an injective function $g: X \backslash \{a\} \to X \backslash \{a\}$ such that $g(X \backslash \{a\})$ is a proper subset of $X \backslash \{a\}$. Case (1) is considered first. Define $g: X \backslash \{a\} \to X$ by $g(x) = f(x)$ for all $x \in X \backslash \{a\}$. By its definition, $g = f|(X \backslash \{a\})$, from which it follows that g is injective because f was injective. Also, since $a \notin f(X)$, the element $f(a) \in f(X)$ belongs to $X \backslash \{a\}$ but $f(a) \notin g(X \backslash \{a\})$. Therefore, we have an injection $g: X \backslash \{a\} \to X \backslash \{a\}$

such that $g(X\backslash\{a\})$ is not all of $X\backslash\{a\}$. Finally, $X\backslash\{a\} \neq \phi$ because the assumption $X\backslash\{a\} = \phi$ implies either $X = \phi$ or $X = \{a\}$, which are both finite. Therefore, $g(X\backslash\{a\}) \neq \phi$. This concludes the proof that $X\backslash\{a\}$ is infinite under the conditions of case (1).

In case (2), $a \in f(X)$ so that under the injection f there is a single point $b \in X$ such that $f(b) = a$. If we now select any point c in the nonempty set $X\backslash f(X)$, a function $h: X \to X$ may be defined as follows:

$$h(x) = f(x) \text{ for all } x \in X \text{ where } x \neq b$$
$$h(b) = c.$$

Since $h(b) = c \notin f(X)$ and h agrees with f elsewhere on X, h is injective. Also, $a \in X\backslash h(X)$ which implies $h(X)$ is a proper subset of X. Therefore, we have the same situation as in case (1) where h now plays the role of f. It follows that $X\backslash\{a\}$ is infinite.

There is a rather easy way of telling whether a subset of N is finite or not. This is probably not surprising since N itself has many special properties. We shall state and prove our condition, which is again intuitively plausible, after a definition and supporting theorem.

4.6. Definition: For each $n \in N$, let us define $I_n = \{k \in N : k \leq n\}$.

4.7. THEOREM: For each $n \in N$, I_n is finite.

Proof: Our proof is by mathematical induction. We know $I_1 = \{1\}$ is finite. Now let k be any natural number such that I_k is finite and consider the set I_{k+1}. If I_{k+1} were infinite, then $I_{k+1}\backslash\{k + 1\} = I_k$ would be infinite by Theorem 4.5, which contradicts the inductive hypothesis. Therefore, if I_k is finite, I_{k+1} is finite, and by the Principle of Mathematical Induction, I_n is finite for each $n \in N$.

4.8. THEOREM: *A subset A of N is finite iff there exists a natural number m such that $A \subset I_m$.*

Proof: First, suppose there exists a natural number m such that $A \subset I_m$. Since I_m is finite by Theorem 4.7, A is finite by Theorem 4.4. Now let A be finite and we shall show there exists a natural number m such that $A \subset I_m$ by the indirect method. Therefore, let us assume that for each $n \in N$ there is an $a \in A$ such that $a > n$. Knowing this, we may now define a function $f: N \to A$, using the fact that N is well ordered, as follows:

$$f(1) = \text{smallest natural number in } A$$

$$f(n + 1) = \text{smallest natural number in } A \backslash \bigcup_{k=1}^{n} f(k).$$

We may make such a definition because $A \backslash \bigcup_{k=1}^{n} f(k) \neq \phi$ for all $n \in N$.

For if $A \backslash \bigcup_{k=1}^{n_0} f(k) = \phi$ for some $n_0 \in N$, then $a \leq f(n_0)$ for every $a \in A$, this being contrary to our assumption. The function f is surjective since $a \in A$ is the smallest natural number in $N \backslash \{k \in N : k < a\}$ and is, therefore, the image of some $n \in N$ under f. From its definition, f is also injective. If we define $g : N \to N$ by $g(n) = n + 1$, then a function $h : A \to A$ may be defined as follows:

$$h(a) = f(g(f^{-1}(a))) \text{ for each } a \in A.$$

These facts are observed about h: The function h is injective because f and g are injective; since $h(2) \in h(A)$, $h(A) \neq \phi$; the smallest element of A does not belong to $h(A)$ so that $h(A) \neq A$. Consequently, A is an infinite set and this contradicts our hypothesis. This means our assumption was false and there does exist a natural number m such that $A \subset I_m$.

Actually, there is a close association between any given finite set which is nonempty and a subset I_m of N for some $m \in N$. The association we have in mind is the fact that $X \neq \phi$ is finite iff there exists an $m \in N$ and a bijective function $f : X \to I_m$. The remaining theorems in this section are ones which point toward the proof of this fact, while the exercises complete the job. Once the proof is complete, the elements of any nonempty finite set may be indexed with a finite set of consecutive natural numbers. In reality, we have already been doing this due to the great intuitive appeal of this fact.

Perhaps we should point out that a different approach to finite and infinite sets can be taken rather than our particular route. We can define a set X to be finite iff it is empty or there exists an $m \in N$

and a bijective function $f: X \to I_m$. An infinite set can then be defined as one which is not finite. From this approach we can prove that a set X is infinite iff there exists an injective function $f: X \to X$ where $f(X)$ is a proper subset of X. The amount of work in doing this, however, is roughly equivalent to that in our present treatment.

4.9. THEOREM: *If $f: X \to Y$ is injective and X is infinite, then Y is infinite.*

Proof: First show the subset $f(X)$ of Y is infinite, and then use the logically equivalent contrapositive of Theorem 4.4: If $X \subset Y$ and X is infinite, then Y is infinite. The details of showing $f(X)$ infinite are left as an exercise.

The proof of the next theorem is based on a fundamental axiom that states that a point may be selected or chosen from a nonempty subset of a given set. This axiom is called the *Axiom of Choice* and has many equivalent forms. Because of the importance of this axiom here, as well as in mathematics in general, we state it now in a form compatible with our present subject matter.

4.10. The Axiom of Choice: Let X be a set and $\mathscr{A}$ a nonempty collection of nonempty subsets of X. Then there exists a function $g: \mathscr{A} \to X$ such that $g(A) \in A$ for each $A \in \mathscr{A}$.

4.11. THEOREM: *If $f: X \to Y$ is surjective and X is finite, then Y is finite.*

Proof: If $X = \phi$, $f: X \to Y$ surjective implies Y must also be empty from which it follows Y is finite. For the remainder of the proof we consider $X \neq \phi$. If we let $\mathscr{A}$ be the collection of all nonempty subsets of X, the Axiom of Choice states there is a function $g: \mathscr{A} \to X$ such that $g(A) \in A$ for every $A \in \mathscr{A}$. We will use the function g after recalling f is surjective so that $f^{-1}(y) \neq \phi$ for each $y \in Y$ and observing that for any two points $y_1 \neq y_2$ in Y, $f^{-1}(y_1) \cap f^{-1}(y_2) = \phi$. For each $y \in Y$ we use g to select a point from the nonempty set $f^{-1}(y)$ and thereby define a function $h: Y \to X$ as follows:

$$h(y) = g(f^{-1}(y)) \text{ for each } y \in Y.$$

The function h is injective. To see this, let $y_1 \neq y_2$ be points in Y. Then $f^{-1}(y_1) \cap f^{-1}(y_2) = \phi$ and, therefore, $h(y_1) = g(f^{-1}(y_1)) \neq g(f^{-1}(y_2)) = h(y_2)$. Now if we assume Y is infinite, Theorem 4.9 gives X infinite, contrary to our hypothesis. It follows that Y is finite as asserted.

For a slightly different perspective on Theorem 4.11, we see that any function $f: X \to Y$ has the property that $f: X \to f(X)$ is surjective. Therefore, if X finite, $f(X) \subset Y$ is finite.

The final theorem that we prove in this section also uses the Axiom of Choice.

4.12. THEOREM: *If X is finite, there exists an injective function $f: X \to N$.*

Proof: If $X = \phi$, then $\phi: \phi \to N$ is injective. For $X \neq \phi$, the proof is accomplished by showing that if no injective $f: X \to N$ exists, then X is infinite. Since $X \neq \phi$, we first use the Axiom of Choice to obtain a function $g: \mathscr{A} \to X$ such that $g(A) \in A$ for each A in the collection $\mathscr{A}$ of all nonempty subsets of X. We then define a function $f: X \to N$ inductively as follows:

Let $g(X) = x_1$ and define $f(x_1) = 1$.

Let $g(X \setminus \bigcup_{k=1}^{n-1} x_k) = x_n$ and define $f(x_n) = n$ for each $n \in N$.

For each $x \in X \setminus \bigcup_{n \in N} x_n$, define $f(x) = 1$.

For each $n \in N$, the set $X \setminus \bigcup_{k=1}^{n} x_k \neq \phi$, or else $f: X \to N$ would be injective, contrary to our assumption. This means that f is indeed a well-defined function. Now define a function $h: X \to X$ given by

$$h(x_n) = x_{n+1} \text{ for each } n \in N.$$
$$h(x) = x \text{ for each } x \in X \setminus \bigcup_{n \in N} x_n.$$

The function h is injective, $h(X) \neq \phi$, and since $x_1 \notin h(X)$, $h(X) \neq X$. We conclude X is infinite.

Exercises:

1. Prove R^1 is infinite.

2. Prove Theorem 4.9.

3. (a) If $\{X_\alpha : \alpha \in \Delta\}$ is an indexed family of sets, each of which is infinite, prove $\bigcup_{\alpha \in \Delta} X_\alpha$ is infinite.

   (b) If $\{X_\alpha : \alpha \in \Delta\}$ is an indexed family of sets, each of which is finite, prove $\bigcap_{\alpha \in \Delta} X_\alpha$ is finite.

4. Prove that if there exists an injective $f : N \to X$, then X is infinite.

5. Prove that if there exists an $m \in N$ and a bijective function $f : X \to I_m$, then X is finite.

6. Prove that a nonempty subset A of N is finite iff there exists an $m \in N$ and a bijective function $f : X \to I_m$.

7. If $X \neq \phi$ is finite, prove there exists an $m \in N$ and a bijective function $f : X \to I_m$.

8. (a) If X and Y are both finite, prove $X \cup Y$ is finite.
   (b) Use part (a) and mathematical induction to prove that a finite union of finite sets is a finite set.

9. If X is infinite and $X = A \cup B$, prove that A or B must be infinite.

10. Prove that if X is infinite and $A \subset X$ is finite, then $X \backslash A$ is infinite.

5. Countable Sets

For a final application of functions in this chapter, we will define and study the concept of a *countable set*. As the term "countable" suggests, we want to include in our definition all nonempty finite sets since there exists a bijective function from a subset I_n of N, for some $n \in N$, onto the set in question. In other words, we can actually count the elements of such sets in the usual sense of our counting process. However, in keeping with the usual practice in mathematics, we want to broaden our outlook somewhat so as to also call countable any set whose elements can be indexed with N. In so doing, there are certain infinite sets which are to be classified as countable along with the finite ones. Since the process of indexing the elements of a given set with N is precisely the process of defining a surjective function from N onto the set, functions are still the unifying concept in thinking about countable sets. In view of this, and our knowledge about finite sets, the definition of a countable set may now be made precise.

5.1. Definition: The set X is *countable* iff $X = \phi$ or there exists a surjective function $f: N \rightarrow X$. If X is not countable, then X is called *uncountable*.

In terms of an indexing set, a nonempty countable set is one whose elements may be indexed with N, whereas for an uncountable set there is no such indexing possible. To justify making such a statement, recall that when we first described an indexed family of sets we did not demand that each set in the family be indexed with only one element from the indexing set. Some examples will now illustrate our definition.

5.2. Example: The set $X = \{0, 1, 2\}$ is countable. Define $f(1) = 0$, $f(2) = 1$ and for $n \geq 3$, $f(n) = 2$. Then according to Definition 5.1, X is countable.

5.3. Example: The set $X = \{0, 1, 2, \ldots\}$ is countable. We exhibit one of the many surjective functions $f: N \rightarrow X$ as follows: $f(n) = n - 1$ for each $n \in N$.

The next theorem shows a familiar subset of R^1 to be uncountable.

5.4. THEOREM: *The set* $X = \{x \in R^1 : 0 < x < 1\}$ *is uncountable.*

Proof: The proof is of the contradictory type. Assume there is a surjective function $f: N \rightarrow X$. Upon this assumption we shall show there is an element of X which is not in $f(N)$. For each natural number $n \in N$, $f(n)$ is a number between 0 and 1 and, therefore, has a decimal representation denoted by $. x_{n1} x_{n2} x_{n3} \ldots$. The function f is illustrated as follows where each x_{ij} is an integer between 0 and 9 inclusive.

$$f(1) = . x_{11} x_{12} x_{13} \ldots$$
$$f(2) = . x_{21} x_{22} x_{23} \ldots$$
$$f(3) = . x_{31} x_{32} x_{33} \ldots$$
$$\ldots$$

Let us consider the number between 0 and 1 described as follows: The number will be denoted by $. y_1 y_2 y_3 \ldots$ where $y_1 = 7$ if $x_{11} \neq 7$ and $y_1 = 3$ if $x_{11} = 7$; $y_2 = 7$ if $x_{22} \neq 7$ and $y_2 = 3$ if $x_{22} = 7$; etc. Then the number $y = . y_1 y_2 y_3 \ldots$ is such that $0 < y < 1$ does not have two decimal representations because for each $n \in N$, y_n is never zero or nine, and furthermore, y is different from $f(1)$ in the first decimal position, from $f(2)$ in the second decimal position and from $f(n)$ in the nth decimal position for each $n \in N$. It follows that y is not in $\{f(n): n \in N\}$. This implies that f is not surjective and contradicts our assumption that f is surjective. Consequently, there is no surjective function $f: N \rightarrow X$ which means that X is uncountable.

Frequent use will be made of the following result in establishing the countability or uncountability of sets.

5.5. THEOREM: *Let $X \subset Y$. If Y is countable, then X is countable.*

Proof: If $X = \phi$, then X is countable by definition. If $X \neq \phi$, we need to exhibit a surjective function $g : N \rightarrow X$. Since Y is countable by hypothesis, we know that there exists a surjective function $f: N \rightarrow Y$. Now our plan is to define a surjective function $h : Y \rightarrow X$, and then let $g = hf$ so as to have $g : N \rightarrow X$ a surjection. To this end, define $h : Y \rightarrow X$ as follows:

If $x \in X$, let $h(x) = x$

If $x \in Y \backslash X$, let $h(x) = a \in X$ where a is a fixed element in X.

There is such a point $a \in X$ because $X \neq \phi$. As defined, $h : Y \rightarrow X$ is surjective. If we form the composition $hf = g$ and use the fact that $f(N) = Y$ and $h(Y) = X$, it follows that $g(N) = h(f(N)) = h(Y) = X$. This shows g is surjective and implies that X is countable.

The contrapositive of Theorem 5.5 is worth emphasizing: Let $X \subset$

Y. If X is uncountable, then Y is uncountable. This fact will help us decide the uncountability of certain sets as is illustrated next.

5.6. THEOREM: *The set R^1 is uncountable.*

Proof: Theorem 5.4 and the contrapositive of Theorem 5.5.

If an infinite set X were under consideration, it would probably be no surprise that X contains an infinite subset that is also countable. This is the essence of the next theorem. The proof of this fact involves selecting a set of points from X indexed by the natural numbers and, therefore, relies on the Axiom of Choice. The heart of the proof is given while the details are an easy test of the use of the Axiom of Choice.

5.7. THEOREM: *Let X be any infinite set. There exists an injective function $f: N \to X$. That is, every infinite set contains a countable infinite subset.*

Proof: Let $\mathscr{A}$ be the collection of all nonempty subsets of X and $g: \mathscr{A} \to X$ the function given by the Axiom of Choice. Now define $f: N \to X$ by

$$f(1) = g(X) \text{ and}$$

$$f(n+1) = g(X \setminus \bigcup_{k=1}^{n} f(k)) \text{ for each } n \in N.$$

The remaining details may be supplied by the reader.

A natural question to raise is whether the product of two countable sets is countable. Theorem 5.8 shows the answer to be yes for the special case of $N \times N$, while the exercises give an affirmative answer in general. To help us prove this general assertion, we may make use of Theorem 5.9, which in turn makes use of Theorem 5.8.

5.8. THEOREM: *The set $N \times N$ is countable.*

Proof: Consider the subset $K = \{2^m 3^n \in N : (m, n) \in N \times N\}$ of N. Since N is countable and $K \subset N$, then K is countable by Theorem 5.5 and, therefore, there exists a surjective function $h: N \to K$. Now define a function $f: K \to N \times N$ given by $f(2^m 2^n) = (m, n)$. The function f is evidently surjective, and if we then form the composition

$fh : N \to N \times N$, we have a surjective function from N onto the set $N \times N$ in question. This gives the conclusion that $N \times N$ is countable.

5.9. THEOREM: *The union of a countable collection of countable sets is countable. That is, if A_n, $n \in N$, is a countable set, then $\bigcup_{n \in N} A_n$ is a countable set.*

Proof: If the union is empty, it is countable by definition. If the union is not empty, at least one A_n, $n \in N$, is a nonempty set. For each nonempty A_n in the countable collection $\{A_n : n \in N\}$, let us set $A_n = B_n$. Then the subcollection $\{B_n\}$ of nonempty sets is also countable by Theorem 5.5 and may, therefore, be indexed by the natural numbers. It follows that $\bigcup_{n \in N} A_n = \bigcup_{n \in N} B_n$. Now, for each $n \in N$, there exists a surjective function $f_n : N \to B_n$ because B_n is a countable set. With this information, we can complete the proof by defining a function $g : N \times N \to \bigcup_{n \in N} B_n$ as follows:

$$g(m, n) = f_m(n) \text{ for each } (m, n) \in N \times N.$$

Since each element $b \in \bigcup_{n \in N} B_n$ belongs to B_m for at least one $m \in N$, it follows that $b \in f_m(N)$, due to the fact that $f_m : N \to B_m$ is surjective. This means $b \in g (N \times N)$, or that g is surjective. At this point, we use Theorem 5.8 to give the existence of a surjective function $h : N \to N \times N$ and then to form the composition $gh : N \to \bigcup_{n \in N} B_n$. This composition is surjective because both g and h are surjective and this in turn implies $\bigcup_{n \in N} B_n = \bigcup_{n \in N} A_n$ is a countable set.

We also make use of the last theorem in our concluding remarks about countable sets. This will be to show the rational numbers are countable, while the irrational numbers are uncountable. These facts will be quite useful to us in the future when studying the topology of R^1.

5.10. THEOREM: *The positive rational numbers are countable.*

Proof: Let us begin by forming the following sets of rational numbers:

$$A_1 = \left\{ \frac{1}{1}, \frac{2}{1}, \frac{3}{1}, \frac{4}{1}, \ldots \right\}$$

$$A_2 = \left\{ \frac{1}{2}, \frac{2}{2}, \frac{3}{2}, \frac{4}{2}, \ldots \right\}$$

$$A_3 = \left\{ \frac{1}{3}, \frac{2}{3}, \frac{3}{3}, \frac{4}{3}, \ldots \right\}$$

. . .

$$A_n = \left\{ \frac{1}{n}, \frac{2}{n}, \frac{3}{n}, \frac{4}{n}, \ldots \right\}$$

We next observe that for each $n \in N$, A_n is countable because the function $f_n : N \to A_n$ given by $f_n(m) = m/n$ is surjective. Theorem 5.9 then tells us that $\bigcup_{n \in N} A_n$ is a countable set. Now, if we let P represent the set of positive rationals and show $P = \bigcup_{n \in N} A_n$, our proof will be complete. To see this set equality we use the fact that each positive rational number has the form m/n, where $m, n \in N$, so that set inclusion easily follows both ways to give $P = \bigcup_{n \in N} A_n$.

5.11. THEOREM: *The rational numbers are countable.*

Proof: A bijection $f : P \to T$ from the positive rationals P onto the negative rationals, denoted by T, may be defined by $f(m/n) = -(m/n)$ where m and n belong to N. Since P is countable by Theorem 5.10, a surjection $g : N \to P$ exists so that $fg : N \to T$ is a surjection which shows T is countable. The set consisting of the single rational number zero is also countable, and the entire set of rationals may be expressed as $Q = P \cup T \cup \{0\}$. Thus, we have Q expressed as the union of a countable number (three, to be exact!) of sets, each of which is countable, so that by Theorem 5.9, Q is countable.

5.12. THEOREM: *The irrational numbers are uncountable.*

Proof: Denote the set of all irrationals by A. Then $R^1 = Q \cup A$ where Q represents the rationals. If we suppose that A is countable, then Theorem 5.11 together with Theorem 5.9 gives R^1 countable. But this contradicts Theorem 5.6 and, therefore, the irrationals must be uncountable.

Exercises:

1. Prove that the set of all integers is countable.

2. (a) Give an example of a countable collection of sets, each of which is finite, but such that their union is infinite.
   (b) Does there exist a countable infinite collection of countable subsets of R^1 such that each pair of these sets are disjoint?

3. Prove Theorem 5.7.

4. Prove that the union of two uncountable sets is uncountable.

5. Prove that if both X and Y are countable, then $X \times Y$ is countable.

6. Prove that the set of all points in R^2 having both coordinates rational is countable.

7. Prove that the collection of all circles in R^2 having rational radii and centers at points having both coordinates rational is countable.

8. Prove that if X is countable and $f: X \to Y$ is surjective, then Y is countable.

9. (a) Prove that every finite set is countable.
   (b) Prove that every uncountable set is infinite.

10. Prove the converse to Theorem 5.7. That is, if there exists an injection $f: N \to X$, then X is infinite.

3

Topological Spaces

1. Defining a Topology

We are all familiar with investigating properties of an algebraic structure induced by one or more binary operations on a set X. To name but a few, these properties include commutativity, associativity, identity elements, etc., which constitute some of the basic concepts in beginning studies of algebra. From this groundwork, a host of other ideas are introduced and included in the subject of algebra. The beginning concepts of topology come about in a similar manner. In this case, however, we are going to induce a *structure* on a set X by use of a collection of subsets of X. For this, we shall soon set forth a set of axioms which a collection of subsets must obey in order to fall within the realm of our studies. Any collection of subsets of X satisfying these axioms will be called a *topology* of X. After studying these axioms, you will find that there are usually many different topologies for a given set and that the fundamental questions to be asked about a topology are set theoretical in nature. For instance, among some of the most basic properties to be investigated are the boundary of a set, the interior of a set, the exterior of a set, etc. Again, a diversity of ideas flow from the basic ones and constitute the subject of topology. As we shall see, a given set $A \subset X$ may have its boundary change, for example, if the topology or structure on X is changed. From prior experience with subsets of the real line or the plane, you may think that this seems strange indeed! This comes about from the fact that many of the examples and objects studied in topology are ones with which we have had geometric experience in a Euclidean setting where distance and shape are of prime importance. As our work unfolds, we will begin to see how topology is sometimes loosely described as a geometry of configurations that are independent of size, shape, and location. In our present work, however, try to think in terms of the given definitions only when studying examples. In all cases, these definitions conform to our usual Euclidean geometry ideas if the proper topology is defined on the set.

1.1. Definition: Let $X \neq \phi$ be a set. Then a *topology* on X is a subset $\mathcal{T}$ of $\mathcal{P}(X)$ obeying the following axioms:
  (a) X and ϕ belong to $\mathcal{T}$.
  (b) If U_1 and U_2 belong to $\mathcal{T}$, then $U_1 \cap U_2$ belongs to $\mathcal{T}$.
  (c) If $\{U_\alpha : \alpha \in \Delta\}$ is an indexed family of sets, each of which belongs to $\mathcal{T}$, then $\bigcup_{\alpha \in \Delta} U_\alpha$ belongs to $\mathcal{T}$.

Notice that axiom (b), along with the Principle of Mathematical Induction, actually tells us that any *finite* intersection of elements from $\mathcal{T}$ is again an element of $\mathcal{T}$. We could then say that a topology for a set X is a collection $\mathcal{T}$ of subsets of X, including ϕ and X, such that finite intersections and arbitrary unions of elements of $\mathcal{T}$ again are elements of $\mathcal{T}$.

For any nonempty set X a topology may be defined as $\mathcal{T} = \{X, \phi\}$. This topology is called the *trivial* or *indiscrete* topology for X. Another way of defining a topology on X is to let $\mathcal{T} = \mathcal{P}(X)$. This latter topology is called the *discrete* topology for X. That these collections of subsets of X actually do form topologies is a trivial exercise in verifying the axioms of Definition 1.1. It follows that every set containing more than one point always has at least two topologies.

1.2. Example: Let $X = \{a, b, c\}$. Among the several topologies on X is $\mathscr{T}_1 = \{\phi, X, \{a\}, \{a, b\}\}$. Another is $\mathscr{T}_2 = \{\phi, X, \{a\}, \{b\}, \{a, b\}\}$. That these collections of subsets satisfy the axioms of Definition 1.1 is easily verified. The collection of subsets $\{\phi, X, \{a\}, \{b\}, \{a, c\}\}$ is not a topology for X, since $\{a\} \cup \{b\}$ is not among the original elements in the collection.

1.3. Example: (The left ray topology for R^1.) Let the set in question be R^1 and for each real number a, define $L_a = \{x : x < a\}$ to be an *open left ray* of real numbers. Notice that $a \notin L_a$. The point a is called a *right end point* of the left ray L_a. Now consider the collection of subsets $\mathscr{T} = \{L_a : a \in R^1\} \cup \{R^1\} \cup \{\phi\}$. That is, $\mathscr{T}$ consists of all possible open left rays in R^1 together with R^1 and ϕ. The set $\mathscr{T} \subset \mathscr{P}(R^1)$ forms a topology for R^1. To verify this, first note that (a) of Definition 1.1 is satisfied. Next let L_a and L_b be any two of the open left rays in $\mathscr{T}$. Then $L_a \cap L_b = \{x : x < \text{minimum of } a \text{ and } b\} = L_{\min\{a,b\}}$ is also an open left ray of the type in $\mathscr{T}$. Furthermore, $L_a \cap R^1 = L_a$, $L_a \cap \phi = \phi$ and $R^1 \cap \phi = \phi$. Therefore, the intersection of any two elements of $\mathscr{T}$ is again an element of $\mathscr{T}$. Finally, consider an arbitrary union of elements in $\mathscr{T}$. If R^1 is among the elements, the union will be R^1 which is in $\mathscr{T}$, and if the union is empty, we again have an element of $\mathscr{T}$. If the union is formed with a collection $\{L_a : a \in \Delta\}$ of open left rays only, then two cases arise: (1) The union is all of R^1, which is in $\mathscr{T}$, or (2) the union is not all of R^1. In the latter case, the right end points of the open left rays $\{L_a : a \in \Delta\}$ form a set which has an upper bound and, therefore, a least upper bound which we shall call b. It is then an easy exercise to verify that $L_b = \bigcup_{a \in \Delta} L_a$ so that $\bigcup_{a \in \Delta} L_a$ belongs to $\mathscr{T}$. We have now established that $\mathscr{T}$ is a topology for R^1, which will henceforth be known as the *left ray topology* for R^1.

The open left rays used in defining the left ray topology are but one example of a large class of subsets of R^1 called *intervals*. Since topologies on R^1 provide such a rich source of examples during our studies, and intervals of one type or another are often useful in describing those topologies, the definitions of these intervals are given next.

1.4. Definition: A set of real numbers having one of the following forms is called an *open interval:*
    (a) $(a, b) = \{x : a < x < b\}$, where a and b are real numbers.
    (b) $\{x : x > a\}$, where a is a real number.
    (c) $\{x : x < a\}$, where a is a real number.
    (d) R^1.

A set of real numbers having one of the following forms is called a *closed interval:*

(a) $[a, b] = \{x : a \leq x \leq b\}$, where a and b are real numbers.
(b) $\{x : x \geq a\}$, where a is a real number.
(c) $\{x : x \leq a\}$, where a is a real number.
(d) R^1.

An *interval* of real numbers is a subset of R^1 which is either an open interval, a closed interval, has the the form $\{x : a \leq x < b\} = [a, b)$, or has the form $\{x : a < x \leq b\} = (a, b]$.

Observe that R^1 qualifies as both an open and closed interval. A closed interval of the form $[a, b]$ where $a = b$ consists of a single point, but otherwise every interval contains more than one point. As a matter of terminology, it is convenient to call any set consisting of a single point, whether it be an interval or not, a *degenerate set*. Sets having more than one point may be referred to as *nondegenerate sets*.

In mathematical studies where topologies are involved, there always seem to be certain ones which stand out as the most natural and are, therefore, perhaps the most important. The reason for this varies with the subject matter at hand, how helpful the topology is in applications of the subject matter, and whether the topology is useful as a unifying concept with other areas of mathematics. That the real numbers are important in all of mathematics goes without saying. For this set R^1, a fair assessment to make is that there is one topology which stands above all others in importance. This topology is so natural for R^1 that it will be called the *standard topology for R^1*, and when one speaks of *the* topology on R^1, this is invariably the one in mind. The reasons for this naturalness will be seen as its properties unfold. For a sketchy preview, it is this standard topology on R^1 that is used in the study of calculus and elementary analysis, and makes the boundary, interior, and exterior, to name only a few properties of subsets of R^1, coincide with what they should be from the viewpoint of Euclidean geometry. The standard topology will now be defined.

1.5. Example: (The standard topology for R^1.) Let the set in question be R^1 and let $\mathscr{T}$ be the collection of subsets of R^1 consisting of ϕ, R^1 and all sets U having the following property: For each $x_0 \in U$ there exists an open interval of the form $(a, b) = \{x : a < x < b\}$ containing x_0 such that $(a, b) \subset U$. To see that $\mathscr{T}$ is a topology for R^1, we first note that R^1 and ϕ belong to $\mathscr{T}$. Consider next the intersection of any two elements U_1 and U_2 of $\mathcal{T}$. If either of these elements is ϕ or R^1 or it happens that $U_1 \cap U_2 = \phi$, the resulting intersection belongs to $\mathcal{T}$. Otherwise, let $x_0 \in U_1 \cap U_2$. Since $x_0 \in U_1$, there exists an open interval (a, b) containing x_0 such that $(a, b) \subset U_1$. Similarly, there exists an open interval (c, d) containing x_0 such that $(c, d) \subset U_2$. The open interval (maximum $\{a, c\}$, minimum $\{b, d\}$) contains x_0 and is a subset of $U_1 \cap U_2$ and is, therefore, an open interval of the proper form to put $U_1 \cap U_2$ in $\mathcal{T}$. It follows that axiom (b) of Definition 1.1 is satisfied. Finally, let $\{U_\alpha \in \mathcal{T} : \alpha \in \Delta\}$ be a collection of elements of $\mathcal{T}$. If $U_\alpha = R^1$ for at least one $\alpha \in \Delta$, or the union is empty, then $\bigcup_{\alpha \in \Delta} U_\alpha \in \mathcal{T}$. Otherwise, let $x_0 \in \bigcup_{\alpha \in \Delta} U_\alpha$. Then $x_0 \in U_\alpha$ for some $\alpha \in \Delta$, and there exists an open interval (a, b) containing x_0 such that $(a, b) \subset U_\alpha$. Consequently, $(a, b) \subset \bigcup_{\alpha \in \Delta} U_\alpha$, implying the union belongs to $\mathcal{T}$. Thus, $\mathcal{T}$ is a topology for R^1 and henceforth will be called the *standard topology for* R^1. (The standard topology is also known as the *Euclidean* topology for R^1.)

Some examples of sets belonging to the standard topology for R^1 are as follows: $\{x : 0 < x < 1\}$, $\{x : 1 < x < 3\} \cup \{x : 5 < x\}$, $\{x : x < 0\}$, and $R^1 \backslash \{0\}$. Some sets which do not belong to the standard topology for R^1 are $\{0\}$, $\{x : 1 \leq x < 3\}$, $\{x : x \geq 5\}$ and N.

To see some connection between intervals of reals and the standard topology for R^1 we have the following result.

1.6. THEOREM: *Every open interval of real numbers belongs to the standard topology for* R^1.

Proof: The easy proof is accomplished by showing that each of the four types of open intervals belongs to the standard topology.

We have already encountered the standard topology for R^1 many times in calculus, whether explicitly mentioned or not. The study of continuous functions and convergence of sequences name but two of such instances. There, open intervals of the form $a - \epsilon < x < a + \epsilon$, where ϵ is a positive real number, were of key importance. While it is not true that every element of the standard topology is an open interval, we will show in Section 3 of this chapter that every nonempty subset of R^1 belonging to the standard topology is the union of a countable collection of disjoint open intervals of reals. This important relationship, along with Theorem 1.6, means that open intervals can be used to fully describe the subsets of R^1 which make up the standard topology. Thus, the word "open" is a rather natural adjective to carry forth in describing the elements of a topology. With this in mind, we shall henceforth call elements of a topology on *any set X, open subsets of X.* This will be done whether or not the elements of the topology are actually "open" in any intuitive sense. For instance, in Example 1.2 each element of $\mathscr{T}_1$, namely, $\phi, X, \{a\}$ and $\{a, b\}$, would be called an open subset of X. In view of this terminology, to define a topology for a set X is to describe a collection of *open sets*, each a member of $\mathscr{P}(X)$, such that the collection obeys the axioms of Definition 1.1. For future notation and terminology, we have the following definition.

1.7. Definition: A *topological space* is a set X together with a topology $\mathscr{T}$ on X. The notation $(X, \mathscr{T})$ will often be used for a topological space, but the shortened notation *the space X* will also be used when no confusion arises concerning the topology on X. When referring to a topological space $(X, \mathscr{T})$, we shall always assume $X \neq \phi$. When R^1 is the set, the standard topology will *always* be assumed, unless otherwise explicitly stated.

To help build a collection of spaces for use in illustrating various topological concepts, the following two examples are offered.

1.8. Example: (The cofinite topology.) Let X be a nonempty set. Define $\mathscr{T}$ as the collection of subsets of X consisting of ϕ, X and all sets U such that $X\backslash U$ is finite. We shall show only that if U_1 and U_2 belong to $\mathscr{T}$, then $U_1 \cap U_2$ also belongs to $\mathscr{T}$, while the reader may complete the verification that $\mathscr{T}$ is a topology. According to the definition of $\mathscr{T}$, if $U_1 \in \mathscr{T}$, then $U_1 = \phi, U_1 = X$ or $X\backslash U_1$ is a finite set. The same holds true for $U_2 \in \mathscr{T}$. Thus, if either U_1 or U_2 is empty, $U_1 \cap U_2 = \phi \in \mathscr{T}$. If $U_1 = X$, then $U_1 \cap U_2 = U_2 \in \mathscr{T}$, and if $U_2 = X$, then $U_1 \cap U_2 = U_1 \in \mathscr{T}$. Finally, consider the remaining case where neither U_1 nor U_2 is the empty set and neither is X. Then $X\backslash(U_1 \cap U_2) = X\backslash U_1 \cup X\backslash U_2$ by Demorgan's Theorem, and both $X\backslash U_1$ and $X\backslash U_2$ are finite sets. Therefore, their union is finite which makes $X\backslash(U_1 \cap U_2)$ finite, which is precisely what is needed to insure $U_1 \cap U_2 \in \mathscr{T}$. We have now completed the verification that if U_1 and U_2 belong to $\mathscr{T}$, then $U_1 \cap U_2 \in \mathscr{T}$. The topology $\mathscr{T}$ just described will henceforth be known as the *cofinite topology*.

1.9. Example: (The co-countable topology.) Let X be a nonempty set. Define $\mathcal{T}$ as the collection of subsets of X consisting of ϕ, X and all sets U such that $X \backslash U$ is countable. The reader may supply the verification that $\mathcal{T}$ is a topology which will henceforth be known as the *co-countable topology*.

If we consider a given set X and two topologies $\mathcal{T}_1$ and $\mathcal{T}_2$ on X, it may happen that some of the members of $\mathcal{T}_1$ besides ϕ and X also belong to $\mathcal{T}_2$. Such a situation leads to a way of comparing the topologies. If, for instance, each member of $\mathcal{T}_1$ also belongs to $\mathcal{T}_2$, then we would write $\mathcal{T}_1 \subset \mathcal{T}_2$ to describe the subset relationship between them, since $\mathcal{T}_1$ and $\mathcal{T}_2$ are themselves subsets of $\mathcal{P}(X)$. In this case, $\mathcal{T}_2$ has at least as many elements as $\mathcal{T}_1$ which allows us to say $\mathcal{T}_2$ is "larger than" $\mathcal{T}_1$ or that $\mathcal{T}_1$ is "smaller than" $\mathcal{T}_2$ if we think of comparing the number of elements in the two topologies. Under these same circumstances, some would prefer to say $\mathcal{T}_2$ is "finer than" $\mathcal{T}_1$ and that $\mathcal{T}_1$ is "coarser than" $\mathcal{T}_2$. Of course, if $\mathcal{T}_1 \subset \mathcal{T}_2$ and $\mathcal{T}_2 \subset \mathcal{T}_1$, then the two topologies are equal and we would write $\mathcal{T}_1 = \mathcal{T}_2$. There is also the distinct possibility that $\mathcal{T}_1$ is not a subset of $\mathcal{T}_2$ nor $\mathcal{T}_2$ a subset of $\mathcal{T}_1$. In this event, we simply say that $\mathcal{T}_1$ and $\mathcal{T}_2$ are *not comparable*. The future importance of having terminology such as described in this paragraph prompts us to set forth our thoughts in a formal definition.

1.10. Definition: Let $\mathcal{T}_1$ and $\mathcal{T}_2$ be two topologies for X. If $\mathcal{T}_1 \subset \mathcal{T}_2$, i.e., if each member of $\mathcal{T}_1$ is also a member of $\mathcal{T}_2$, then $\mathcal{T}_1$ is said to be smaller than $\mathcal{T}_2$, and $\mathcal{T}_2$ is larger than $\mathcal{T}_1$. If neither $\mathcal{T}_1 \subset \mathcal{T}_2$ nor $\mathcal{T}_2 \subset \mathcal{T}_1$ is true, then $\mathcal{T}_1$ and $\mathcal{T}_2$ are said to be *not comparable*.

For R^1, the left ray topology is smaller than the standard topology, and the standard topology is larger than the left ray topology. In Example 1.2, the first of the two given topologies is smaller than the second. The topology $\mathcal{T} = \{\phi, X, \{c\}\}$ on $X = \{a, b, c\}$ is not comparable with either of the two given in Example 1.2 since it is not a subset of, nor does it contain, either of the topologies. For any topology $\mathcal{T}$ on X, it always follows that the trivial topology on X is smaller than $\mathcal{T}$, which in turn is smaller than the discrete topology on X. That is, the trivial topology is the smallest topology possible on X while the discrete is the largest topology that may be put on X.

By the use of the next theorem, we shall see that a collection of topologies on a set X leads to a topology that is smaller than all of the original ones in a rather natural way, yet is the largest topology contained in all of the original ones.

1.11. THEOREM: *Let X be a nonempty set and $\{\mathcal{T}_\alpha : \alpha \in \Delta\}$ be an indexed family of topologies each defined on X. Then $\bigcap_{\alpha \in \Delta} \mathcal{T}_\alpha$ is a topology for X. Furthermore, $\bigcap_{\alpha \in \Delta} \mathcal{T}_\alpha$ is the largest topology on X that is contained in $\mathcal{T}_\alpha$ for each $\alpha \in \Delta$.*

Proof: The empty set and X belong to every $\mathcal{T}_\alpha$ and, therefore, to $\bigcap_{\alpha \in \Delta} \mathcal{T}_\alpha$. Now let U_1 and U_2 belong to $\bigcap_{\alpha \in \Delta} \mathcal{T}_\alpha$. Then U_1 and U_2 each belong to $\mathcal{T}_\alpha$ for every $\alpha \in \Delta$, so that $U_1 \cap U_2$ belongs to $\mathcal{T}_\alpha$ for every $\alpha \in \Delta$, since each $\mathcal{T}_\alpha$ is a topology. Thus $U_1 \cap U_2 \in \bigcap_{\alpha \in \Delta} \mathcal{T}_\alpha$. Finally, let $\{U_\beta : \beta \in \Omega\}$ be a family of sets where each $U_\beta \in \bigcap_{\alpha \in \Delta} \mathcal{T}_\alpha$. Then, since $\mathcal{T}_\alpha$, $\alpha \in \Delta$, is a topology for X, $\bigcup_{\beta \in \Omega} U_\beta$ belongs to $\mathcal{T}_\alpha$ for every $\alpha \in \Delta$ so that $\bigcup_{\beta \in \Omega} U_\beta \in \bigcap_{\alpha \in \Delta} \mathcal{T}_\alpha$. The conclusion is that $\bigcap_{\alpha \in \Delta} \mathcal{T}_\alpha$ is a topology for X.

Since $\bigcap_{\alpha \in \Delta} \mathcal{T}_\alpha \subset \mathcal{T}_\alpha$ for every $\alpha \in \Delta$, $\bigcap_{\alpha \in \Delta} \mathcal{T}_\alpha$ is smaller than all of the original topologies on X. Now, if we let $\mathcal{T}$ be any topology such that $\bigcap_{\alpha \in \Delta} \mathcal{T}_\alpha \subset \mathcal{T}$ and $\mathcal{T} \subset \mathcal{T}_\alpha$ for every $\alpha \in \Delta$, it follows that each member of $\mathcal{T}$ belongs to $\mathcal{T}_\alpha$ for every $\alpha \in \Delta$, which implies each member of $\mathcal{T}$ belongs to $\bigcap_{\alpha \in \Delta} \mathcal{T}_\alpha$. Therefore, $\mathcal{T} \subset \bigcap_{\alpha \in \Delta} \mathcal{T}_\alpha$ so that $\mathcal{T} = \bigcap_{\alpha \in \Delta} \mathcal{T}_\alpha$. In other words, there is no topology larger than $\bigcap_{\alpha \in \Delta} \mathcal{T}_\alpha$ that is also contained in $\mathcal{T}_\alpha$ for every $\alpha \in \Delta$. This allows us to say that $\bigcap_{\alpha \in \Delta} \mathcal{T}_\alpha$ is the largest topology on X that is contained in each of the original topologies $\mathcal{T}_\alpha$, $\alpha \in \Delta$, under consideration.

Exercises:

1. List all topologies for a set containing three distinct elements.

2. Prove the collection of subsets described in Example 1.8 actually forms a topology.

3. Prove the collection of subsets described in Example 1.9 actually forms a topology.

4. (a) Is there a set upon which the discrete and indiscrete topologies are equal?

(b) Give an example of a topology on an infinite set which has only a finite number of elements. (Do not use the indiscrete topology.)

5. Give three examples of topologies for R^1 other than the standard, the left ray, the cofinite, co-countable, trivial or discrete.

6. Use (b) of Definition 1.1 and mathematical induction to prove that the intersection of a finite number of open sets in a topological space is always an open set in the space.

7. Give an example of a set X and two topologies $\mathscr{T}_1$ and $\mathscr{T}_2$ for X such that $\mathscr{T}_1 \cup \mathscr{T}_2$ is not a topology for X.

8. Prove that cofinite topology on a finite set X is the same as the discrete topology on X.

9. Prove that $\mathscr{T}$ is the discrete topology for X iff every point in X is an open set.

10. Let the set under consideration be N. For each $n \in N$, define $U_n = \{n, n + 1, n + 2, \ldots\}$ and let $\mathscr{T}$ consist of ϕ, N and all subsets U_n of N. Prove that $\mathscr{T}$ is a topology for N.

11. Verify the statement $L_b = \bigcup_{a \in \Delta} L_a$ made in Example 1.3.

12. For R^1, how does the cofinite topology compare with the standard topology? With the left ray topology? With the co-countable topology?

13. Give an example of a collection of open sets in a space $(X, \mathscr{T})$ whose intersection is not open.

2. Closed Sets

Now that we have some feeling for what a topology is, we are ready to begin investigating the structure of topological spaces. Our first investigation concerns closed sets.

2.1. Definition: Let $(X, \mathscr{T})$ be a topological space. A set $A \subset X$ is *closed* iff $X \backslash A$ is an open subset of X.

2.2. Example: For the space R^1, the set $A = \{x : 0 \leq x \leq 1\}$ is closed because $R^1 \backslash A = \{x : x < 0\} \cup \{x : x > 1\}$ is open in R^1. The set $\{x : 0 \leq x < 1\}$ is neither open nor closed. The sets ϕ and R^1 are both open and closed.

If R^1 has the left ray topology, the set $A = \{x : 0 \leq x \leq 1\}$ is not closed because $R^1 \backslash A$ is not one of the elements of the topology. The given set is not open either, of course. It would be instructive to investigate which sets in this space qualify as closed.

Example 2.2 shows how properties of sets may change when the topology is changed. It also illustrates the fact that a subset of a space may be both open and closed, or it may be neither. This means that we should beware of the following fallacy: If $A \subset X$ is not closed, then A is open.

By definition of a topological space, the union of any number of open sets is again an open set, while the intersection of a finite number of open sets is always an open set. Examples easily show the intersection of an infinite number of open sets may not result in an open set. Contrast this situation with Theorem 2.3.

2.3. THEOREM: *Let $(X, \mathcal{T})$ be a topological space. Then*
    (a) The intersection of any family of closed sets is closed.
    (b) The union of any finite number of closed sets is closed.

Proof: Let $\{A_\alpha : \alpha \in \Delta\}$ be a family of closed subsets of X. To show the intersection is closed, we must show $X \backslash (\bigcap\limits_{\alpha \in \Delta} A_\alpha)$ is open. By DeMorgan's Theorem, $X \backslash (\bigcap\limits_{\alpha \in \Delta} A_\alpha) = \bigcup\limits_{\alpha \in \Delta} (X \backslash A_\alpha)$. The fact that $X \backslash A_\alpha$ is open in X for every $\alpha \in \Delta$ completes the proof of (a). The remaining details are left to the reader.

Theorem 2.3 shows the duality of the roles played by open and closed subsets of a space. These results, and the fact that ϕ and X are always closed subsets of X, show that a topology could be described on X in terms of which subsets are closed, rather than which are open.

Even though a subset A of a space X may not be closed, A is always contained in at least one closed set in X, namely X itself, and possible many more. This observation makes the next definition meaningful.

2.4. Definition: Let $(X, \mathcal{T})$ be a topological space and $A \subset X$. Then the *closure* of A is the intersection of all closed sets in X which contain A and is denoted by $\bar{A}$.

From Theorem 2.3(a) and the definition, $\bar{A}$ is always a closed set containing A regardless of the nature of A. Furthermore, if B is any closed set containing A, then $\bar{A} \subset B$, thus proving the next theorem.

2.5. THEOREM: *Let A be a subset of the space X. Then $\bar{A}$ is the smallest closed set containing A.*

Another observation that may be easily made is contained in the following theorem.

2.6. THEOREM: *A subset A of a space X is closed iff $\bar{A} = A$.*

Proof: The proof is left to the reader.

2.7. Example: For R^1, if $A = \{x : 0 < x \leq 1\}$, then $\bar{A} = \{x : 0 \leq x \leq 1\}$. If B is the set of all rationals, then $\bar{B} = R^1$. If $C = \{x : 0 \leq x \leq 1\}$, then $\bar{C} = C$.

2.8. Example: Consider R^1 with the left ray topology. If $A = \{x : 0 \leq x \leq 1\}$, then $\bar{A} = \{x : x \geq 0\}$. To understand this, notice that any closed set containing A must have as its complement a left ray L_a, and that the intersection of all such closed sets yields the set $\{x : x \geq 0\}$.

In conclusion, some properties of the closure are now considered.

2.9. THEOREM: *Let A and B be subsets of the space X. Then*
    (a) $\bar{\phi} = \phi$.
    (b) $\overline{A \cup B} = \bar{A} \cup \bar{B}$.
    (c) $\bar{\bar{A}} = \bar{A}$.
    (d) If $A \subset B$, then $\bar{A} \subset \bar{B}$.

Proof: Only the proof of (b) will be given here. The smallest closed set containing A is certainly a subset of the smallest closed set containing both A and B. Therefore, $\bar{A} \subset \overline{A \cup B}$. Likewise, $\bar{B} \subset \overline{A \cup B}$ so that $\bar{A} \cup \bar{B} \subset \overline{A \cup B}$. To see the reverse subset implication, suppose $\overline{A \cup B}$ is not a subset of $\bar{A} \cup \bar{B}$. Then there is a point x belonging to $\overline{A \cup B}$ that does not belong to $\bar{A} \cup \bar{B}$. Now $\bar{A} \cup \bar{B}$ is a closed set containing $A \cup B$ but not x, and, therefore, the intersection of all closed sets containing $A \cup B$ does not contain x. This implies $x \notin$

$\overline{A \cup B}$ and contradicts the fact that $x \in \overline{A \cup B}$. It follows that $\overline{A \cup B} \subset \bar{A} \cup \bar{B}$. The final conclusion is $\bar{A} \cup \bar{B} = \overline{A \cup B}$.

Exercises:

1. Prove that each closed interval is a closed set in the space R^1.

2. List all closed subsets of X for each of the spaces $(X, \mathscr{T}_1)$ and $(X, \mathscr{T}_2)$ described in Example 1.2 of the previous section.

3. Prove Theorem 2.3.

4. Give an example of a collection of closed sets in a topological space whose union is not closed.

5. Let R^1 have the left ray topology. Describe what kind of sets are closed in this space. Do the same for the cofinite topology on R^1.

6. In R^1, do the rationals form an open set? Closed set? Neither? Both? Prove your assertion.

7. Consider the set of real numbers $A = \{x : 0 < x < 1\} \cup \{2\}$ in R^1. Describe $\bar{A}$ for the following topologies on R^1:
   (a) Standard.
   (b) Cofinite.
   (c) Left ray.
   (d) Discrete.

8. Give an example to show $\overline{A \cap B} \neq \bar{A} \cap \bar{B}$. Does the subset implication apply one way? If so, which way. Prove your assertion.

9. Prove parts (a), (c) and (d) of Theorem 2.9.

10. Prove Theorem 2.6.

11. Prove that the subset A of the space X is open iff $X \backslash A$ is closed.

12. Let $A \subset R^1$ have an upper bound. Prove that the least upper bound of A belongs to $\bar{A}$.

3. A Closer Look at the Standard Topology on R^1

We now delve more deeply into the basic structure of open sets which make up the standard topology on R^1. After a few preliminary theorems, we shall see that the main result of this section is that a nonempty subset of R^1 is open iff it is the union of a countable collection of disjoint open intervals.

3.1. THEOREM: *Between any two distinct real numbers a and b there is a rational number. In particular, every open interval contains a rational number.*

Proof: Assume a is the smaller of the two, and let $b - a = p$ where p is a positive real number. Then there exists a natural number n such that $n(b - a) = np > 10$ by use of the Archemedian Principle of real numbers. That is, na and nb are at least 10 units apart, so that between na and nb there is an *integer* which we shall denote by k. The resulting inequality $na < k < nb$ may be rewritten as $a < k/n < b$, so that the rational number k/n lies between a and b.

3.2. THEOREM: *For R^1, if $\{U_\alpha : \alpha \in \Delta\}$ is any family of disjoint open intervals, then $\{U_\alpha : \alpha \in \Delta\}$ is countable.*

Proof: The set of all rationals Q is countable and hence the subset $A \subset Q$ of all rationals contained in $\bigcup_{\alpha \in \Delta} U_\alpha$ is also countable according to Theorem 5.5 of Chapter 2. This means that there exists a surjection $f : N \to A$. Now let us define a function $g : A \to \{U_\alpha : \alpha \in \Delta\}$ as follows: For each $x \in A$, let $g(x)$ be the unique member of $\{U_\alpha\}$ to which x belongs. As defined, g is a surjection since each open interval contains a rational by Theorem 3.1. Therefore, $gf : N \to \{U_\alpha : \alpha \in \Delta\}$ is a surjection which shows the collection $\{U_\alpha : \alpha \in \Delta\}$ is countable.

3.3. THEOREM: *Let R^1 have the standard topology and let $\{A_\alpha : \alpha \in \Delta\}$ be an indexed family of open intervals each containing the point p. Then $\bigcup_{\alpha \in \Delta} A_\alpha$ is an open interval containing p.*

Proof: We first note that since $p \in A_\alpha$ for all $\alpha \in \Delta$, then $p \in \bigcup_{\alpha \in \Delta} A_\alpha$. Also, since each A_α is an open subset of R^1, then $\bigcup_{\alpha \in \Delta} A_\alpha$ is an open subset of R^1. We need to establish that $\bigcup_{\alpha \in \Delta} A_\alpha$ has one of the forms given in Definition 1.4. To do this, let us examine the cases where $\bigcup_{\alpha \in \Delta} A_\alpha$ (1) has an upper bound, (2) has no upper bound, (3) has a lower bound, or (4) has no lower bound. Then considering case (1) with each of (3) and (4), and then (2) with each of (3) and (4), we reach the desired conclusion of the theorem.

Considering case (1), if the nonempty set $\bigcup_{\alpha \in \Delta} A_\alpha$ has an upper bound, then it has a least upper bound which we shall call b. Now $b \notin \bigcup_{\alpha \in \Delta} A_\alpha$, because every open interval (x_1, x_2) containing b contains a point x such that $b < x < x_2$, which means $x \notin \bigcup_{\alpha \in \Delta} A_\alpha$, and this implies b cannot belong to the open set $\bigcup_{\alpha \in \Delta} A_\alpha$. Therefore, $p < b$.

Furthermore, if $x \in R^1$ has the property that $p < x < b$, then $x \in \bigcup_{\alpha \in \Delta} A_\alpha$. The reason for this is that b is the least upper bound of $\bigcup_{\alpha \in \Delta} A_\alpha$ and hence there is at least one interval A_α containing p such that the right end point b_α of A_α has the property $x < b_\alpha < b$. It follows that $x \in A_\alpha \subset \bigcup_{\alpha \in \Delta} A_\alpha$.

If $\bigcup_{\alpha \in \Delta} A_\alpha$ has a lower bound, the $\bigcup_{\alpha \in \Delta} A_\alpha$ has a greatest lower bound a. The reasoning of the previous paragraph lets us see that $a \notin \bigcup_{\alpha \in \Delta} A_\alpha$, and if $a < x < p$, then $x \in \bigcup_{\alpha \in \Delta} A_\alpha$.

The reader may now consider the four cases previously listed to reach the conclusion of the theorem. Now, for the main theorem of this section, we have the following.

3.4. THEOREM: *A nonempty subset $U \subset R^1$ belongs to the standard topology on R^1 iff U is the union of a countable collection of disjoint open intervals.*

Proof: If U is the union of a collection (countable or otherwise) of disjoint open intervals, then U is open since, by Theorem 1.6, each open interval is a member of the standard topology for R^1. Now if we suppose U is any open subset of R^1, we can show that U is the union of a countable collection of disjoint open intervals. Since U is open, for each point $p \in U$ there exists an open interval containing p and lying entirely in U. Therefore, define for each $p \in U$, the set I_p to be the union of all open intervals containing p and lying in U. The sets I_p have properties (a), (b) and (c) as follows:

(a) Each I_p is an open interval containing p and lying in U. This fact comes from Theorem 3.3 and the definition of I_p.

(b) If $I_p \cap I_q \neq \phi$, then $I_p = I_q$. To see this, first let $x \in I_p$ and let $z \in I_p \cap I_q$. Then there exists an open interval (a, b) which contains both x and p, an open interval (c, d) which contains both p and z, and an open interval (e, f) containing both z and q, each of these intervals being a subset of U. Thus, $(a, b) \cup (c, d) \cup (e, f)$ is an interval, a subset of U, and contains both x and q and, therefore,

$x \in I_q$. Consequently, $I_p \subset I_q$. Similarly, it may be shown that $I_q \subset I_p$, so that $I_p = I_q$.

(c) If x and y are points in U, then either $I_x = I_y$ or $I_x \cap I_y = \phi$. By definition, $I_x \subset U$ for each $x \in U$ and conversely, for each $x \in U$ there is defined a set I_x so that $U \subset \bigcup_{x \in U} I_x$, which implies $U = \bigcup_{x \in U} I_x$. Thus, U is expressed as the union of a collection of open disjoint intervals which by Theorem 3.2 is countable.

3.5. Corollary: A nonempty subset A of R^1 is closed iff A is the complement of a countable number of disjoint open intervals. In particular, every finite subset of R^1 is closed.

Proof: The definition of a closed set gives this result.

Exercises:

1. Prove that every open interval in R^1 contains infinitely many distinct points. (This shows no nonempty finite set is open in R^1.)

2. Give an example of a collection of closed subsets of R^1 whose union is not closed.

3. Complete the proof of Theorem 3.3.

4. Let U be any open subset of R^1 and $A \subset R^1$ any finite set. Prove $U \backslash A$ is open.

5. Give an example of two subsets A and B of R^1 such that A and $A \backslash B$ are both open but B is not closed.

6. If a and b are two distinct points in R^1, prove there exist disjoint open sets containing a and b, respectively.

7. (a) Give an example of a countable set in R^1 that is not closed.
   (b) Give an example of an infinite collection of closed sets in R^1 whose union is closed.

8. Let $A \subset R^1$. Prove that $x \in \bar{A}$ iff every open set containing x also contains at least one point of A.

9. A set $A \subset R^1$ is called *nowhere dense* iff for every open interval I of R^1 there exists an open interval U of R^1 such that $U \subset I$ and $U \cap A = \phi$.
   (a) Prove that the set of rational numbers is not nowhere dense.
   (b) Prove that the set $\{1/n : n \in N\}$ is nowhere dense.

10. Use Exercise 8 to prove that if a closed set $M \subset R^1$ contains no nondegenerate interval, then M is nowhere dense.

4. Topologies Induced by Functions

Our studies in this section begin by showing how any function $f: X \to Y$ may be used to define a topology on Y if one is given for X, and on X if one is given for Y. Once this is done, there is an important application which we wish to make concerning the topologizing of subsets of a topological space.

4.1. THEOREM: Let $f: X \to Y$ be any function, and suppose X has a topology $\mathscr{T}_X$. Then the collection $\mathscr{T}_Y = \{V : V \subset Y \text{ and } f^{-1}(V) \in \mathscr{T}_X\}$ of subsets of Y is a topology for Y.

Proof: The sets ϕ and Y belong to $\mathscr{T}_Y$ because $f^{-1}(\phi) = \phi$ and $f^{-1}(Y) = X$ both belong to $\mathscr{T}_X$. We next establish that if V_1 and V_2 belong to $\mathscr{T}_Y$, then $V_1 \cap V_2$ belongs to $\mathscr{T}_Y$. According to the definition of $\mathscr{T}_Y$, if we can show $f^{-1}(V_1 \cap V_2)$ belongs to $\mathscr{T}_X$, then $V_1 \cap V_2$ will belong to $\mathscr{T}_Y$. Pursuing this line of thought, V_1 and V_2 in $\mathscr{T}_Y$ means $f^{-1}(V_1)$ and $f^{-1}(V_2)$ both belong to $\mathscr{T}_X$. Since $\mathscr{T}_X$ is given as a topology for X, we know that $f^{-1}(V_1) \cap f^{-1}(V_2)$ is a member of $\mathscr{T}_X$. Now we see the fact that $f^{-1}(V_1) \cap f^{-1}(V_2) = f^{-1}(V_1 \cap V_2)$ gives $f^{-1}(V_1 \cap V_2)$ as a member of $\mathscr{T}_X$, so that $V_1 \cap V_2$ belongs to $\mathscr{T}_Y$ as desired. Finally, using a similar argument based on the union of members of $\mathscr{T}_X$ being in $\mathscr{T}_X$, we have that if $\{V_\alpha : \alpha \in \Delta\}$ is a collection of subsets each belonging to $\mathscr{T}_Y$, then $f^{-1}(\bigcup_{\alpha \in \Delta} V_\alpha) = \bigcup_{\alpha \in \Delta} f^{-1}(V_\alpha)$ belongs to $\mathscr{T}_X$, implying $\bigcup_{\alpha \in \Delta} V_\alpha$ is in $\mathscr{T}_Y$. The collection $\mathscr{T}_Y$ then meets the requirements for a topology on Y.

With the results of Theorem 4.1 behind us, let us next proceed to some terminology.

4.2. Definition: The topology $\mathscr{T}_Y$ described in Theorem 4.1 is called the *topology induced on Y* by f and $(X, \mathscr{T}_X)$.

4.3. Example: Let $X = Y = R^1$ and $f: X \to Y$ be a function described as

$$f(x) = \begin{cases} 0 \text{ if } x < 0 \\ x \text{ if } x \geq 0 \end{cases}.$$

Consider the domain X of f to have the standard topology. Our problem will be to describe what kind of sets belong to the topology induced on $Y = R^1$ by f and X. That is, what kind of sets $V \subset Y$ have the property that $f^{-1}(V)$ is open in $X = R^1$? Notice first that for each $y < 0, f^{-1}(y) = \phi$ which is open in R^1. This means each point $y < 0$ is an element of the induced topology and, consequently, any subset of $\{y : y < 0\}$ belongs to this topology. Single points y where $y \geq 0$, however, are not open since $f^{-1}(y) = y$ and points are not open in R^1 with the standard topology. If $V \subset \{y : y \geq 0\}$, it follows that $f^{-1}(V) = V$ is open in R^1 iff V belongs to the standard topology for R^1. Thus, the topology induced on $Y = R^1$ contains all subsets of R^1 which belong to the standard topology, together with every subset of $\{y \in Y : y < 0\}$.

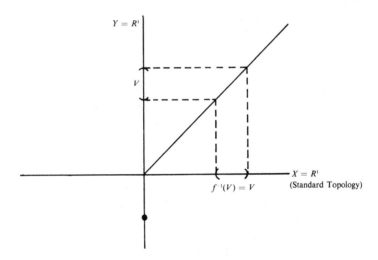

Figure 12

Perhaps it is worthwhile to observe that if $f : X \to Y$ is not surjective, then each point in $Y \backslash f(X)$ is a member of the topology induced on Y by f and $(X, \mathcal{T}_x)$. The reason for this is that $f^{-1}(y) = \phi$ for all $y \in Y \backslash f(X)$.

Now suppose we consider the possibility that $f: X \to Y$ is a given function, and Y has a topology $\mathscr{T}_Y$. The next theorem shows how a topology may be defined on X and, when done in this fashion, will be called the topology induced on X by f and $(Y, \mathscr{T}_Y)$.

4.4. THEOREM: *Let $f: X \to Y$ be any function, and let Y have a topology $\mathscr{T}_Y$. Then the collection $\mathscr{T}_X = \{f^{-1}(U): U \in \mathscr{T}_Y\}$ of subsets of X is a topology for X.*

Proof: The proof is left as an easy exercise.

4.5. Example: Let $X = \{a, b, c\}$ and $Y = \{1, 3, 5, 7\}$ where $\mathscr{T}_Y = \{\phi, Y, \{1, 5\}, \{5, 7\}, \{5\}, \{1, 5, 7\}\}$. Let $f = \{(a, 5), (b, 7), (c, 7)\}$. Then the topology $\mathscr{T}_X$ induced on X by f and $\mathscr{T}_Y$ is $\{\phi, X, \{a\}\}$.

Suppose we now turn to the application we mentioned earlier. For this, let us consider a topological space $(X, \mathscr{T})$ and any nonempty subset A of X. Then we may form the inclusion function $f: A \to X$, given by $f(x) = x$ for each $x \in A$, and since X has a topology $\mathscr{T}$, the function f induces a topology $\mathscr{T}_A$ on A in the manner described in Theorem 4.4. It follows that $(A, \mathscr{T}_A)$ is always a topological space.

4.6. Definition: Let $(X, \mathscr{T})$ be a space and A a nonempty subset of X. Then the topology induced on A by the inclusion function $f: A \to X$ and $\mathscr{T}$ is called the *subspace topology* for A. Under these conditions, the subspace topology will be denoted by $\mathscr{T}_A$.

According to Definition 4.6, we see that every nonempty subset of a topological space may be made into a topological space in its own right by "inheriting" its topology from the original space. This point is made clearer by considering the very workable characterization of the subspace topology given next.

4.7. THEOREM: *Let $A \neq \phi$ be a subset of the space $(X, \mathscr{T})$. Then $U \subset A$ belongs to the subspace topology $\mathscr{T}_A$ iff there exists an element V of $\mathscr{T}$ such that $U = A \cap V$.*

Proof: First, let us observe that for the inclusion function $f: A \to X$, if B is any subset of X, then $f^{-1}(B) = \{x \in A: f(x) = x \in B\} = B \cap A$. Now let $U \subset A$ belong to the subspace topology $\mathscr{T}_A$ on A. Then, according to the definition of $\mathscr{T}_A$, $U = f^{-1}(V)$ for some $V \in \mathscr{T}$ so that $U = f^{-1}(V) = A \cap V$. Conversely, suppose $U \subset A$ and there exists an open set $V \in \mathcal{T}$ such that $U = A \cap V$. Then $U = A \cap V = f^{-1}(V)$, which means $U \in \mathcal{T}_A$. This completes the proof.

4.8. Example: Let $A = \{x \in R^1 : 0 < x \leq 1\}$ be a subset of R^1 with the standard topology. Consider the subspace topology $\mathcal{T}_A$ on A. Then set $U = \{x : \frac{1}{2} < x \leq 1\}$ belongs to the subspace topology $\mathcal{T}_A$ because $V = \{x : \frac{1}{2} < x < 10\} \in \mathcal{T}$ is an open set in R^1 such that $U = A \cap V$. This means $F = \{x : 0 < x \leq \frac{1}{2}\}$ is closed in the subspace $(A, \mathcal{T}_A)$. If $I \subset R^1$ represents the integers, then the subspace topology $\mathcal{T}_I$ is the discrete topology.

4.9. Example: Let R^1 have the left ray topology. Suppose $A = \{x : 0 \leq x \leq 1\}$. Then the elements of the subspace topology $\mathcal{T}_A$ have the form $L_a \cap A = \{x : 0 \leq x < a\}$ where $a \in R^1$. Of course, ϕ and A are open in A.

For the special case when A is an open subset of the space X, we have the following result.

4.10. THEOREM: Let A be an open subset of the space $(X, \mathcal{T})$. Then $\mathcal{T}_A \subset \mathcal{T}$. In other words, every open subset of an open subspace A of X is open in X.

Proof: Let $U \subset A$ be open in the subspace A. Then $U = A \cap V$ where V is open in X, according to Theorem 4.7. But A is also open in X by hypothesis and, therefore, $A \cap V$ is open in X. It follows that U is open in X.

Exercises:

1. (a) Compare the topology on Y described in Example 4.3 with the standard topology on the reals.
   (b) Give an example to show that if Y has the discrete topology and $f : X \rightarrow Y$, the topology induced on X by f and the space Y need not be discrete.

2. Prove Theorem 4.4.

3. Suppose $X = Y = R^1$ and $f : X \rightarrow Y$ is given by $f(x) = 1$ if $x \geq 0$ and $f(x) = -1$ if $x < 0$. Let X have the left ray topology. Describe the topology induced on Y by f and the space $X = R^1$ with the left ray topology.

4. Consider the same function f as in Exercise 3. Suppose $Y = R^1$ has the cofinite topology. Describe the topology induced on the domain $X = R^1$ by f and $Y = R^1$ with the cofinite topology.

5. Let $(A, \mathcal{T}_A)$ be a subspace of $(X, \mathcal{T})$. Prove that F is closed in A iff there exists a closed set M in X such that $F = A \cap M$.

6. Prove that every closed subset of a closed subspace is closed in X.

7. (a) Prove that the rational numbers Q, considered as a subspace of R^1, do not have the discrete topology.
   (b) Prove that the set $A = \{1/n : n \in N\} \subset R^1$ has the discrete subspace topology.

8. (a) Let X have the discrete topology and $f : X \to Y$ any function. Prove that the topology induced on Y by f and the space X is the discrete topology.
   (b) Let Y be a set and $\mathcal{T}$ a topology for Y with the property that for each pair of distinct points a and b in Y, there exist open sets U and V in Y such that $a \in U, b \in V$, and $U \cap V = \phi$. If $f : X \to Y$ is injective, prove the topology induced on X by f and $(Y, \mathcal{T})$ has the same property described for $\mathcal{T}$.

9. Let $\mathcal{T}_Y$ be the topology induced on Y by the function $f : X \to Y$ and the space $(X, \mathcal{T})$. Prove that for $U \in \mathcal{T}, f(U) \in \mathcal{T}_Y$ iff $(f^{-1}f)(U) \in \mathcal{T}$.

10. Let $(X, \mathcal{T})$ be a space. Prove that a subspace of a subspace is a subspace of $(X, \mathcal{T})$.

5. The Interior, Exterior and Boundary of a Set

As we stated at the beginning of this chapter, the concepts of the interior, exterior, and boundary of a set are among the most basic for investigation in a topological space. We are now ready for such an investigation. Our first step will be to define these concepts and look at some examples of each.

5.1. Definition: Let $(X, \mathcal{T})$ be a topological space and $A \subset X$. A point $x \in A$ is an *interior point* of A iff there exists an open set $U \in \mathcal{T}$ containing x such that $U \subset A$. The set of interior points of A is called the *interior* of A and is denoted by Int (A). A point $x \in X$ is an *exterior point* of A iff there exists an open set $U \in \mathcal{T}$ containing x such that $U \cap A = \phi$. The set of exterior points of A is called the *exterior* of A and is denoted by Ext (A). A point $x \in X$ is a *boundary point* of A iff every open set in X containing x contains at least one point of A, and at least one point of $X \backslash A$. The set of boundary points of A is called the *boundary* of A and is denoted by Bd(A).

The definition tells us directly that Int $(A) \subset A$, Ext $(A) \subset X \backslash A$ while boundary points of A may lie in either A or $X \backslash A$. It also tells us that these three sets are pairwise disjoint.

5.2. THEOREM: *Let $(X, \mathcal{T})$ be a topological space and $A \subset X$. Then each point in X belongs to one and only one of the sets Int(A), Ext(A), or* Bd(A).

Proof: The proof is left as an exercise.

The next examples to be considered, as well as some of the exercises for this section, show that the same set may have drastically different interiors, exteriors, and boundaries, depending upon the topology for X. These concepts do, however, conform to our usual intuitive notions in R^1 with the standard topology.

5.3. Example: Let $A = \{x : 0 \leq x \leq 1\}$ be a subset of R^1. For the standard topology on R^1 it is easy to see by the definitions that Int$(A) = \{x : 0 < x < 1\}$, Ext$(A) = \{x : x < 0\} \cup \{x : x > 1\}$, and Bd$(A) = \{0, 1\}$. Suppose we now change to the cofinite topology on R^1 and ask which points of A are interior points of A. Asking this question for the specific point $x = \frac{1}{2}$, for instance, is to ask if there is an open set U containing $x = \frac{1}{2}$ and lying entirely in A. Since every open set U in the cofinite topology is of the form $U = R^1 \backslash \{x_1, x_2, \ldots, x_n\}$ and $R^1 \backslash A$ is an infinite set, then $U \cap (R^1 \backslash A) \neq \phi$. Consequently, no open set containing $x = \frac{1}{2}$ lies entirely in A, which implies $x = \frac{1}{2}$ is not an interior point. In the same fashion, it may be argued that no point of A is an interior point and, hence, Int$(A) = \phi$. Similar investigations show Ext$(A) = \phi$, while Bd$(A) = R^1$.

Let us digress for a moment from our study of interiors, exteriors, and boundaries to prove a fundamental result about open sets which the reader may have already observed. We will find this result has many helpful applications in this section as well as the ones that lie ahead.

5.4. THEOREM: *The set U is open in the space $(X, \mathcal{T})$ iff for each $x \in U$ there exists an open set $V(x)$ containing x such that $V(x) \subset U$.*

Proof: Suppose first that U is open in X. Then for each $x \in U$, $V(x) = U$ is an open set containing x such that $V(x) \subset U$.

Conversely, suppose that for each $x \in U$ there exists an open set $V(x)$ containing x such that $V(x) \subset U$. It follows that $\bigcup_{x \in U} V(x) \subset U$. On the other hand, if $x \in U$, then x belongs to one of the open sets $V(x)$ for which $V(x) \subset U$. Consequently, $x \in \bigcup_{x \in U} V(x)$, and we have shown $U = \bigcup_{x \in U} V(x)$. Now each $V(x)$ is open so that $\bigcup_{x \in U} V(x)$ is open, which shows $U = \bigcup_{x \in U} V(x)$ is open.

The first application of this theorem will occur in proving the next result.

5.5. THEOREM: *If A is any subset of the space X, then $Int(A)$ is an open subset of X.*

Proof: The definition of interior point and Theorem 5.4 give $Int(A)$ open.

5.6. THEOREM: *The set $Int(A)$ is the largest open set contained in A.*

Proof: Certainly $Int(A) \subset A$ and is open by Theorem 5.5. Now let U be any open set which is contained in A. Then for each $x \in U$, U is an open set containing x and lying entirely in A. Thus, each $x \in U$ is an interior point of A. It follows that $U \subset Int(A)$ and, consequently, $Int(A)$ must be the largest open set in A.

5.7. Corollary: The set $A \subset X$ is open iff $A = Int(A)$.

A reexamination of Example 5.3 for the case where R^1 has the standard topology will suggest why the next theorem holds. Later on, when we have defined the standard topology on R^2 and R^3, the intuitiveness of this theorem will be further reinforced. Its proof for all topological spaces, however, follows from the definitions and theorems we have developed thus far.

5.8. THEOREM: *Let $A \subset X$ be any subset of the space X. Then*

    (a) *A is open iff A contains none of its boundary points.*
    (b) *A is closed iff A contains all of its boundary points.*

Proof: Only the proof of (a) is given. Let A be open in X. Then $Int(A) = A$, according to Corollary 5.7, so that $Bd(A) \cap A = \phi$, which means A contains none of its boundary points.

    Next, let A be a set containing none of its boundary points and it will be shown that A is open. To this end, let $x \in A$. Then $x \notin Bd(A)$, and, since $x \notin Ext(A)$, it follows that $x \in Int(A)$. This shows that $A \subset Int(A)$, and, since it is always true that $Int(A) \subset A$, we have $A = Int(A)$. Corollary 5.7 then says A is open.

    Suppose we now consider any subset A of the space X and let $x \in Bd(A \cup Bd(A))$. We can then show that $x \in Bd(A)$. To do this, first observe that if U is any open set containing x, then $U \cap (A \cup Bd(A)) \neq \phi$ and $U \cap (X \backslash (A \cup Bd(A))) \neq \phi$. Rewriting these two expressions, we have $(U \cap A) \cup (U \cap Bd(A)) \neq \phi$ and $U \cap (X \backslash A \cap X \backslash Bd(A)) \neq \phi$. From the first of these expressions, either $U \cap A \neq \phi$ or $U \cap Bd(A) \neq \phi$, and from the second it is always true that $U \cap X \backslash A \neq \phi$. Therefore, our considerations reduce to the following two cases: (1) $U \cap A \neq \phi$ and $U \cap X \backslash A \neq \phi$, or (2) $U \cap Bd(A) \neq \phi$ and $U \cap X \backslash A \neq \phi$. If (1) holds, it is evident that $x \in Bd(A)$. If (2) holds and $z \in U \cap Bd(A)$, then U is an open set containing $z \in Bd(A)$ so that $U \cap A \neq \phi$ and $U \cap X \backslash A \neq \phi$, and again $x \in Bd(A)$. Our conclusion is that if $x \in Bd(A \cup Bd(A))$, then $x \in Bd(A)$. As a consequence, if $x \in Bd(A \cup Bd(A))$, then $x \in A \cup Bd(A)$ so that $A \cup Bd(A)$ contains all of its boundary points and is, therefore, closed by Theorem 5.8(b). Knowing this, we can now express the closure of a set in terms of the boundary of a set.

5.9. THEOREM: *Let $A \subset X$ be any subset of the space X. Then $\bar{A} = A \cup Bd(A) = Int(A) \cup Bd(A)$.*

Proof: Only the first equality will be proved. The set $A \cup Bd(A)$ is closed according to the remarks preceding the statement of the theorem. It will now be shown that $A \cup Bd(A)$ is the smallest closed set containing A, from which it follows that $\bar{A} = A \cup Bd(A)$. To prove this assertion, suppose there exists a closed set F containing A such that F is a proper subset of $A \cup Bd(A)$. Then there is a point $x \in A \cup Bd(A)$ such that $x \notin F$. This means that $x \in A$ or $x \in Bd(A)$, but since $A \subset F$, it follows that $x \in Bd(A)$ must hold. However, F closed implies $X \backslash F$ is an open set containing x but no point of

$A \subset F$. This implies that $x \in \text{Ext}(A)$. We now have $x \in \text{Ext}(A)$ and $x \in \text{Bd}(A)$, which is a contradiction according to Theorem 5.2. Thus, the supposition of the existence of the closed set F smaller than $A \cup \text{Bd}(A)$ and containing A is false, making $A \cup \text{Bd}(A)$ the smallest closed set containing A.

To close this section, we show how the boundary of a set may be written as the intersection of two closed sets. From our knowledge of closed sets, it follows that the boundary of a set is always closed.

5.10. THEOREM: *Let A be a subset of the space X. Then* $\text{Bd}(A) = \bar{A} \cap \overline{X \backslash A}$.

Proof: Take $x \in \bar{A} \cap \overline{X \backslash A}$ and show every open set in X which contains x also contains a point of A and of $X \backslash A$. This may be done indirectly. A similar procedure shows the other subset inclusion. Details are left as an exercise.

Exercises:

1. Prove Theorem 5.2.

2. Consider R^1 and the set $A = \{x \in R^1 : 0 < x \leq 1\}$. Describe $\text{Int}(A)$, $\text{Ext}(A)$, and $\text{Bd}(A)$ if the topology on R^1 is the
   (a) Standard topology.
   (b) Left ray topology.
   (c) Discrete topology.
   (d) Indiscrete topology.
   (e) The topology induced on R^1 in Example 4.3.

3. Let $X = \{a, b, c\}$ with topology $\mathscr{T} = \{\phi, X, \{a\}, \{a, b\}\}$. If $A = \{a\}$, what is $\text{Int}(A)$? $\text{Ext}(A)$? $\text{Bd}(A)$?

4. Prove part (b) of Theorem 5.8.

5. Prove Theorem 5.10.

6. Prove the second equality in Theorem 5.9.

7. Prove that $\text{Int}(A) = A \backslash \text{Bd}(A)$.

8. Let A be a subset of the space X. Prove that $\text{Bd}(A) = \bar{A} \backslash \text{Int}(A)$.

9. Let A and B be subsets of the space X.
   (a) Prove that $(\text{Int}(A) \cup \text{Int}(B)) \subset \text{Int}(A \cup B)$
   (b) Give an example to show that $\text{Int}(A \cup B)$ need not be a subset of $\text{Int}(A) \cup \text{Int}(B)$.

10. Let A and B be subsets of the space X, and suppose $\text{Bd}(A) \cap \text{Bd}(B) = \phi$. Prove that $\text{Int}(A \cup B) = \text{Int}(A) \cup \text{Int}(B)$.

11. Let A be a subset of the space X.
    (a) Prove that $\text{Bd}(\text{Int}(A)) \subset \text{Bd}(A)$.
    (b) Give an example to show the reverse inclusion in part (a) need not hold.

12. Let A and B be subsets of the space X.
    (a) Prove that $\text{Int}(A \backslash B) \subset \text{Int}(A) \backslash \text{Int}(B)$.
    (b) Give an example to show the reverse inclusion in part (a) need not hold.

13. Let A and B be subsets of the space X.
    (a) Prove that $\text{Bd}(A \cup B) \subset \text{Bd}(A) \cup \text{Bd}(B)$.
    (b) Give an example to show the reverse inclusion in part (a) need not hold.

14. Let A and B be subsets of the space X.
    (a) Prove that $\text{Ext}(A \cup B) = \text{Ext}(A) \cap \text{Ext}(B)$.
    (b) Prove that $\text{Ext}(A) = \text{Ext}(X \backslash \text{Ext}(A))$.

6. Cluster Points

If A is a subset of the space $(X, \mathcal{T})$ and $x \in \text{Ext}(A)$, there exists an element $U \in \mathcal{T}$ containing x such that $U \cap A = \phi$. When we can find such a set U, we are in a sense using the topology $\mathcal{T}$ to "separate" the point x from the set A. From the definition of an exterior point, the larger the topology, the better the chance that x belongs to $\text{Ext}(A)$. By contrast, no point in $A \cup \text{Bd}(A)$ can be separated from A by $\mathcal{T}$ in the sense just mentioned. In other words, for each point in $A \cup \text{Bd}(A)$, every element of $\mathcal{T}$ containing the point has a nonempty intersection with the set A. Among the points in X that cannot be separated from A by $\mathcal{T}$, there is a particularly useful class which we shall study in this section. A point in this class is one about which the points of A cluster or accumulate.

6.1. Definition: Let X be a topological space, $x \in X$, and $A \subset X$. Then x is a *cluster point* (or *accumulation point*, or *limit point*) of A iff every open set containing x contains at least one point of A different from x.

6.2. Example: Let R^1 have the standard topology where $A = \{x \in R^1 : 0 < x \leq 1\}$. The point $x = 0$ is a cluster point of A, but does not belong to A. In fact, if x is any point in $\{x \in R^1 : 0 \leq x \leq 1\}$, it readily follows from Definition 6.1 that x is a cluster point of A. These are the only cluster points of A because if $x \in R^1$ does not belong to the closed set $\{x \in R^1 : 0 \leq x \leq 1\}$, then there is an open set U containing x such that $U \cap A = \phi$. Thus, x would not be a cluster point of A. In the same space R^1, if $B = \{x \in R^1 : 0 < x \leq 1\} \cup \{2\}$, then each point of $\{x \in R^1 : 0 \leq x \leq 1\}$ is a cluster point of B, and these are the only cluster points of B. The point $x = 2$ belongs to B, is a boundary point of B, but is not a cluster point of B because the open set $U = \{x \in R^1 : \frac{3}{2} < x < \frac{5}{2}\}$ contains $2 \in B$, but no other point of B. For the set $N \subset R^1$, there are no cluster points.

6.3. Example: Let $A = \{x \in R^1 : 0 \leq x \leq 1\}$ be a subset of R^1 with the left ray topology. We first consider the point $x = 5$ and ask whether this point is a cluster point of A. This means asking whether every open left ray containing $x = 5$ also contains a point of A different from $x = 5$. The answer is clearly yes. Therefore, $x = 5$ qualifies as a cluster point of A. The point $x = -2$ is not a cluster point of A because the open left ray L_{-1} contains $x = -2$, but no point of A. An examination will show that the set of cluster points for A is $\{x \in R^1 : x \geq 0\}$.

These examples show that a cluster point of a set may or may not belong to the set and that boundary points of a set may or may not be cluster points of that set. The definitions of cluster point and boundary point tell us that if $x \in \text{Bd}(A)$ and $x \notin A$, then x is a cluster point of A. Therefore, any point in $A \cup \text{Bd}(A)$ that is not a cluster point of A, must necessarily belong to A. The points of A that are not cluster points of A are called isolated points of A according to the next definition.

6.4. Definition: Let A be any subset of the space X. Then $x \in A$ is an *isolated* point of A iff there exists an open set U of X containing x such that $U \cap A = \{x\}$.

It follows from our definition that for any subset A of the space $(X, \mathscr{T})$, a given point $x \in X$ belongs to one and only one of the sets $\text{Ext}(A)$, the set of all cluster points of A, or the set of all isolated points of A. It also follows that the smaller that $\mathscr{T}$ is, then the better the chance that a point $x \in X$ is a cluster point of A, and that the larger that $\mathscr{T}$ is, then the better the chance that a point $a \in A$ is an isolated point of A.

Suppose we now turn to showing the usefulness of the definition of a cluster point.

6.5. THEOREM: *Let A be a subset of the space X. Then A is closed iff A contains all of its cluster points.*

Proof: First, let A be closed and let x be a cluster point of A. It will be shown that $x \in A$. Suppose $x \notin A$. Then x belongs to the open set $X \backslash A$ and, hence, there exists an open set containing x but no point of A. This implies x is not a cluster point of A, which contradicts the fact that x was taken to be a cluster point of A. The supposition $x \notin A$ is then false and $x \in A$.

Now let A be a set which contains all of its cluster points. It will be shown that A is closed by showing $X \backslash A$ is open. Let $x \in X \backslash A$. Then $x \notin A$, and from our hypothesis this means that x is not a cluster point of A. Thus, there is some open set U containing x such that $U \cap A = \phi$, which means $U \subset X \backslash A$. According to Theorem 5.4, $X \backslash A$ is open and this implies A is closed.

6.6. Definition: For any set A in the space X, the set of all cluster points of A is called the *derived set* of A. The derived set of A is denoted by A'.

From this definition comes a characterization of the closure of a set which, in many cases, is easier to use than the original definition.

6.7. THEOREM: *Let A be a subset of the space X. Then $\bar{A} = A \cup A'$.*

Proof: The instructive proof is left to the reader.

Theorem 6.7 is a handy tool to use in proving the following theorem. Contrast part (b) with Theorem 2.9(b).

6.8. THEOREM: *Let $\{A_\alpha : \alpha \in \Delta\}$ be an indexed family of subsets of the space $(X, \mathscr{T})$. Then*

(a) $\overline{\bigcap\limits_{\alpha \in \Delta} A_\alpha} \subset \bigcap\limits_{\alpha \in \Delta} \bar{A}_\alpha$ *and*

(b) $\bigcup\limits_{\alpha \in \Delta} \bar{A}_\alpha \subset \overline{\bigcup\limits_{\alpha \in \Delta} A_\alpha}$.

Proof: The proof uses only definitions and is left as an exercise. A study of Exercise 6 at the end of this section shows why the inclusions in parts (a) and (b) cannot be reversed.

We are now going to generalize to arbitrary topological spaces a fact that is true about the real numbers with the standard topology. The fact in mind is that for each real number x there is a rational number *arbitrarily close* to x. By arbitrarily close we mean that for each real number x every open set which contains x also contains a rational number. Stated in different terms, if Q represents the rational numbers, then $\bar{Q} = R^1$. We say that the rationals are *dense* in R^1.

6.9. Definition: Let X be a space. Then $A \subset X$ is *dense* in X iff $\bar{A} = X$.

6.10. Example: In R^1, the set of all rationals Q is dense in R^1. The set of all irrationals is also dense in R^1. The set of all integers is not dense in R^1. However, if the left ray topology is used on R^1, then the integers are dense in R^1 since every point of R^1 is a cluster point of the set of integers and, hence, the closure of the integers is R^1.

As a final theorem for this chapter, we give three conditions which are equivalent to saying $A \subset X$ is dense in the space X.

6.11. THEOREM: *Let A be a subset of the space $(X, \mathcal{T})$. Then the following four conditions are equivalent:*

(a) *The set A is dense in X.*

(b) *If B is any closed subset of X and $A \subset B$, then $B = X$.*

(c) *For each $x \in X$, every open set in X containing x has a nonempty intersection with A.*

(d) *$Int(X \backslash A) = \phi$.*

Proof: (a) implies (b). From (a), we known $\bar{A} = X$. Now let B be a closed subset of X such that $A \subset B$. Then using Theorem 2.9(d) and the fact that B is closed, we have $X = \bar{A} \subset \bar{B} = B$.

(b) implies (c). Let $x \in X$ and U be a nonempty open set in X containing x. Assume $U \cap A = \phi$. Then $A \subset X \backslash U$. The fact that $X \backslash U$ is a closed set in X allows us to use (b) to infer that $X \backslash U = X$.

But, on the other hand, $U \neq \phi$ implies $X \backslash U \neq X$. This contradiction means our assumption is false, so that $U \cap A \neq \phi$.

(c) implies (d). Assume $\text{Int}(X \backslash A) \neq \phi$. Then $\text{Int}(X \backslash A)$ is a non-empty set which is open by Theorem 5.5. However, $(X \backslash A) \cap A = \phi$ and since $\text{Int}(X \backslash A) \subset X \backslash A$, we have $\text{Int}(X \backslash A) \cap A = \phi$. This contradicts (c) and means $\text{Int}(X \backslash A) = \phi$.

(d) implies (a). First consider $x \in X \backslash A$. Since $\text{Int}(X \backslash A) = \phi$ by (d), then no open set containing x can be a subset of $X \backslash A$ so that every open set containing x must also contain a point of A. It follows that $x \in A'$ Therefore, if $x \in X$, then $x \in A \cup A'$, and we conclude that $X = A \cup A'$. Now using Theorem 6.7, $\bar{A} = A \cup A' = X$, showing that A is dense in X.

Exercises:

1. Let R^1 have the cofinite topology. What are the cluster points of $A = \{x \in R^1 : 0 < x < 1\}$? In general, which subsets do and which subsets do not have cluster points in this space? Give an example of a set A that is dense in this space, but is not dense if the standard topology is used on R^1.

2. Prove Theorem 6.7.

3. Prove Theorem 6.8.

4. Prove that a set A in a space X is not open
   (a) iff $X \backslash A$ is not closed.
   (b) iff there exists a point $x \in A$ such that x is a cluster point of $X \backslash A$.
   (c) iff A contains at least one of its boundary points.

5. Prove that a set F in a space X is not closed
   (a) iff $X \backslash A$ is not open.
   (b) iff there exists a cluster point of A that does not belong to A.
   (c) iff A fails to contain all of its boundary points.

6. (a) Give an example of an indexed family $\{A_\alpha : \alpha \in \Delta\}$ of subsets of a space X for which there exists a cluster point x of $\bigcup_{\alpha \in \Delta} A_\alpha$, but x is not a cluster point of A_α for every $\alpha \in \Delta$.
   (b) Give an example to show the reverse inclusion in Theorem 6.8(a) need not hold.

7. Let X be a space having no isolated points. Prove that every open subset of X can have no isolated point.

8. For any two subsets A and B of the space X, if $A \subset B$, prove that $A' \subset B'$.

9. If A and B are subsets of the space X,
   (a) Prove that $(A \cup B)' = A' \cup B'$. (This shows that if x is a cluster point of $A \cup B$, then x is a cluster point of A or of B.)
   (b) Give an example of show $(A \cap B)' \neq A' \cap B'$.

10. Let A be a subset of the space X. Prove that $x \in A'$ iff $x \in \overline{A \setminus \{x\}}$.

11. Prove that if x is a cluster point of $A \subset R^1$ and U is any open set in R^1 containing x, then $U \cap A$ is an infinite set.

12. Give an example of a subset A of a space X for which A' is not closed.

13. The set A is called *perfect* iff $A = A'$. Prove that a set A is perfect iff it is closed and has no isolated points.

14. Give an example of a dense subset A which has an isolated point in a space X which itself has no isolated points.

15. (a) Let A and B be subsets of the space X. Prove that $\bar{A} \setminus \bar{B} \subset \overline{A \setminus B}$.
    (b) Give an example to show the reverse inclusion in part (a) need not hold.

4

Bases, Subbases and Products

1. Bases

For a topological space $(X, \mathscr{T})$, it is often useful to describe $\mathscr{T}$ by defining a collection $\mathscr{B}$ of subsets of X which, in turn, yield the nonempty elements of $\mathscr{T}$ by forming all possible unions of members of $\mathscr{B}$. In so doing, we are able to investigate certain problems in the reduced collection $\mathscr{B}$ without dealing directly with the topology, as well as to give some useful classifications of topologies as we shall see later. On the other hand, we shall see what restrictions must be placed on a collection of subsets of a set X in order that the set of all possible unions of these subsets, together with ϕ, form a topology for X. From this viewpoint, there is an important application of defining a topology on the finite product of topological spaces.

Suppose we now turn to defining the collection $\mathscr{B}$ we have in mind.

1.1. Definition: Let $(X, \mathscr{T})$ be a space. A *base* or *basis* for $\mathscr{T}$ is a collection $\mathscr{B}$ of subsets of X such that

    (a) each member of $\mathscr{B}$ is also a member of $\mathscr{T}$ and

    (b) if $U \in \mathscr{T}$ and $U \neq \phi$, then U is the union of sets belonging to $\mathscr{B}$.

Because each element of $\mathscr{B}$ belongs to $\mathscr{T}$, we sometimes may refer to the elements of $\mathscr{B}$ as *basic open sets* in X. In view of part (a) and the fact that any union of members of $\mathscr{T}$ is again a member of $\mathscr{T}$, part (b) could be restated to say "if $U \neq \phi$, then $U \in \mathscr{T}$ iff U is the union of sets belonging to $\mathscr{B}$." As a consequence of this, a base for $\mathscr{T}$ completely determines $\mathscr{T}$. We also see that any topology is a base for itself, which means a given topology always has at least one base. It is not necessary, however, that a base for a topology be a topology as illustrated in our first example.

1.2. Example: The collection $\mathscr{B}$ of all open intervals in R^1 serves as a base for the standard topology on R^1. This follows from Definition 1.1 and Theorem 3.4 of Chapter 3.

1.3. Example: Let $X = \{a, b, c\}$ with topology $\mathscr{T} = \{\phi, X, \{a\}, \{b\}, \{a, b\}\}$. Other than $\mathscr{T}$ itself, the only bases for $\mathscr{T}$ are $\mathscr{B}_1 = \{\phi, X, \{a\}, \{b\}\}$, and $\mathscr{B}_2 = \{X, \{a\}, \{b\}\}$.

Our first theorem shows how a base for a topology may be used in a rather natural way to determine whether or not a set is open.

1.4. THEOREM: *Let $(X, \mathscr{T})$ be a space and $\mathscr{B}$ a base for $\mathscr{T}$. Then for $U \neq \phi$, $U \subset X$ is open iff for each $x \in U$ there is an element $B \in \mathscr{B}$ such that $x \in B \subset U$.*

Proof: Let $U \subset X$ be open and $x \in U$. Then by Definition 1.1, $U = \bigcup_{\alpha \in \Delta} B_\alpha$ where $B_\alpha \in \mathscr{B}$ and Δ is some indexing set. Therefore, $x \in U = \bigcup_{\alpha \in \Delta} B_\alpha$ implies $x \in B_\alpha$ for at least one $\alpha \in \Delta$ and since $B_\alpha \subset U$, half of our theorem is proved.

For the converse, suppose $U \subset X$ is a set such that for each $x \in U$

there is a $B \in \mathscr{B}$ such that $x \in B \subset U$. Since each $B \in \mathscr{B}$ is open, Theorem 5.4 of Chapter 3 gives U open. This concludes the proof.

Theorem 1.4 actually gives a characterization of the kinds of collections that may serve as a base for $\mathscr{T}$. Stated in theorem form, we have the following result.

1.5. THEOREM: *For a space $(X, \mathscr{T})$, a collection $\mathscr{B} \subset \mathscr{T}$ is a base for $\mathscr{T}$ iff for each nonempty $U \subset \mathscr{T}$ and each $x \in U$ there is a $B \in \mathscr{B}$ such that $x \in B \subset U$.*

Proof: The easy proof is left to the reader.

A base for the subspace topology on a set $A \subset X$ may be readily obtained from a base for the topology on X, as the next theorem shows.

1.6. THEOREM: *Let $(X, \mathscr{T})$ be a topological space and $\mathscr{B}$ a base for $\mathscr{T}$. If $A \subset X$, then $\{B \cap A : B \in \mathscr{B}\}$ is a base for the subspace topology $\mathscr{T}_A$ on A.*

Proof: The instructive proof is left as an exercise.

Suppose we now consider a nonempty set X and an arbitrary collection $\mathscr{B}$ of subsets of X. Examples easily show that if we form all possible unions of members of $\mathscr{B}$ and then adjoin ϕ, the resulting collection may not be a topology for X having $\mathscr{B}$ as a base. However, we show next that if certain restrictions are placed on $\mathscr{B}$, then the set of all possible unions of members of $\mathscr{B}$, along with ϕ, always forms a topology on X for which $\mathscr{B}$ is a base. In this case, the topology on X is said to be "generated" by $\mathscr{B}$. In other words, we are topologizing the set X by use of the collection $\mathscr{B}$. We now set forth the restrictions on $\mathscr{B}$ and prove our assertions about generating a topology on X.

1.7. THEOREM: *Let $\mathscr{B}$ be a collection of subsets of a nonempty set X. Then there exists a topology on X for which $\mathscr{B}$ is a base iff*
    (a) for each $x \in X$ there is at least one $U \in \mathscr{B}$ such that $x \in U$ and
    (b) if $U \in \mathscr{B}$, $V \in \mathscr{B}$, and $x \in U \cap V$, then there exists a $W \in \mathscr{B}$ such that $x \in W \subset U \cap V$.

Proof: First, let $\mathscr{B}$ be a base for a topology $\mathscr{T}$ on X. Since each member of $\mathscr{B}$ is also an element of $\mathscr{T}$, conditions (a) and (b) follow readily.

Conversely, suppose $\mathscr{B}$ is a collection of subsets of X for which (a) and (b) hold. Define $\mathscr{T}(\mathscr{B})$ as a set consisting of all possible unions of members of $\mathscr{B}$ together with ϕ. We shall establish that $\mathscr{T}(\mathscr{B})$ is a topology for X having $\mathscr{B}$ as a base. By (a) there is a $U(x) \in \mathscr{B}$ containing x for each $x \in X$ and, therefore, the definition of $\mathscr{T}(\mathscr{B})$ gives $\bigcup_{x \in X} U(x) = X \in \mathscr{T}(\mathscr{B})$. Since each element of $\mathscr{T}(\mathscr{B})$ is the union of members of $\mathscr{B}$, an arbitrary union of elements in $\mathscr{T}(\mathscr{B})$ is again a union of members of $\mathscr{B}$ and, thus, belongs to $\mathscr{T}(\mathscr{B})$. To complete the proof we need only show that if U and V belong to $\mathscr{T}(\mathscr{B})$, then $U \cap V$ belongs to $\mathscr{T}(\mathscr{B})$. To this end, let $U = \bigcup_{\alpha \in \Delta} B_\alpha$ and $V = \bigcup_{\gamma \in \Omega} C_\gamma$ belong to $\mathscr{T}(\mathscr{B})$. Then $U \cap V = \bigcup_{\alpha \in \Delta} B_\alpha \cap \bigcup_{\gamma \in \Omega} C_\gamma = \bigcup_{\substack{\alpha \in \Delta \\ \gamma \in \Omega}} (B_\alpha \cap C_\gamma)$ according to Exercise 4(a), Section 3 of Chapter 1. Using (b), it is easy to see that each $B_\alpha \cap C_\gamma$ is the union of members of $\mathscr{B}$ or is empty. Therefore, $U \cap V$ is an element of $\mathscr{T}(B)$, and this completes the proof that $\mathscr{T}(\mathscr{B})$ is a topology for X. From the construction of $\mathscr{T}(\mathscr{B})$, $\mathscr{B}$ is a base for $\mathscr{T}(\mathscr{B})$.

Not only is the collection $\mathscr{T}(\mathscr{B})$, as defined in the proof of Theorem 1.7, a topology for X, but it is also unique.

1.8. THEOREM: *Let $\mathscr{B}$ be a collection of subsets of a nonempty set X obeying properties (a) and (b) of Theorem 1.7, and define $\mathscr{T}(\mathscr{B})$ as in the proof of Theorem 1.7. Then $\mathscr{T}(\mathscr{B})$ is the unique topology on X having $\mathscr{B}$ as a base.*

Proof: We know from the proof of Theorem 1.7 that $\mathscr{T}(\mathscr{B})$ is a topology on X having $\mathscr{B}$ as a base. Now let $\mathscr{T}$ be any topology on X having $\mathscr{B}$ as a base. Then $\mathscr{T}$ contains all members of $\mathscr{B}$ and hence all unions of these members. From this it follows that $\mathscr{T}(\mathscr{B}) \subset \mathscr{T}$. On the other hand, if $U \in \mathscr{T}$, then U may be expressed as the union of elements of $\mathscr{B}$ since $\mathscr{B}$ is a base for $\mathscr{T}$. This means $U \in \mathscr{T}(\mathscr{B})$ by the construction of $\mathscr{T}(\mathscr{B})$, so that $\mathscr{T} \subset \mathscr{T}(\mathscr{B})$. Consequently, $\mathscr{T} = \mathscr{T}(\mathscr{B})$.

In summary, the proof of Theorem 1.7 gives a way of constructing or generating a topology from certain collections $\mathscr{B}$ of subsets of a given set X that obey conditions (a) and (b) of that theorem. The topology $\mathcal{T}(\mathcal{B})$ is formed by simply taking all possible unions of members of $\mathcal{B}$ and adjoining the empty set. Of course, each member of the original collection $\mathcal{B}$ belongs to the topology $\mathcal{T}(\mathcal{B})$ and, furthermore, $\mathcal{B}$ is a base for $\mathcal{T}(\mathcal{B})$. Theorem 1.8, then, says that there is one and only one topology, $\mathcal{T}(\mathcal{B})$, on X having $\mathcal{B}$ as a base. From this, we can now make the following definition:

1.9. Definition: Let $\mathcal{B}$ be a collection of subsets of a set X which obey conditions (a) and (b) of Theorem 1.7. Define $\mathcal{T}(\mathcal{B})$ as the set consisting of all possible unions of members of $\mathcal{B}$ together with the empty set. Then $\mathcal{T}(\mathcal{B})$ is called the topology *generated* by $\mathcal{B}$ and having $\mathcal{B}$ as a base.

For a collection $\mathcal{B}$ of subsets of X obeying (a) and (b) of Theorem 1.7, there may be many topologies for X which contain $\mathcal{B}$, but not having $\mathcal{B}$ as a base. It is an easy exercise to show that $\mathcal{T}(\mathcal{B})$ is the smallest topology on X containing $\mathcal{B}$.

Example 1.3 shows that a topology may have more than one base. In this case, we shall use the terminology defined next.

1.10. Definition: Two collections $\mathcal{B}_1$ and $\mathcal{B}_2$ of subsets of X are *equivalent* bases iff there exists a topology $\mathcal{T}$ for X such that $\mathcal{B}_1$ and $\mathcal{B}_2$ are both bases for $\mathcal{T}$.

Our last two theorems for this section give useful information about equivalent bases.

1.11. THEOREM: *Let $(X, \mathcal{T})$ be a space, $\mathcal{B}_1$ a base for $\mathcal{T}$ and $\mathcal{B}_2$ a collection of subsets of X. Then $\mathcal{B}_1$ and $\mathcal{B}_2$ are equivalent bases iff*
 (a) *for each $U_1 \in \mathcal{B}_1$ and $x \in U_1$, there is a $U_2 \in \mathcal{B}_2$ such that $x \in U_2 \subset U_1$ and*
 (b) *for each $U_2 \in B_2$ and $x \in U_2$, there is a $U_1 \in \mathcal{B}_1$ such that $x \in U_1 \subset U_2$.*

Proof: We leave the proof to the reader.

1.12. THEOREM: *Let $\mathcal{B}_1$ and $\mathcal{B}_2$ be collections of subsets of X both satisfying (a) and (b) of Theorem 1.7. Then $\mathcal{B}_1$ and $\mathcal{B}_2$ are equivalent bases iff $\mathcal{T}(\mathcal{B}_1) = \mathcal{T}(\mathcal{B}_2)$.*

Proof: The proof is again left for the reader.

Exercises:

1. Give a proof for Theorem 1.5.

2. Prove Theorem 1.6.

3. (a) In the proof of Theorem 1.7, give the details that conditions (a) and (b) hold if $\mathscr{B}$ is a base for a topology $\mathscr{T}$ on X.
   (b) In the proof of Theorem 1.7, verify that $B_\alpha \cap C_\gamma$ is the union of elements of $\mathscr{B}$.

4. Prove Theorem 1.11.

5. Prove Theorem 1.12.

6. In R^1, consider the collection $\mathscr{B}$ of all open intervals with rational end points.
   (a) Show $\mathscr{B}$ is a base for some topology on R^1.
   (b) Show that $\mathscr{B}$ is a base equivalent to the collection of all open intervals in R^1.

7. Find a base for the left ray topology on R^1 which is different from the topology itself.

8. Let $X \neq \phi$ be a set and $\mathscr{B}$ a collection of subsets of X obeying (a) and (b) of Theorem 1.7. Prove that $\mathscr{T}(\mathscr{B})$ is the smallest topology on X containing $\mathscr{B}$.

9. Let $(X, \mathscr{T})$ be a topological space and $\mathscr{B}$ a base for $\mathscr{T}$. Prove that $A \subset X$ is dense in the space iff each nonempty element of $\mathscr{B}$ contains a point of A.

10. Let $(X, \mathscr{T})$ be a space such that there exists a countable base $\mathscr{B}$ for $\mathscr{T}$. Prove there exists a countable dense subset of X.

11. (a) Prove that R^1 with the cofinite topology does not have a countable base.
    (b) Use part (a) to give an example showing the converse of Exercise 10 need not hold.

2. Finite Products of Topological Spaces

For a finite family of topological spaces $(X_1, \mathscr{T}_1), (X_2, \mathscr{T}_2), \ldots$ $(X_n, \mathscr{T}_n)$, there is a rather natural way to topologize the product $X_1 \times X_2 \times \ldots \times X_n$ by essentially producting the given topologies. This new topology is called the *product topology*. We shall discuss the details of this topology in this section.

2.1. THEOREM: *Let* $(X_1, \mathcal{T}_1), (X_2, \mathcal{T}_2), \ldots, (X_n, \mathcal{T}_n)$ *be a finite collection of topological spaces, and let* $X = X_1 \times X_2 \times \ldots \times X_n$. *If* $\mathcal{B}$ *is the collection of all sets in* X *of the form* $U_1 \times U_2 \times \ldots \times U_n$, *where* U_k *is an open set in* X_k *for* $k = 1, 2, \ldots, n$, *then* $\mathcal{B}$ *is the base for a topology on* X.

Proof: We verify that conditions (a) and (b) of Theorem 1.7 hold for the collection $\mathcal{B}$. Conditions (a) follows from the fact that each coordinate x_k of $(x_1, x_2, \ldots, x_n) \in X$ lies in some open set U_k of X_k for $k = 1, 2, \ldots, n$ and, therefore, $(x_1, x_2, \ldots, x_n) \in U_1 \times U_2 \times \ldots \times U_n \in \mathcal{B}$. To see that condition (b) holds, let $U = U_1 \times U_2 \times \ldots \times U_n$ and $V = V_1 \times V_2 \times \ldots \times V_n$ be any two elements of $\mathcal{B}$, and suppose $(x_1, x_2, \ldots, x_n) \in U \cap V$. According to Theorem 3.7(a) of Chapter 1, we may now express $U \cap V$ an follows:

$$U \cap V = (U_1 \times U_2 \times \ldots \times U_n) \cap (V_1 \times V_2 \times \ldots \times V_n)$$
$$= (U_1 \cap V_1) \times (U_2 \cap V_2) \times \ldots \times (U_n \cap V_n).$$

If we then let $W = (U_1 \cap V_1) \times (U_2 \cap V_2) \times \ldots \times (U_n \cap V_n)$, it follows that $x \in W = U \cap V$, so we may write $x \in W \subset U \cap V$ and, furthermore, $W \in \mathcal{B}$ because each of the sets U_k and V_k are open in X_k giving $U_k \cap V_k$ open for $k = 1, 2, \ldots, n$. Condition (b) is now satisfied and by Theorem 1.7, there exists a topology $\mathcal{T}(\mathcal{B})$ on $X = X_1 \times X_2 \times \ldots \times X_n$ for which $\mathcal{B}$ is a base.

If the collection $\mathcal{B}$ described in Theorem 2.1 is used and if we form the topology $\mathcal{T}(\mathcal{B})$ on X, Theorem 1.8 tells us that $\mathcal{T}(\mathcal{B})$ is unique. This fact allows us to make the next definition.

2.2. Definition: Using the notation of Theorem 2.1, the topology $\mathcal{T}(\mathcal{B})$ generated by $\mathcal{B}$ and having $\mathcal{B}$ as a base is called the *product topology* on $X = X_1 \times X_2 \times \ldots \times X_n$. The space consisting of the set X together with the product topology is called the *topological product*, or simply the *product space*, of $(X_1, \mathcal{T}_1), (X_2, \mathcal{T}_2), \ldots, (X_n, \mathcal{T}_n)$. For each $1 \leq k \leq n$, the space $(X_k, \mathcal{T}_k)$ is called the kth *factor space* of the topological product.

Theorem 1.4 gives us a convenient way of describing the nonempty elements of the product topology on $X = X_1 \times X_2 \times \ldots \times X_n$.

According to that theorem, a nonempty set $U \subset X$ is open iff for each $x = (x_1, x_2, \ldots, x_n) \in U$ there exists a set $B = U_1 \times U_2 \times \ldots \times U_n$, where U_k is open in the space X_k for each $k = 1, 2, \ldots, n$, such that $x \in B \subset U$. In other words, $U \subset X$ is open iff for each $x = (x_1, x_2, \ldots, x_n) \in U$, there exist open sets $U_k \subset X_k$ such that $x_k \in U_k$, for each $k = 1, 2, \ldots, n$, and $U_1 \times U_2 \times \ldots \times U_n \subset U$.

Since R^n is the product of R^1 with itself n times, an immediate application of the product topology is to introduce a standard topology on R^n for $n \geq 2$.

2.3. Definition: Let R^1 have the standard topology. Then the product topology on R^n is called the *standard* or *Euclidean* topology for R^n. The topological space consisting of R^n together with the standard topology is sometimes known as an *n-dimensional Euclidean space*. In all of our future work, the standard topology on R^n will be assumed unless otherwise stated.

In the plane, R^2, for instance, the collection of all *open rectangles* form a base for the standard topology. By the term "open rectangle"

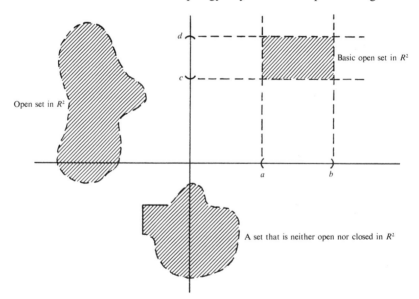

Figure 13

we mean the product of two open intervals in R^1. In R^3, open rectangular parallelepipeds are basic open sets.

To learn more about the product topology, we will now investigate the product of two closed subsets and the product of two subspaces. As we shall prove later, the results we will obtain hold for the product of any number (finite or infinite) of spaces. Even though this will make our present theorem and its proof redundant, we hope that a study of the proof in this special case will make the general one more understandable.

2.4. THEOREM: *Let $(X_1, \mathscr{T}_1)$ and $(X_2, \mathscr{T}_2)$ be spaces and consider the product topology on $X_1 \times X_2$. Let $A_1 \subset X_1$ and $A_2 \subset X_2$.*
  (a) Then $\overline{A_1 \times A_2} = \overline{A_1} \times \overline{A_2}$.
  (b) Let A_1 and A_2 have the subspace topologies $\mathscr{T}_{A_1}$ and $\mathscr{T}_{A_2}$, respectively. Then the product topology on $A_1 \times A_2$ is the same as the subspace topology on $A_1 \times A_2 \subset X_1 \times X_2$.

Proof: (a). Let $x = (x_1, x_2) \in \overline{A_1 \times A_2}$. Then x is either a point or a cluster point of $A_1 \times A_2$ so that every open set U in $X_1 \times X_2$ containing x has the property $U \cap (A_1 \times A_2) \neq \phi$. Now let U_1 and U_2 be any open sets in X_1 and X_2, respectively, such that $x_1 \in U_1$ and $x_2 \in U_2$. Since $U_1 \times U_2$ is open in $X_1 \times X_2$ and contains $x = (x_1, x_2)$, then $\phi \neq (U_1 \times U_2) \cap (A_1 \times A_2) = (U_1 \cap A_1) \times (U_2 \cap A_2)$ by use of Theorem 3.7(a) of Chapter 1. The product $(U_1 \cap A_1) \times (U_2 \cap A_2)$ $\neq \phi$ means $U_1 \cap A_1 \neq \phi$ and $U_2 \cap A_2 \neq \phi$ so that x_1 and x_2 are, respectively, points or cluster points of A_1 and A_2. In either case $x_1 \in \overline{A_1}$ and $x_2 \in \overline{A_2}$ and, therefore, $x = (x_1, x_2) \in \overline{A_1} \times \overline{A_2}$. We have now shown that $\overline{A_1 \times A_2} \subset \overline{A_1} \times \overline{A_2}$.

For the reverse subset inclusion, let $x = (x_1, x_2) \in \overline{A_1} \times \overline{A_2}$, from which it follows that $x_1 \in \overline{A_1}$ and $x_2 \in \overline{A_2}$. If we assume $x \notin \overline{A_1 \times A_2}$, there exist open sets U_1 and U_2 of X_1 and X_2, respectively, such that $x_1 \in U_1, x_2 \in U_2$ and $(U_1 \times U_2) \cap (A_1 \times A_2) = \phi$. Thus, $(U_1 \times U_2)$ $\cap (A_1 \times A_2) = (U_1 \cap A_1) \times (U_2 \cap A_2) = \phi$, which means $U_1 \cap A_1 = \phi$ or $U_2 \cap A_2 = \phi$. In other words, either $x_1 \notin \overline{A_1}$ or $x_2 \notin \overline{A_2}$, both of which are impossible. We conclude that $x \in \overline{A_1 \times A_2}$. This gives $\overline{A_1} \times \overline{A_2} \subset \overline{A_1 \times A_2}$ and finishes the proof to (a).

Part (b) is left as an exercise.

We turn now to the interior, boundary, and derived set of subsets in the product of two spaces. Actually, the following results hold for any finite product of spaces, but we defer the more general inductive proof until a theorem concerning the associativity of the product has been proved in the next chapter. In view of this, the next theorem is the first step of the inductive proof.

2.5. THEOREM: *Let $(X_1, \mathcal{T}_1)$ and $(X_2, \mathcal{T}_2)$ be spaces and consider the product topology on $X_1 \times X_2$. If $A_1 \subset X_1$ and $A_2 \subset X_2$, then*

    (a) $\text{Int}(A_1 \times A_2) = \text{Int}(A_1) \times \text{Int}(A_2)$.

    (b) $(A_1 \times A_2)' = (A_1' \times \overline{A_2}) \cup (\overline{A_1} \times A_2')$.

    (c) $\text{Bd}(A_1 \times A_2) = (\text{Bd}(A_1) \times \overline{A_2}) \cup (\overline{A_1} \times \text{Bd}(A_2))$.

Proof: Only the proof of (b) will be given, while those of (a) and (c) are left an exercises. The point $x = (x_1, x_2)$ belongs to $(A_1 \times A_2)'$ iff $(x_1, x_2) \in \overline{(A_1 \times A_2) \backslash \{(x_1, x_2)\}}$, according to Exercise 10, Section 6 of Chapter 3. Now the definition of the product of two sets implies that $(A_1 \times A_2) \backslash \{(x_1, x_2)\} = (A_1 \backslash \{x_1\} \times A_2) \cup (A_1 \times A_2 \backslash \{x_2\})$, so that we may write $(x_1, x_2) \in (A_1 \times A_2)'$ iff $(x_1, x_2) \in \overline{((A_1 \backslash \{x_1\}) \times A_2) \cup (A_1 \times (A_2 \backslash \{x_2\}))}$. Using Theorem 2.9(b) of Chapter 3 and then Theorem 2.4(a) of this chapter, we have

$$\overline{((A_1 \backslash \{x_1\}) \times A_2) \cup (A_1 \times (A_2 \backslash \{x_2\}))} = \overline{((A_1 \backslash \{x_1\}) \times A_2)} \cup \overline{(A_1 \times (A_2 \backslash \{x_2\}))}$$
$$= ((A_1 \backslash \{x_1\}) \times \overline{A_2}) \cup (\overline{A_1} \times \overline{(A_2 \backslash \{x_2\})})$$

from which it follows that $(x_1, x_2) \in (A_1 \times A_2)'$ iff $(x_1, x_2) \in (A_1' \times \overline{A_2}) \cup (\overline{A_1} \times A_2')$. This proves part (b).

Exercises:

1. Let $X_1 = \{a, b, c\}$ with topology $\mathcal{T}_1 = \{\phi, X_1, \{a\}\}$ and $X_2 = \{x, y, z\}$ with topology $\mathcal{T}_2 = \{\phi, X_2, \{y\}, \{x, z\}\}$. List all elements in the product topology on $X_1 \times X_2$.

2. Let R^2 have the standard topology and suppose $M = \{(x, y) : 0 \leq x < 1, 2 < y \leq 3\}$. Describe $\text{Int}(M)$, $\text{Ext}(M)$ and $\text{Bd}(M)$. Do the same if R^2 has the cofinite topology.

3. Let R^1 have the left ray topology. Describe some of the open sets in the product space formed by the topological product of this space with itself. Give examples of sets which are closed and of sets which have nonempty interiors in this product.

4. List two bases, other than open rectangles, for the standard topology on R^2.

5. Let $\{(X_k, \mathcal{T}_k) : k = 1, 2, \ldots, n\}$ be a finite family of topological spaces and let $\mathcal{B}_k$ be a base for $\mathcal{T}_k$, $k = 1, 2, \ldots, n$. If we denote $\mathcal{B} = \{B_1 \times B_2 \times \ldots \times B_n : B_1 \in \mathcal{B}_1, B_2 \in \mathcal{B}_2, \ldots, B_n \in \mathcal{B}_n\}$, prove that the topology $\mathcal{T}(\mathcal{B})$ on the product $X = X_1 \times X_2 \times \ldots \times X_n$ is the same as the product topology on X.

6. Prove part (b) of Theorem 2.4.

7. Prove part (a) of Theorem 2.5.

8. Prove part (c) of Theorem 2.5.

9. Let $(X_1, \mathcal{T}_1)$ and $(X_2, \mathcal{T}_2)$ be spaces, and consider the product topology on $X_1 \times X_2$. Prove that $A_1 \times A_2$ is dense in the space $X_1 \times X_2$ iff A_1 is dense in X_1 and A_2 is dense in X_2.

10. Let $\{(X_k, \mathcal{T}_k) : k = 1, 2, \ldots, n\}$ be a finite family of spaces and consider the product topology on $X = X_1 \times X_2 \times \ldots \times X_n$. Prove that $A \subset X$ is closed iff A is the intersection of a collection of sets $\{B_\alpha : \alpha \in \Delta\}$ where each B_α is a finite union of sets of the form $A_1 \times A_2 \times \ldots \times A_n$ for which A_k is closed in X_k, $k = 1, 2, \ldots, n$.

3. Subbases

We have seen that under certain conditions, a set $\mathcal{B}$ of subsets of a set X will generate a topology on X having $\mathcal{B}$ as a base. We now define the idea of a subbase for a topology and then show that *any* set $\mathcal{S}$ of subsets of X may be used to construct a topology on X having $\mathcal{S}$ as a subbase.

3.1. Definition: Let $(X, \mathcal{T})$ be a topological space. A subbase for $\mathcal{T}$ is a collection of subsets $\mathcal{S}$ of X having the following properties:
    (a) Each element of $\mathcal{S}$ is also a member of $\mathcal{T}$.
    (b) The collection of all finite intersections of elements of $\mathcal{S}$, together with X, forms a base for $\mathcal{T}$.

Some examples will help clarify the definition.

3.2. Example: For R^1, the collection $\mathcal{S}$ of all open left rays together with all open right rays forms a subbase for the standard topology.

This is because each element of $\mathscr{S}$ belongs to the standard topology and each basic open set (an open interval) other than R^1 is either an element of $\mathscr{S}$ or can be expressed as the intersection of two elements of $\mathscr{S}$.

3.3. Example: A subbase for the standard topology on R^2 is the set of all $U = \{(x, y) \in R^2 : x$ belongs to an open interval in $R^1\}$ together with all $V = \{(x, y) \in R^2 : y$ belongs to an open interval in $R^1\}$. In other words, the collection of all open horizontal and vertical strips in the plane form a subbase for the standard topology. Typical elements of this subbase are shown in Figure 14.

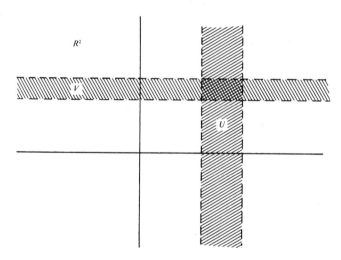

Figure 14

Using the definition of subbase and our knowledge about a base for a topology, we may characterize open sets in terms of a subbase.

3.4. THEOREM: Let $(X, \mathscr{T})$ be a space with subbase $\mathscr{S} = \{S_\alpha : \alpha \in \Delta\}$. Then $U \subset X$, $U \neq \phi$, is open iff for each $x \in U$ there is a finite intersection $\bigcap_{k=1}^{n} S_{\alpha_k}$ such that $x \in \bigcap_{k=1}^{n} S_{\alpha_k} \subset U$.

Proof: The proof is an exercise.

For any topology $\mathscr{T}$ on X, $\mathscr{T}$ itself would qualify as a subbase so that every topology has at least one subbase. It would be natural to ask, on the other hand, if a given collection $\mathscr{S}$ of subsets of a nonempty set X is a subbase for some topology on X. Herein lies the importance of the concept of a subbase. If $\mathscr{S}$ is *any* collection of subsets of X, we shall see next that $\mathscr{S}$ may be used to construct a topology on X such that each member of $\mathscr{S}$ belongs to the topology, $\mathscr{S}$ is a subbase for the topology and, furthermore, this topology is unique. Verification of these facts is made in following two theorems.

3.5. THEOREM: *Let $X \neq \phi$ be a set, and let $\mathscr{S} = \{S_\alpha : \alpha \in \Delta\}$ be any collection of subsets of X. Define $\mathscr{B}[\mathscr{S}]$ to be the collection of all possible finite intersections of members of $\mathscr{S}$ together with X. Then $\mathscr{B}[\mathscr{S}]$ is a base for a topology on X.*

Proof: We show conditions (a) and (b) of Theorem 1.7 are satisfied by $\mathscr{B}[\mathscr{S}]$. Since $X \in \mathscr{B}[\mathscr{S}]$, condition (a) is clearly satisfied. To see condition (b), let $U = \bigcap_{k=1}^{n} S_{\alpha_k}$ and $V = \bigcap_{j=1}^{m} S_{\beta_j}$ be two of the finite intersections in $\mathscr{B}[\mathscr{S}]$, and let $x \in U \cap V = \bigcap_{k=1}^{n} S_{\alpha_k} \cap \bigcap_{j=1}^{m} S_{\beta_j}$. Then x belongs to each S_{α_k} and also to each S_{β_j} and, therefore, to the finite intersection of the sets $W = S_{\alpha_1} \cap S_{\alpha_2} \ldots \cap S_{\alpha_n} \cap S_{\beta_1} \cap S_{\beta_2} \cap \ldots \cap S_{\beta_m} = \bigcap_{k=1}^{n} S_{\alpha_k} \cap \bigcap_{j=1}^{n} S_{\beta_j}$. Consequently, we have demonstrated the existence of an element $W \in \mathscr{B}[\mathscr{S}]$ such that $x \in W \subset U \cap V$, which shows that (b) holds. We have now shown that $\mathscr{B}[\mathscr{S}]$ satisfies (a) and (b) of Theorem 1.7. According to Theorem 1.8, there is a unique topology $\mathscr{T}(\mathscr{B}[\mathscr{S}])$ for X generated by $\mathscr{B}[\mathscr{S}]$ and having $\mathscr{B}[\mathscr{S}]$ as a base.

3.6. THEOREM: *The topology $\mathscr{T}(\mathscr{B}[\mathscr{S}])$ generated by $\mathscr{B}[\mathscr{S}]$ and having $\mathscr{B}[\mathscr{S}]$ as a base is the unique topology on X having $\mathscr{S}$ as a subbase.*

Proof: The instructive proof is left to the reader.

Thus, we have seen that for $X \neq \phi$ and $\mathscr{S}$ any collection of subsets of X, we may construct a topology $\mathscr{T}(\mathscr{B}[\mathscr{S}])$ on X for which each member of $\mathscr{S}$ belongs to the topology. Furthermore, $\mathscr{T}(\mathscr{B}[\mathscr{S}])$ is the unique topology on X having $\mathscr{S}$ as a subbase. Of course, there may be many other topologies on X which contain $\mathscr{S}$ but do not have $\mathscr{S}$

as a subbase. The next theorem states that $\mathscr{T}(\mathscr{B}[\mathscr{S}])$ is the smallest topology on X containing $\mathscr{S}$.

3.7. THEOREM: *Let $X \neq \phi$ and $\mathscr{S}$ any collection of subsets of X. Then $\mathscr{T}(\mathscr{B}[\mathscr{T}])$ is the smallest topology on X containing $\mathscr{S}$.*

Proof: The proof is again left to the reader.

For our purposes, the real usefulness of the subbase will come in defining the product topology for an arbitrary product of topological spaces. This will be done in the next section. As a final theorem for this section, we state when two collections of subsets of a set X are subbases for the same topology on X. We have already seen an analogous theorem for bases.

3.8. THEOREM: *Let $\mathscr{S}_1$ and $\mathscr{S}_2$ be collections of subsets of a set X such that the following conditions hold:*
   (a) If $x \in S_1 \in \mathscr{S}_1$, there exists an $S_2 \in \mathscr{S}_2$ such that $x \in S_2 \subset S_1$.
   (b) If $x \in S_2 \in \mathscr{S}_2$, there exists an $S_1 \in \mathscr{S}_1$ such that $x \in S_1 \subset S_2$. Then $\mathscr{S}_1$ and $\mathscr{S}_2$ subbases for the same topology on X.

Proof: The proof is left as an exercise.

Exercises:

1. Describe the topologies $\mathscr{T}(\mathscr{B}[\mathscr{S}])$ on R^2 where $\mathscr{S}$ consists of
   (a) All open half planes.
   (b) All straight lines.
   (c) All horizontal straight lines.

2. Find a subbase with as few elements as possible for the topology $\mathscr{T} = \{\phi, X, \{a\}, \{b, c\}\}$ on the set $X = \{a, b, c\}$.

3. Let $X = \{a, b, c\}$ and let $\mathscr{S} = \{\{a, b\}, \{b, c\}\}$. List all elements in the topology $\mathscr{T}(\mathscr{B}[\mathscr{S}])$.

4. Prove Theorem 3.4.

5. Prove Theorem 3.6.

6. Prove Theorem 3.7.

7. Prove Theorem 3.8.

8. Let $\mathscr{S}$ be a collection of subsets of X. Let T be the intersection of all topologies on X containing $\mathscr{S}$. Prove that $T = \mathscr{T}(\mathscr{B}[\mathscr{S}])$.

9. Let $(A, \mathcal{T}_A)$ be a subspace of $(X, \mathcal{T})$. Let $\mathcal{S} = \{S_\alpha : \alpha \in \Delta\}$ be a subbase for $\mathcal{T}$. Prove that $\{A \cap S_\alpha : \alpha \in \Delta\}$ is a subbase for $\mathcal{T}_A$.

10. Let X be a set for which each of the topologies $\{\mathcal{T}_\alpha : \alpha \in \Delta\}$ is a topology for X. Define $\bigvee_{\alpha \in \Delta} \mathcal{T}_\alpha$ to be the topology having $\bigcup_{\alpha \in \Delta} \mathcal{T}_\alpha$ as a subbase. Prove that $\bigvee_{\alpha \in \Delta} \mathcal{T}_\alpha$ is the smallest topology on X which is larger than every $\mathcal{T}_\alpha$, $\alpha \in \Delta$.

11. Prove that the collection of all sets of the form $\{x : x < b\}$ together with those of the form $\{x : x > a\}$, where a and b are *rational* numbers, is a subbase for the standard topology on R^1.

4. General Products

In this section we want to define and discuss the topological product of an arbitrary (not necessarily finite) family of topological spaces. To do this we first need to define the product of a family of sets which is indexed by any set whatever. In so doing, we will want our new definition to be the same as the previous one if the indexing set is finite. In addition, we will need a more general definition for the projection function from the product to each of the factor sets. We begin our studies with the general definition of the product of sets. The viewpoint of our definition is that the product is a set of functions.

4.1. Definition: Let $\{X_\alpha : \alpha \in \Delta\}$ be an indexed family of sets and let $X = \bigcup_{\alpha \in \Delta} X_\alpha$. Then the *product* of the family $\{X_\alpha : \alpha \in \Delta\}$ is defined as $\prod_{\alpha \in \Delta} X_\alpha = \{f : f : \Delta \to X, \text{ where } f(\alpha) \in X_\alpha \text{ for all } \alpha \in \Delta\}$. The point $f(\alpha) \in X_\alpha$ is called the αth coordinate of the point f.

In other words, the product of $\{X_\alpha : \alpha \in \Delta\}$ is the set of all possible functions, each of which has the indexing set Δ for a domain, and such that for each $\alpha \in \Delta$, $f(\alpha) \in X_\alpha$. Consequently, we are thinking of each "point" in the product as a function. Figure 15 shows our former concept of a point in R^2 as an ordered pair of real numbers, as well as the new concept in which the same point is labeled as a function.

For a better understanding of Definition 4.1, the reader should consider an example such as $X_1 = \{a, b, c\}$ and $X_2 = \{x, y\}$. First form $X_1 \times X_2$ as a set of ordered pairs, and then form $X_1 \times X_2$ as a set of

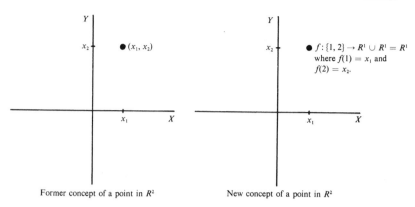

Former concept of a point in R^2 New concept of a point in R^2

Figure 15

functions from the indexing set $\Delta = \{1, 2\}$ to $X_1 \cup X_2$ where $f(1) \in X_1$ and $f(2) \in X_2$. For instance, the point (a, y) would be labeled by the function f_1 where $f_1(1) = a$ and $f_1(2) = y$. The point (b, c) would be labeled with the function f_2 where $f_2(1) = b$ and $f_2(2) = c$. There are six such functions in $X_1 \times X_2$.

These comparisons should clearly illustrate that Definition 4.1 is a generalization of the previous definition for finite products and that the two definitions indeed agree (except for viewpoint) when the indexing set is finite. Thus, if a finite product is involved, we may use the definition of our choice. If, however, arbitrary products are involved, the functional notation and viewpoint of Definition 4.1 must be used. Many of the theorems that we have already proved concerning finite products may also be proved in the present more general setting. Some of these will be extended in this section or in the exercises of this section. When dealing with topological spaces, there are many important results which hold for arbitrary products, but which are rather difficult to prove as far as the scope of this book is concerned. As a consequence, we shall from time to time prove a result for a finite product when, in fact, it holds for more general indexing sets. In all cases, we will try to point out specifically what limitations, if any, are on the indexing set.

Theorem 2.5 of Chapter 1 and Exercise 5 of Section 3, Chapter 1, are results concerning finite products that may now be generalized to arbitrary indexing sets. Recall that we have agreed to use only non-empty indexing sets. One implication of this is that the empty function ϕ cannot be an element of $\prod_{\alpha \in \Delta} X_\alpha$. However, $\prod_{\alpha \in \Delta} X_\alpha$ may be the empty set.

4.2. THEOREM: *For the family of sets* $\{X_\alpha : \alpha \in \Delta\}$, $\prod_{\alpha \in \Delta} X_\alpha \neq \phi$
iff for each $\alpha \in \Delta$, $X_\alpha \neq \phi$.

Proof: Suppose first that for each $\alpha \in \Delta$, $X_\alpha \neq \phi$. We then use
the Axiom of Choice to ensure the existence of a function $f : \Delta \rightarrow$
$\bigcup_{\alpha \in \Delta} A_x$ such that $f(\alpha) \in A_x$. Therefore, $f \in \prod_{\alpha \in \Delta} X_\alpha \neq \phi$. On the other
hand, if $\prod_{\alpha \in \Delta} X_\alpha \neq \phi$, there exists a function $f : \Delta \rightarrow \bigcup_{\alpha \in \Delta} X_\alpha$, such that
$f(\alpha) \in X_\alpha$ for each $\alpha \in \Delta$. Since $\Delta \neq \phi$, $X_\alpha \neq \phi$ for each $\alpha \in \Delta$.

We now see how Theorem 2.4 (a) and (b) of Chapter 1 may be
generalized.

4.3. THEOREM: *Let* $\{X_\alpha : \alpha \in \Delta\}$ *be an indexed family of sets
and* Y *any set. Then*

    (a) $(\bigcup_{\alpha \in \Delta} X_\alpha) \times Y = \bigcup_{\alpha \in \Delta} (X_\alpha \times Y)$.

    (b) $(\bigcap_{\alpha \in \Delta} X_\alpha) \times Y = \bigcap_{\alpha \in \Delta} (X_\alpha \times Y)$.

Proof: The proof of (a) will be given here. On the left side of (a)
we are dealing with the product of two sets and may, therefore, con-
sider them indexed with the set $\{1, 2\}$. Now if $f \in (\bigcup_{\alpha \in \Delta} X_\alpha) \times Y$, then
$f(1) \in \bigcup_{\alpha \in \Delta} X_\alpha$ and $f(2) \in Y$, from which it follows that $f(1) \in X_\alpha$
for at least one $\alpha \in \Delta$. Consequently, $f \in X_\alpha \times Y$ for at least one
$\alpha \in \Delta$, implying that $f \in \bigcup_{\alpha \in \Delta} (X_\alpha \times Y)$.

Conversely, let $f \in \bigcup_{\alpha \in \Delta} (X_\alpha \times Y)$. Then $f \in X_\alpha \times Y$ for at least
one $\alpha \in \Delta$, and because we are again dealing with the product of two
sets, $f(1) \in X_\alpha$ and $f(2) \in Y$ for some $\alpha \in \Delta$. Since $f(1) \in X_\alpha$,
$f(1) \in \bigcup_{\alpha \in \Delta} X_\alpha$ and, therefore, $f \in (\bigcup_{\alpha \in \Delta} X_\alpha) \times Y$.

A detailed proof of the next theorem will further help the reader
become acquainted with the general definition of the product of sets.
The theorem is a generalization of Theorem 3.7, Chapter 1.

4.4. THEOREM: *Let* $\{X_\alpha : \alpha \in \Delta\}$ *and* $\{Y_\alpha : \alpha \in \Delta\}$ *be two families of sets each indexed by the same nonempty set* Δ. *Then*

    (a) $\prod_{\alpha \in \Delta} X_\alpha \cap \prod_{\alpha \in \Delta} Y_\alpha = \prod_{\alpha \in \Delta} (X_\alpha \cap Y_\alpha)$.

    (b) $\prod_{\alpha \in \Delta} X_\alpha \cup \prod_{\alpha \in \Delta} Y_\alpha \subset \prod_{\alpha \in \Delta} (X_\alpha \cup Y_\alpha)$.

Proof: The proof uses only the definitions of union, intersection and
the general product.

With the general product now defined, we next give the general definition for the projection functions.

4.5. Definition: Let $\prod_{\alpha \in \Delta} X_\alpha$ be the product of a family of sets $\{X_\alpha :$ $\alpha \in \Delta\}$. For each $\alpha \in \Delta$, define a function $p_\alpha : \prod_{\alpha \in \Delta} X_\alpha \rightarrow X_\alpha$ by $p_\alpha(f) = f(\alpha)$. The function p_α is called the αth *projection function*.

4.6. Example: Let $X_\alpha = R^1$ for $\alpha \in \Delta = \{1, 2\}$. Suppose $f \in$ $\prod_{\alpha \in \Delta} X_\alpha$ represents the point $(3, 2)$, i.e., $f(1) = 3$ and $f(2) = 2$. Then $p_1(f) = f(1) = 3$ and $p_2(f) = f(2) = 2$. If $U = \{x : 1 < x < 2\}$ is a subset of $X_1 = R^1$, then $p_1^{-1}(U) = \{f : f \in \prod_{\alpha \in \Delta} X_\alpha$ and $f(1) \in U\}$ is an infinite vertical strip in R^2 whose points have first coordinate between 1 and 2.

With the generalized definitions given thus far in this section, we are now in a position to define the topological product of an arbitrary family of topological spaces. Recall that we have defined topologies on nonempty sets only so that in view of Theorem 4.2, whenever a topological product is mentioned it will always be assumed to be nonempty.

4.7. Definition: Let $\{(X_\alpha, \mathcal{T}_\alpha) : \alpha \in \Delta\}$ be a family of topological spaces. Define the collection $\mathcal{S}$ of subsets of $\prod_{\alpha \in \Delta} X_\alpha$ to be $\mathcal{S} =$ $\{p_\alpha^{-1}(U_\alpha) : \alpha \in \Delta$ and U_α is open in $X_\alpha\}$. That is, $\mathcal{S} = \{f \in \prod_{\alpha \in \Delta} X_\alpha :$ $p_\alpha(f) = f(\alpha) \in U_\alpha, \alpha \in \Delta$, and U_α open in $X_\alpha\}$. Using $\mathcal{S}$ as a subbase, the unique topology $\mathcal{T}(\mathcal{B}[\mathcal{S}])$ on $\prod_{\alpha \in \Delta} X_\alpha$ generated by $\mathcal{B}[\mathcal{S}]$ and having $\mathcal{B}[\mathcal{S}]$ as a base is called the *product topology* on $\prod_{\alpha \in \Delta} X_\alpha$. The set $\prod_{\alpha \in \Delta} X_\alpha$ together with the product topology is called the *topological product* of the family $\{(X_\alpha, \mathcal{T}_\alpha) : \alpha \in \Delta\}$ or simply the *product space*. The space X_α is called the αth factor space of the product.

4.8. Example: Consider R^1 with the standard topology and form $R^1 \times R^1 = R^2$. According to Definition 4.7, the elements of $\mathcal{S}$ in R^2 will be open vertical or horizontal strips. Finite intersections of these yield open rectangles as a base and hence the product topology is precisely the standard topology on R^2.

In forming the topological product according to Definition 4.7, we must first take all possible *finite* intersections of members of the subbase collection $\mathscr{S}$ and then adjoin $\prod_{\alpha \in \Delta} X_\alpha$ to obtain a base for the product topology. That is, aside from the entire space, basic open sets have the appearance $p_{\alpha_1}^{-1}(U_{\alpha_1}) \cap P_{\alpha_2}^{-1}(U_{\alpha_2}) \cap \ldots \cap p_{\alpha_n}^{-1}(U_{\alpha_n})$ where n is an arbitrary natural number. Thus, basic open sets in $\prod_{\alpha \in \Delta} X_\alpha$ restrict only a *finite* number of coordinates, $\alpha_1, \alpha_2, \ldots, \alpha_n$, while for all other $\alpha \in \Delta$, *all* points of X_α are used. Consequently, basic open sets are all of those having the form $\bigcap_{k=1}^{n} p_{\alpha_k}^{-1}(U_{\alpha_k}) = U_{\alpha_1} \times U_{\alpha_2} \times \ldots \times U_{\alpha_n} \times \prod_{\alpha \in \Delta} \{X_\alpha : \alpha \neq \alpha_1, \alpha_2, \ldots, \alpha_n\}$ where U_{α_k} is an open set in the space X_{α_k} for $k = 1, 2, \ldots, n$. It is an easy exercise to verify that this equality holds. For a finite indexing set Δ having n elements, Definition 4.7 gives the product topology precisely as we had earlier defined it. Notice, however, that if Δ is infinite and U_α is open in X_α, $U_\alpha \neq X_\alpha$, then $\prod_{\alpha \in \Delta} U_\alpha$ is never open in the product space since no basic open set in the product (remember they restrict only a finite number of coordinates) can be a subset of $\prod_{\alpha \in \Delta} U_\alpha$.

The following theorem is a generalization of Theorem 2.4(a) to an arbitrary topological product.

4.9. THEOREM: *Suppose* $\{(X_\alpha, \mathscr{T}_\alpha) : \alpha \in \Delta\}$ *is a family of topological spaces and consider the product topology on* $\prod_{\alpha \in \Delta} X_\alpha$. *Let* $A_\alpha \subset X_\alpha$ *for each* $\alpha \in \Delta$. *Then* $\prod_{\alpha \in \Delta} \overline{A_\alpha} = \overline{\prod_{\alpha \in \Delta} A_\alpha}$. *That is, the product of closed sets is always closed.*

Proof: Let $f \in \prod_{\alpha \in \Delta} \overline{A_\alpha}$. Then for each α, $p_\alpha(f) = f(\alpha) \in \overline{A_\alpha}$ by the previous definitions of p_α and f. Now let W be any open set in the product topology containing f. Then there exists a basic open set V such that $f \in V = U_{\alpha_1} \times U_{\alpha_2} \times \ldots \times U_{\alpha_n} \times \prod_{\alpha \in \Delta} \{X_\alpha : \alpha \neq \alpha_1, \alpha_2, \ldots, \alpha_n\} \subset W$. Since $p_\alpha(f) \in \overline{A_\alpha}$ for every $\alpha \in \Delta$, $p_{\alpha_k}(f) \in U_{\alpha_k}$ implies U_{α_k} contains a point of A_{α_k} for $k = 1, 2, \ldots, n$. Thus, $V \cap \prod_{\alpha \in \Delta} A_\alpha \neq \phi$, which implies $W \cap \prod_{\alpha \in \Delta} A_\alpha \neq \phi$ so that f is a point or a cluster point of $\prod_{\alpha \in \Delta} A_\alpha$ and, therefore, $f \in \overline{\prod_{\alpha \in \Delta} A_\alpha}$. It follows that $\prod_{\alpha \in \Delta} \overline{A_\alpha} \subset \overline{\prod_{\alpha \in \Delta} A_\alpha}$.

Now let $f \in \overline{\prod_{\alpha \in \Delta} A_\alpha}$ and note that any open set W in the product topology which contains f has the property that $W \cap \prod_{\alpha \in \Delta} A_\alpha \neq \phi$. For each $\alpha \in \Delta$ consider $p_\alpha(f) = f(\alpha) \in X_\alpha$ and let $U_\alpha \subset X_\alpha$ be any open set containing $p_\alpha(f)$. Then $p_\alpha^{-1}(U_\alpha)$ is a subbasic open set in the product topology and contains f. Thus $p_\alpha^{-1}(U_\alpha) \cap \prod_{\alpha \in \Delta} A_\alpha \neq \phi$ and, therefore, $U_\alpha \cap A_\alpha \neq \phi$. Consequently, $p_\alpha(f) \in \overline{A_\alpha}$ for each $\alpha \in \Delta$ so that $f \in \prod_{\alpha \in \Delta} \overline{A_\alpha}$. We now have $\overline{\prod_{\alpha \in \Delta} A_\alpha} \subset \prod_{\alpha \in \Delta} \overline{A_\alpha}$ so that by using the reverse inclusion of the previous paragraph, our theorem is proved.

Our final theorem generalizes Theorem 2.4(b). Its proof is left as an exercise.

4.10. THEOREM: *Let $\{(X_\alpha, \mathcal{T}_\alpha) : \alpha \in \Delta\}$ be a family of spaces and consider the product topology on $\prod_{\alpha \in \Delta} X_\alpha$. If $A_\alpha \subset X_\alpha$ for each $\alpha \in \Delta$, then $\prod_{\alpha \in \Delta} A_\alpha$ (where each A_α has the subspace topology) has the same topology as if considered as a subspace of $\prod_{\alpha \in \Delta} X_\alpha$.*

Proof: The proof is an exercise.

Exercises

1. Prove Theorem 4.3(b).

2. Prove Theorem 4.4.

3. Write out a proof for the equality $\bigcap_{k=1}^{n} p_{\alpha_k}^{-1}(U_{\alpha_k}) = U_{\alpha_1} \times U_{\alpha_2} \times \ldots \times U_{\alpha_n} \times \prod_{\alpha \in \Delta} \{X_\alpha : \alpha \neq \alpha_1, \alpha_2, \ldots, \alpha_n\}$.

4. Let $\{X_\alpha : \alpha \in \Delta\}$ and $\{Y_\alpha : \alpha \in \Delta\}$ be indexed families of sets.
   (a) If for each $\alpha \in \Delta$, $X_\alpha \subset Y_\alpha$, then prove $\prod_{\alpha \in \Delta} X_\alpha \subset \prod_{\alpha \in \Delta} Y_\alpha$.
   (b) If for each $\alpha \in \Delta$, $X_\alpha \neq \phi$ and $\prod_{\alpha \in \Delta} X_\alpha \subset \prod_{\alpha \in \Delta} Y_\alpha$, then prove $X_\alpha \subset Y_\alpha$.

5. For each $\alpha \in \Delta$, let $A_\alpha \subset X_\alpha$ and consider $\prod_{\alpha \in \Delta} X_\alpha$. Prove that if p_α is the αth projection function, then
   (a) $\prod_{\alpha \in \Delta} A_\alpha = \bigcap_{\alpha \in \Delta} p_\alpha^{-1}(A_\alpha)$.
   (b) $(\prod_{\alpha \in \Delta} X_\alpha) \backslash p_\alpha^{-1}(A_\alpha) = p_\alpha^{-1}(X_\alpha \backslash A_\alpha)$.

(c) $(\prod_{\alpha \in \Delta} X_\alpha)\backslash(\prod_{\alpha \in \Delta} A_\alpha) = \bigcup_{\alpha \in \Delta} p_\alpha^{-1}(X_\alpha \backslash A_\alpha).$

6. Let Δ be infinite and each $X_\alpha, \alpha \in \Delta$, be a nondegenerate set having the discrete topology. Prove that the product topology on $\prod_{\alpha \in \Delta} X_\alpha$ is not discrete.

7. If Δ is infinite and $\prod_{\alpha \in \Delta} X_\alpha$ has the product topology, prove that if $U \subset \prod_{\alpha \in \Delta} X_\alpha$ is open, then $p_\alpha(U) = X_\alpha$ for all $\alpha \in \Delta$ except a finite number.

8. Let $\prod_{\alpha \in \Delta} X_\alpha$ have the product topology. Prove that $\prod_{\alpha \in \Delta} A_\alpha$ is dense in $\prod_{\alpha \in \Delta} X_\alpha$ iff A_α is dense in $(X_\alpha, \mathcal{T}_\alpha)$ for each $\alpha \in \Delta$.

9. Let $\{(X_\alpha, \mathcal{T}_\alpha) : \alpha \in \Delta\}$ be an indexed family of spaces and for each $\alpha \in \Delta$, let $\mathcal{B}_\alpha$ be a base for $\mathcal{T}_\alpha$. Let $\mathcal{S}_1 = \{p_\alpha^{-1}(B_\alpha) : \alpha \in \Delta$ and $B_\alpha \in \mathcal{B}_\alpha\}$ where p_α is the αth projection function. Prove that $\mathcal{S}_1$ is a subbase for the product topology on $\prod_{\alpha \in \Delta} X_\alpha$.

10. Prove Theorem 4.10.

11. Let f_0 be a fixed element in the topological product $\prod_{\alpha \in \Delta} X_\alpha$. Define $A \subset \prod_{\alpha \in \Delta} X_\alpha$ as follows: $A = \{f : f(\alpha) \neq f_0(\alpha)$ for only a finite number of $\alpha \in \Delta\}$. Prove A is dense in $\prod_{\alpha \in \Delta} X_\alpha$.

5

Continuous Functions

1. Defining a Continuous Function

We have already considered several special types of functions in our studies thus far. Among them are injective functions, surjective functions, identity functions, etc. Our attention now turns to what is perhaps the most important type of function we shall encounter, the *continuous function*. Not only are continuous functions important here, but their importance in all of mathematics cannot be overemphasized. In order to discuss a continuous function $f: X \to Y$, we always assume that X and Y have topologies and in a moment we shall see what relationship f must have with these topologies in order to be called continuous. First, however, it should be recalled that continuous functions played a major role in the study of calculus. In calculus functions were generally from the R^n spaces into R^n spaces, and in particular from R^1 to R^1, with the standard topology even though the term "topology" might not have been mentioned. To see why the standard topology was involved, remember that the definition of a continuous function $f: R^1 \to R^1$ at a point $a \in R^1$ is usually stated as follows: Given any real number $\epsilon > 0$, there exists a real number $\delta > 0$ such that if $|x - a| < \delta$, then $|f(x) - f(a)| < \epsilon$. The function f is then continuous iff it is continuous at each point of R^1. We recognize immediately that $|x - a| < \delta$ and $|f(x) - f(a)| < \epsilon$ describe the open sets $\{x : a - \delta < x < a + \delta\}$ and $\{f(x): f(a) - \epsilon \ll f(a) + \epsilon\}$, respectively, from the standard topology on R^1.

In defining continuity of a function from one topological space to another, it is desirable to use a definition general enough to apply in all spaces, yet reduce to our previous ideas when applied to familiar spaces. This approach is followed since we have considerable knowledge about topological spaces at this point.

1.1. Definition: Let $(X, \mathcal{T}_x)$ and $(Y, \mathcal{T}_y)$ be topological spaces and $f: X \to Y$. The function f is *continuous* at the point $a \in X$ iff given any open set $V \subset Y$ containing $f(a)$, there exists an open set $U \subset X$ containing a such that $f(U) \subset V$. If f is continuous at every point of X, then f is said to be a *continuous function*.

Notice that the definition implies, and examples will substantiate, that continuity at a point a is strongly dependent upon the topologies involved. The smaller the topology on Y and the larger the topology on X, then the better the chance that a function is continuous. Thus, a given function may be continuous for certain topologies on X and Y, but not continuous if different topologies are defined on these sets. Observe also how the definition conforms to the epsilon-delta definition of calculus of a moment ago. There, the open set $\{f(x) : f(a) - \epsilon < f(x) < f(a) + \epsilon\}$ containing $f(a)$ corresponds to the open set V, while the open set $\{x : a - \delta < x < a + \delta\}$ corresponds to the open set U in Definition 1.1.

Some examples will be helpful at this point.

1.2. Example: Let $f: X \to Y$ be given by $f(x) = x$ where $X = Y = R^1$. We shall show that f is continuous at each point $a \in X$. To do this, let $V \subset Y$ be any open set in Y containing $f(a) = a$. If we choose $U = V$, or any subset of X which is open in X, contains a, and is a subset of V, then $f(U) = U \subset V$ and Definition 1.1 is satisfied. Consequently, f is continuous at each $a \in X$ and is, therefore, a continuous function.

Now suppose the topology on the domain X is changed to the cofinite topology, while that on Y remains the standard topology. We might ask whether f is continuous at the point $a = 1$, for instance. To answer the question, suppose we consider the open set $V = \{y : 0 < y < 2\} \subset Y$ containing $f(a) = f(1) = 1$. The question now asks if there is an open set U belonging to the cofinite topology and containing $a = 1$ such that $f(U) = U \subset V$. The answer is clearly no. Thus, the function fails to be continuous at the point $a = 1$. In fact, the given function $f(x) = x$ is not continuous at any point of the space $X = R^1$ with the cofinite topology.

Definition 1.1, along with a consideration of Example 1.2, leads us to a condition telling when $f : X \to Y$ is not continuous at a point of X.

1.3. A Discontinuity Criterion: The function $f : X \to Y$ is *discontinuous* at $a \in X$ iff there exists some open set $V \subset Y$ containing $f(a)$ such that for every open set $U \subset X$ containing a, $f(U) \not\subset V$.

Many times functions are given whose domains X are proper subsets of larger spaces. For example, X may be a subset of the reals with standard topology. For all of these cases, remember that the topology on X is that of the *subspace topology*.

1.4. Example: Let $X = \{x : 0 \le x \le 2\} \cup \{x : 3 \le x \le 10\}$ be a subset of R^1 and let $Y = R^1$. Define $f : X \to Y$ as follows:

$$f(x) = 2x, \text{ if } 0 \le x \le 2.$$
$$f(x) = 6, \quad \text{if } 3 \le x \le 8.$$
$$f(x) = 10, \text{ if } 8 < x \le 10.$$

As defined, f is continuous at every point of X except $a = 8$. To see that f is not continuous at $a = 8$, let $V = \{y : 5 < y < 7\}$ be an open set containing $f(8) = 6$. Every open set containing $a = 8$ contains a point x such that $x > 8$ and, hence, $f(x) = 10 \notin V$. Therefore, f is not continuous at the point $a = 8$.

To see that f is continuous at the point $x = 2$, for instance, let V be any open set containing $f(2) = 4$. Then there exists an open interval $(a, b) = \{y : a < y < b\} \subset V$ containing the point 4. Furthermore, the left end point a can be chosen so that $a > 0$. Since $f^{-1}(a) = a/2$, consider the open set $(a/2, 2]$ in X. If $a/2 < x \leq 2$, then $f(x) = 2x$ and hence $a < f(x) \leq 4$. Thus, we have found an open set $(a/2, 2]$ in X containing 2 such that $f((a/2, 2]) \subset V$, which shows f is continuous at the point 2. Note that any open set which is a subset of $(a/2, 2]$ and contains 2 would have fulfilled our requirements as well.

Depending upon the individual doing the work and the circumstances at hand, it might be easier to prove that a function is continuous with one method rather than another. This is facilitated by the fact that there are several characterizations of continuous functions and, hence, that any one of them may be used to show continuity of a function. These are given in the next theorem.

1.5. THEOREM: *Let $f : X \to Y$ be a function from one topological space to another. Then the following conditions are equivalent:*
    (a) *The function f is continuous.*
    (b) *For each open set $V \subset Y$, $f^{-1}(V)$ is open in X.*
    (c) *For each basic open set $W \subset Y$, $f^{-1}(W)$ is open in X.*
    (d) *For each subbasic open set $U \subset Y$, $f^{-1}(U)$ is open in X.*
    (e) *For each closed set $M \subset Y$, $f^{-1}(M)$ is closed in X.*
    (f) *For each subset $A \subset X$, $f(\overline{A}) \subset \overline{f(A)}$.*
    (g) *For each subset $B \subset Y$, $\overline{f^{-1}(B)} \subset f^{-1}(\overline{B})$.*

Before considering the proof, notice that the familiar epsilon-delta criterion for continuity is not included above. The reason is that in a general topological space, the expressions $|x - a| < \delta$ and $|f(x) - f(a)| < \epsilon$ have no meaning. A sufficient structure, both topological and algebraic, for using an epsilon-delta characterization of continuity will be at our disposal in a class of spaces called *metric spaces*. The space R^1 is included in this class. When metric spaces are studied, we will also study the epsilon-delta condition. Notice too that the limit characterization of continuity is not listed either. This condition says that $f : X \to Y$ is continuous at $a \in X$ iff for each sequence (x_n) converging to a, the sequence $(f(x_n))$ converges to $f(a)$. This will be studied in the chapter on convergence. This condition is also not valid in general topological spaces.

Proof: (a) implies (b). Let $V \subset Y$ be open and let $x \in f^{-1}(V)$. Then $f(x) \in V$ and V is an open set in Y containing $f(x)$. Since f is continuous at x, there exists an open set $U(x)$ containing x such that $f(U(x)) \subset V$. Thus, $U(x) \subset f^{-1}(V)$. Since each $x \in f^{-1}(V)$ is contained in an open set, which in turn is a subset of $f^{-1}(V)$, $f^{-1}(V)$ is an open subset of X.

(d) implies (e). Let $M \subset Y$ be closed. The set $f^{-1}(M)$ will be shown to be closed by showing $X \backslash f^{-1}(M)$ is open. For any $x \in X \backslash f^{-1}(M)$, $f(x) \in Y \backslash M$. Since M is closed, $Y \backslash M$ is open and, thus, there exists a basic open set W containing $f(x)$ such that $W \subset Y \backslash M$. Furthermore, $W = U_1 \cap U_2 \cap \cdots \cap U_n$, where each $U_k, k = 1, 2, \ldots, n$, is a subbasic open set in Y. That is, $f(x) \in W = U_1 \cap U_2 \cap \ldots \cap U_n \subset Y \backslash M$. Therefore, $x \in f^{-1}(W) = f^{-1}(U_1 \cap U_2 \cap \ldots \cap U_n) = \bigcap_{k=1}^{n} f^{-1}(U_k) \subset f^{-1}(Y \backslash M) = X \backslash f^{-1}(M)$. By (d), each $f^{-1}(U_k)$ is open and, thus, $\bigcap_{k=1}^{n} f^{-1}(U_k)$ is open in X. Consequently, we have shown that for any $x \in X \backslash f^{-1}(M)$, there exists an open set $\bigcap_{k=1}^{n} f^{-1}(U_k)$ such that $x \in \bigcap_{k=1}^{n} f^{-1}(U_k)$ and this open set is a subset of $X \backslash f^{-1}(M)$. This proves $X \backslash f^{-1}(M)$ is open and, thus, $f^{-1}(M)$ is closed.

(f) implies (g). Let $B \subset Y$ and let $A = f^{-1}(B)$. Then $f(A) \subset B$, and by condition (f) and Theorem 2.9(d) of Chapter 3, $f(\bar{A}) \subset \overline{f(A)} \subset \bar{B}$. Thus, $\bar{A} \subset f^{-1}(\bar{B})$. Substituting for A gives $\bar{A} = \overline{f^{-1}(B)} \subset f^{-1}(\bar{B})$.

Proofs of the other implications are left as exercises. Theorem 1.5 leads us to other discontinuity conditions, one of which is given next.

1.6. Another Discontinuity Criterion.   The function $f : X \to Y$ is not continuous iff there is some open set $V \subset Y$ such that $f^{-1}(V)$ is not open in X.

Theorem 1.5 allows several proofs for the following theorem.

1.7. THEOREM: *If $f : X \to Y$ and $g : Y \to Z$ are both continuous functions, then the composition $gf : X \to Z$ is continuous.*

Proof: The easy proof is left as an exercise.

For a function $f : X \to Y$ and a subset $A \subset X$, the restriction of f to A has been defined in Definition 2.5 of Chapter 2. If X is a topological space, then A has the subspace topology, and we may consider the continuity of the restricted function with domain A. Our final theorem of this section tells us that if f is continuous, then f restricted to A is also continuous.

1.8. THEOREM: *Let $f : X \to Y$ be continuous and A a nonempty subset of X. Then $f|A$ is continuous.*

Proof: The instructive proof is left to the reader.

Exercises:

1. Let $X = \{a, b, c\}$, $\mathscr{T}_X = \{\phi, X, \{a\}, \{b, c\}\}$, $Y = \{x, y, z\}$ and $\mathscr{T}_Y = \{\phi, Y, \{y\}, \{z\}, \{y, z\}\}$.
   (a) Let $f = \{(a, x), (b, z), (c, z)\}$. Prove or disprove f is continuous.
   (b) Find all continuous functions from X to Y.

2. Prove that the function in Example 1.4 is continuous at the point $a = 1$. At the point $a = 6$.

3. (a) Let Z be the integers as a subspace of R^1. Let $f : Z \to Z$ be given by $f(x) = 2x$ for each $x \in Z$. Prove or disprove that f is continuous.
   (b) Given an example of a function $f : X \to Y$ and a subset $A \subset X$ such that $f|A$ is continuous but f is not continuous at any point of A.

4. Let $f : X \to Y$ be continuous. If $A \subset X$ has a as a cluster point, prove that $f(a)$ is either a cluster point of $f(A)$ or belongs to $f(A)$. Give an example to show that if a is a cluster point of A, then $f(a)$ need not be a cluster point of $f(A)$.

5. In Theorem 1.5, prove (b) implies (c) and (c) implies (d).

6. In Theorem 1.5, prove (e) implies (f) and (g) implies (a). (This, together with Exercise 5, completes the proof of Theorem 1.5.)

7. Prove or disprove:
   (a) All constant functions are continuous.
   (b) If the domain of the function has the discrete topology, then the function is continuous.
   (c) All injective functions are continuous.

8. Prove Theorem 1.7.

9. Prove Theorem 1.8.

10. Let $X = Y = R^1$ with the standard topology. Prove that $f : X \to Y$ given by $f(x) = x^2$ is continuous.

11. Let $(X, \mathcal{T})$ be a space and R^1 have the standard topology. Prove that $f: X \to R^1$ is continuous iff for each real number $a \in R^1$, both of the sets $\{x \in X : f(x) > a\}$ and $\{x \in X : f(x) < a\}$ are open.

2. Open Functions and Homeomorphisms

If $(X, \mathcal{T}_X)$ and $(Y, \mathcal{T}_Y)$ are topological spaces and $f: X \to Y$ is continuous, then Theorem 1.5(b) shows that the induced function $f^{-1}: \mathcal{P}(Y) \to \mathcal{P}(X)$ preserves open sets. That is, if $V \subset Y$ is open, then $f^{-1}(V)$ is open in X. It is also true that f^{-1} preserves closed sets according to Theorem 1.5(e). It is not true in general even for a continuous function, however, that if $U \subset X$ is open, then $f(U)$ is open in Y. Likewise, for $A \subset X$ closed, $f(A)$ may not be closed in Y. We can verify our last two statements by example.

2.1. Example: Let $X = \{a, b, c\}$, $\mathcal{T}_X = \{\phi, X, \{a\}, \{b, c\}\}$, $Y = \{x, y, z\}$ and $\mathcal{T}_Y = \{\phi, Y, \{y\}\}$ and suppose $f = \{(a, x), (b, y), (c, y)\}$. It is easy to see f is continuous. For the open set $\{a\}$ in X, $f(a) = x$ is not open in Y, and for the closed set $\{b, c\} \subset X$, $f(\{b, c\}) = \{y\}$, which is not a closed set in Y.

For future notation, we attach a special name to functions which preserve open sets and also to those which preserve closed sets.

2.2. Definition: Let $f: X \to Y$ be a given function from the space X to the space Y. (We are not assuming continuity of f in this definition.) Then f is *open* iff for each open set $U \subset X$, $f(U)$ is open in Y. The function f is *closed* iff for each closed set $A \subset X$, $f(A)$ is closed in Y.

Perhaps a word of caution is in order here. The term f *is open* does not refer to f as a subset of $X \times Y$ being open, but rather to the fact that the image under f of each open set in X is open in Y.

It is easy to construct examples of functions which are continuous and open or continuous and closed. One important set of functions that are both continuous and open is the set of projection functions from the topological product to the factor spaces. An exercise at the end of this section will point out that these functions are not necessarily closed, however.

2.3. THEOREM: *Let $\{(X_\alpha, \mathcal{T}_\alpha): \alpha \in \Delta\}$ be a family of topological spaces. Let $p_\alpha: \prod_{\alpha \in \Delta} X_\alpha \to X_\alpha$ be the αth projection function from the topological product $\prod_{\alpha \in \Delta} X_\alpha$ to the space X_α. Then*

(a) *p_α is continuous for each $\alpha \in \Delta$.*
(b) *p_α is open for each $\alpha \in \Delta$.*
(c) *p_α is surjective for each $\alpha \in \Delta$.*

Proof: (a) Let U_α be any open set in X_α. Then $p^{-1}(U_\alpha) = \{f \in \prod_{\alpha \in \Delta} X_\alpha: f(\alpha) \in U_\alpha\}$ is a subbasic open set in $\prod_{\alpha \in \Delta} X_\alpha$. Thus, by Theorem 1.5(b), p_α is continuous.

(b) Let V be an open set in $\prod_{\alpha \in \Delta} X_\alpha$ and it will be shown that $p_\alpha(V)$ is open in X_α. To do this, it will be shown that for each $a \in p_\alpha(V) \subset X_\alpha$, there is an open set S such that $a \in S \subset p_\alpha(V)$. This will establish $p_\alpha(V)$ is open. Thus, let $a \in p_\alpha(V)$. Then there is an $f \in V$ such that $f(\alpha) = a$. Since V is open, there exists an open basic subset S of $\prod_{\alpha \in \Delta} X_\alpha$ such that $f \in S \subset V$ where S has the form $U_{\alpha_1} \times U_{\alpha_2} \times \ldots \times U_{\alpha_n} \times \prod\{X_\alpha: \alpha \neq \alpha_1, \alpha_2, \ldots, \alpha_n\}$, where each U_{α_k} is open in $X_{\alpha_k}, k = 1, 2, \ldots, n$. Therefore, $a = p_\alpha(f) \in p_\alpha(S) \subset p_\alpha(V)$. Now $p_\alpha(S) = U_{\alpha_k}$ if $\alpha = \alpha_k$ for some $k, k = 1, 2, \ldots, n$, and otherwise $p_\alpha(S) = X_\alpha$. Therefore, if $\alpha = \alpha_k$ for some $k = 1, 2, \ldots, n$, then $a = p_\alpha(f) \in p_\alpha(S) = U_{\alpha_k} \subset p_\alpha(V)$. Otherwise $a \in p_\alpha(f) \in p_\alpha(S) = X_\alpha = p_\alpha(V)$. In any case, there is an open set containing the point a which is a subset of $p_\alpha(V)$ thus establishing $p_\alpha(V)$ is open.

(c) By our previous agreement, we are assuming that $X_\alpha \neq \phi$ for all $\alpha \in \Delta$. Let $a \in X_\alpha$. Then by the definition of the general product, there is a function $f \in \prod_{\alpha \in \Delta} X_\alpha$ such that $f(\alpha) = a$. Now by definition of $p_\alpha, p_\alpha(f) = f(\alpha) = a$. Thus, for each $a \in X_\alpha$, there is at least one element f in $\prod_{\alpha \in \Delta} X_\alpha$ such that $p_\alpha(f) = a$, showing that p_α is surjective.

Having the continuity of the projection function, we can now prove a characterization of continuity for functions from a space into a topological product.

2.4. THEOREM: *Let $\{(X_\alpha, \mathcal{T}_\alpha): \alpha \in \Delta\}$ be a family of spaces and $\prod_{\alpha \in \Delta} X_\alpha$ the topological product. Let X be any space $f: X \to \prod_{\alpha \in \Delta} X_\alpha$. Then f is continuous iff $p_\alpha f: X \to X_\alpha$ is continuous for each $\alpha \in \Delta$.*

Proof: If f is continuous, Theorem 1.7 gives $p_\alpha f$ continuous for each $\alpha \in \Delta$. Now suppose $p_\alpha f$ is continuous for each $\alpha \in \Delta$, and consider any subbasic open set $p_\alpha^{-1}(U)$ in $\prod_{\alpha \in \Delta} X_\alpha$. If we can show $f^{-1}(p_\alpha^{-1}(U))$ is open in X, f will be continuous by Theorem 1.5(d). This is established by using Theorem 3.6 of Chapter 2 to obtain $f^{-1}(p_\alpha^{-1}(U)) = (p_\alpha f)^{-1}(U)$, and then using the continuity of $p_\alpha f$ to see that $(p_\alpha f)^{-1}(U)$ is open.

A type of continuous function that is of fundamental importance to the study of topology will be defined at this time. The remainder of this section is devoted to its study.

2.5. Definition: Let $f : X \to Y$ be a bijective function from the space X to the space Y. If f is open and continuous, then f is called a *homeomorphism*. If f is a homeomorphism from X to Y, then the spaces X and Y are said to be *homeomorphic*. This fact will sometimes be denoted by $X \cong Y$.

For characterizations of a function being a homeomorphism we have the following theorem.

2.6. THEOREM: *Let $f : X \to Y$ be bijective. Then the following statements are equivalent:*
   (a) f is a homeomorphism.
   (b) $f : X \to Y$ and $f^{-1} : Y \to X$ are both continuous.
   (c) f is continuous and closed.
   (d) $f(\bar{A}) = \overline{f(A)}$ for each subset $A \subset X$.

Proof: The instructive proof is left to the reader as an exercise.

As a use of Theorem 2.6, consider the following theorem.

2.7. THEOREM: *If $f : X \to Y$ is a homeomorphism and $A \subset X$, then the subspace A is homeomorphic to the subspace $f(A) \subset Y$.*

Proof: Since $f : X \to Y$ is a bijection, certainly $(f|A): A \to f(A)$ is bijective. Furthermore, the restriction of any continuous function is continuous according to Theorem 1.8. Therefore, $f|A$ and $f^{-1}|f(A) = (f|A)^{-1}$ are both continuous so that Theorem 2.6(b) gives $f|A$ as a homeomorphism.

2.8. THEOREM: *On any family of spaces $\{X_\alpha : \alpha \in \Delta\}$, the relation "is homeomorphic to" is an equivalence relation.*

Proof: Again, the straightforward proof is left to the reader.

If $(X, \mathcal{T}_X)$ and $(Y, \mathcal{T}_Y)$ are topological spaces and $f : X \to Y$ is a homeomorphism, a study of Definition 2.5 reveals that not only is f a bijective function from X to Y, but when restricted to $\mathcal{T}_X$ the induced function $f : \mathcal{P}(X) \to \mathcal{P}(Y)$ is also bijective. In other words, f is a bijection from X to Y and, at the same time, a bijection from $\mathcal{T}_X$ to $\mathcal{T}_Y$. This means that anything we say about X or subsets of X which is expressed exclusively in terms of the topology on X, can also be said about Y or the corresponding subsets of Y. Therefore, two homeomorphic spaces have indistinguishable topological structures. (Compare this with *isomorphism* studied in algebra.) Our discussion is now formalized with Definition 2.9.

2.9. Definition: A property of a space X is called a *topological property* iff every space Y homeomorphic to X also has the same property.

For example, the property of having a countable dense subset is a topological property. To see this, let X be a space having a countable dense subset A. That is, A is countable and $\bar{A} = X$. Now let Y be any space homeomorphic to X, and we will establish that Y also has a countable dense subset. If f is the homeomorphism, then $f(A) \subset Y$ is the required countable dense subset. To see this, note that $f(A)$ is countable and by use of Theorem 1.5(f), $\overline{f(A)} \supset f(\bar{A}) = f(X) = Y$, so that $Y \subset \overline{f(A)}$. Of course, $\overline{f(A)} \subset Y$, which gives $Y = \overline{f(A)}$, or that $f(A)$ is a countable dense subset of Y.

We will become acquainted with many other topological properties as our studies progress. Some of immediate importance to us are listed next.

2.10. THEOREM: *Let X and Y be topological spaces and $f : X \to Y$ a homeomorphism. For each subset $A \subset X$, we have*
  (a) *The point $a \in A$ is an interior point of A iff $f(a)$ is an interior point of $f(A) \subset Y$.*
  (b) *The point $a \in X$ is a cluster point of A iff $f(a)$ is a cluster point of $f(A) \subset Y$.*
  (c) *The point $a \in X$ is a boundary point of A iff $f(a)$ is a boundary point of $f(A) \subset Y$.*

Proof: Only the proof of (b) will be given here. We first show that if a is a cluster point of A, then $f(a)$ is a cluster point of $f(A)$. To do this let V be any open set in Y containing $f(a)$. Then $f^{-1}(V)$ is open in X, due to the continuity of f, and contains a. Since a is a cluster point of A, $f^{-1}(V)$ contains a point $x \in A$ such that $x \neq a$ and, therefore, $f(x) \in f(f^{-1}(V)) = V$, $f(x) \in f(A)$ and $f(x) \neq f(a)$. Consequently, every open set V containing $f(a)$ contains a point of $f(A)$ distinct from $f(a)$, implying $f(a)$ is a cluster point of $f(A)$.

Conversely, if $f(a)$ is a cluster point of $f(A)$, let U be any open set containing $a = f^{-1}(f(a))$. Then $f(U)$ is open, because f is a homeomorphism, and contains $f(a)$. Therefore, $f(U)$ contains a point $y \in f(A)$ distinct from $f(a)$ so that the point $f^{-1}(y) \in f^{-1}(f(U)) = U$ and, furthermore, $f^{-1}(y) \neq a$. Our conclusion is that a is a cluster point of A. This proves part (a).

2.11. Example: (Distance is not a topological property.) The sets $X = \{x : 0 \le x \le 1\}$ and $Y = \{x : 0 \le x \le 2\}$ as subspaces of R^1 are homeomorphic. It is not hard to see that $f : X \to Y$, given by $f(x) = 2x$ for each $x \in X$, is a homeomorphism from X to Y. However, X has the property that the distance between any two points in X is less than or equal to 1, while Y does not have this property.

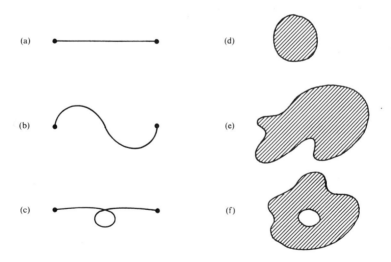

Figure 16

In Figure 16 consider the six figures as subspaces of R^2. The subspaces in (a) and (b) are homeomorphic as a projection shows. However, the subspace in (c) is not homeomorphic to either of those in (a) or (b). The subspaces in (d) and (e) are homeomorphic, while the one in (f) is not homeomorphic to (d) or (e). None of the subspaces in (a), (b), or (c) are homeomorphic to those in (d), (e), or (f).

For each point $(a, b) \in X \times Y$, the subsets $X \times \{b\}$ and $\{a\} \times Y$ of $X \times Y$ are called *slices* through (a, b) and parallel to the X and Y factor spaces, respectively. A slice is homeomorphic to the factor space to which it is parallel.

2.12. THEOREM: *Let $(X, \mathcal{T}_X)$ and $(Y, \mathcal{T}_Y)$ be topological spaces and let (a, b) be any point in the topological product space $X \times Y$. Then the subspace $X \times \{b\}$ is homeomorphic to X, and the subspace $\{a\} \times Y$ is homeomorphic to Y.*

Proof: Only the proof that $X \times \{b\}$ is homeomorphic to X will be given, since the other part of the conclusion is proved in a similar manner. Define the function $h : X \times \{b\} \to X$ as $h(x, b) = x$ for each point $(x, b) \in X \times \{b\}$. If $h(x_1, b) = h(x_2, b)$, then $x_1 = h(x_1, b) = h(x_2, b) = x_2$, which implies that h is injective. Furthermore, h is evidently a surjective function. Theorem 1.8 now tells us h is continuous. The reason is that $h = p_X|(X \times \{b\})$, and p_X is continuous according to Theorem 2.3. (p_X is the projection function $p_X : X \times Y \to X$.) Finally, if we can show that h is open, then h will be a homeomorphism and X will be homeomorphic to $X \times \{b\}$. To help us with this task, observe that since the basic open sets in $X \times Y$ are of the form $U \times V$ where U is open in X and V is open in Y, then the basic open sets of the subspace topology on $X \times \{b\}$ have the form $(U \times V) \cap (X \times \{b\})$ according to Theorem 1.6 of Chapter 4. Now let W be open in $X \times \{b\}$ and we shall show $h(W) = \{x \in X : (x, b) \in W\}$ is open in X. For each $x_0 \in h(W)$ there is a point $(x_0, b) \in W$ such that $h(x_0, b) = x_0$. Therefore, there exists a basic open set in $X \times \{b\}$ containing the point (x_0, b) and which is a subset of W. It follows from our earlier observation that this basic open set has the form $(U(x_0) \times V(b)) \cap (X \times \{b\})$, where $U(x_0)$ is open in X and contains x_0, and $V(b)$ is open in Y and contains b. Using Theorem 3.7(a) of Chapter 1, we have

$$(U(x_0) \times V(b)) \cap (X \times \{b\}) = (U(x_0) \cap X) \times (V(b) \cap \{b\}) =$$
$$U(x_0) \times \{b\},$$

and since $(x_0, b) \in U(x_0) \times \{b\} \subset W$, our conclusion is that $x_0 \in U(x_0)$ $\subset h(W) = \{x \in X : (x, b) \in W\}$. This means $h(W)$ is open.

When we first defined the product of two sets, we pointed out that this operation was not associative, and due to this fact, gave a non-inductive definition for the product of a finite number of sets. (Definition 3.4 of Chapter 1.) Topologically speaking, non-associativity is not a serious problem since such finite product spaces are homeomorphic, as is shown next. [In fact, it may be shown (see, for example, reference 2 of the Bibliography) that for the family $\{(X_\alpha, \mathcal{T}_\alpha) : \alpha \in \Delta\}$ of topological spaces, if $\{\Delta_\beta : \beta \in \Omega\}$ is any partition of Δ whatever and $B_\beta = \prod_{\alpha \in \Delta_\beta} X_\alpha$, then $\prod_{\alpha \in \Delta} X_\alpha \cong \prod_{\beta \in \Omega} B_\beta$.] Topologically speaking, this means for finite products we could have defined $X_1 \times X_2 \times X_3$ $= (X_1 \times X_2) \times X_3$, and inductively in general, that $X_1 \times X_2 \times \ldots$ $\times X_n = (X_1 \times X_2 \times \ldots \times X_{n-1}) \times X_n$, using $=$ rather than $\cong$ with no problems having arisen.

2.13. THEOREM: *Let $\{(X_k, \mathcal{T}_k) : k = 1, 2, \ldots, n\}$ be a finite family of spaces. Then the product spaces $X_1 \times X_2 \times \ldots \times X_n$, $(X_1 \times X_2 \times \ldots \times X_{n-1}) \times X_n$ and $X_1 \times (X_2 \times X_3 \times \ldots \times X_n)$ are homeomorphic.*

Proof: We shall only show that $X_1 \times X_2 \times \ldots \times X_n \cong (X_1 \times X_2 \times \ldots \times X_{n-1}) \times X_n$. To this end, define $h : (X_1 \times X_2 \times \ldots \times X_n)$ $\rightarrow (X_1 \times X_2 \times \ldots \times X_{n-1}) \times X_n$ as $h(x_1, x_2, \ldots, x_n) = ((x_1, x_2, \ldots, x_{n-1}), x_n)$. Then h is clearly a bijection and, furthermore, $h(U_1 \times U_2 \times \ldots \times U_n) = (U_1 \times U_2 \times \ldots \times U_{n-1}) \times U_n$ where U_k is any subset of X_k for $k = 1, 2, \ldots, n$. This last equality should be apparent from the definition of the function h. Since basic open sets in the spaces $X_1 \times X_2 \times \ldots \times X_n$ and $(X_1 \times X_2 \times \ldots \times X_{n-1}) \times X_n$ have the form $U = U_1 \times U_2 \times \ldots \times U_n$ and $V = (U_1 \times U_2 \times \ldots \times U_{n-1}) \times U_n$, respectively, where U_k is an open subset of X_k for $k = 1, 2, \ldots, n$, $h(U) = V$ and $h^{-1}(V) = U$, Theorem 1.5(c) shows both h and h^{-1} are continuous. Therefore, h is a homeomorphism by Theorem 2.6(b).

In view of Theorem 2.13, we know that $R^3 = R^1 \times R^1 \times R^1$ $\cong (R^1 \times R^1) \times R^1 \cong R^2 \times R^1$ and, in general, $R^n \cong R^{n-1} \times R^1$ for each $n \in N$.

Finally, we are now in a position to prove Theorem 2.5 of Chapter 4 for finite product spaces.

2.14. THEOREM: Let $\{(X_k, \mathscr{T}_k) : k = 1, 2, \ldots, n\}$ be a finite family of spaces. If $A_k \subset X_k, k = 1, 2, \ldots, n$, then

(a) $\text{Int}\,(\prod\limits_{k=1}^{n} A_k) = \prod\limits_{k=1}^{n} \text{Int}\,(A_k)$.

(b) $(\prod\limits_{k=1}^{n} A_k)' = (A_1' \times \bar{A}_2 \times \ldots \times \bar{A}_n) \cup$
$(\bar{A}_1 \times A_2' \times \bar{A}_3 \times \ldots \times \bar{A}_n) \ldots \cup$
$(\bar{A}_1 \times \bar{A}_2 \times \ldots \times \bar{A}_{n-1} \times A_n')$.

(c) $\text{Bd}\,(\prod\limits_{k=1}^{n} A_k) = (\text{Bd}(A_1) \times \bar{A}_2 \times \ldots \times \bar{A}_n) \cup$
$(\bar{A}_1 \times \text{Bd}(A_2) \times \bar{A}_3 \times \ldots \times \bar{A}_n) \ldots \cup$
$(\bar{A}_1 \times \bar{A}_2 \times \ldots \times \bar{A}_{n-1} \times \text{Bd}(A_n))$.

Proof: Again, we shall give only the proof to (b), while the others are left as exercises. Our proof is by mathematical induction. Theorem 2.5(b) of Chapter 4 establishes the case for $n = 2$. Now let n be a natural number such that for all $A_k \subset X_k, k = 1, 2, \ldots, n$, it follows that $(A_1 \times A_2 \times \ldots \times A_n)' = (A_1' \times \bar{A}_2 \times \ldots \times \bar{A}_n) \cup (\bar{A}_1 \times A_2' \times \bar{A}_3 \times \ldots \times \bar{A}_n) \cup \ldots \cup (\bar{A}_1 \times \bar{A}_2 \times \ldots \times \bar{A}_{n-1} \times A_n')$ in the product space $\prod\limits_{k=1}^{n} X_k$. Our goal is to prove that for the natural number $n + 1$, if $A_k \subset X_k, k = 1, 2, \ldots, n + 1$, then $(A_1 \times A_2 \times \ldots \times A_{n+1})' = (A_1' \times \bar{A}_2 \times \ldots \times \bar{A}_{n+1}) \cup (\bar{A}_1 \times A_2' \times \bar{A}_3 \times \ldots \times \bar{A}_{n+1}) \cup \ldots \cup (\bar{A}_1 \times \bar{A}_2 \times \ldots \times \bar{A}_n \times A_{n+1}')$. The first step in the proof is to observe that there is a homeomorphism $f : (X_1 \times X_2 \times \ldots \times X_{n+1}) \to (X_1 \times X_2 \times \ldots \times X_n) \times X_{n+1}$ given by $f(x_1, x_2, \ldots, x_{n+1}) = ((x_1, x_2, \ldots, x_n), x_{n+1})$, according to Theorem 2.13. Then we use Theorem 2.10(b) to see that $a \in (A_1 \times A_2 \times \ldots \times A_{n+1})'$ iff $f(a) \in ((A_1 \times A_2 \times \ldots \times A_n) \times A_{n+1})'$. Using the case when $n = 2$, we have $f(a) \in ((A_1 \times A_2 \times \ldots \times A_n) \times A_{n+1})'$ iff $f(a) \in (A_1 \times A_2 \times \ldots \times A_n)' \times \bar{A}_{n+1} \cup \overline{(A_1 \times A_2 \times \ldots \times A_n)} \times A_{n+1}'$. The inductive hypothesis, along with Theorem 4.9 of Chapter 4, states that this is true iff $f(a) \in ((A_1' \times \bar{A}_2 \times \ldots \times \bar{A}_n) \cup (\bar{A}_1 \times A_2' \times \bar{A}_3 \times \ldots \times \bar{A}_n) \cup \ldots \cup (\bar{A}_1 \times \bar{A}_2 \times \ldots \times \bar{A}_{n-1} \times A_n')) \times \bar{A}_{n+1} \cup (\bar{A}_1 \times \bar{A}_2 \times \ldots \times \bar{A}_n) \times A_{n+1}'$. Applying Theorem 4.3(a) of Chapter 4, we now have our last statement as true iff $f(a) \in (A_1' \times \bar{A}_2 \times \ldots \times \bar{A}_n) \times \bar{A}_{n+1} \cup (\bar{A}_1 \times A_2' \times \bar{A}_3 \times \ldots \times \bar{A}_n) \times \bar{A}_{n+1} \cup \ldots \cup (\bar{A}_1 \times \bar{A}_2 \times \ldots \times \bar{A}_{n-1} \times A_n') \times \bar{A}_{n+1} \cup (\bar{A}_1 \times \bar{A}_2 \times \ldots \times \bar{A}_n) \times A_{n+1}'$ and since f is bijective, this is true iff $a \in (A_1' \times \bar{A}_2 \times \bar{A}_3 \times \ldots \times \bar{A}_{n+1}) \cup (\bar{A}_1 \times A_2' \times \bar{A}_3 \times \ldots \times \bar{A}_{n+1}) \cup \ldots \cup (\bar{A}_1 \times \bar{A}_2 \times \ldots \times \bar{A}_n \times A_{n+1}')$. From this, our conclusion is that $(A_1 \times A_2 \times \ldots \times A_{n+1})' = (A_1' \times \bar{A}_2 \times \bar{A}_3 \times \ldots \times \bar{A}_{n+1}) \cup$

$(\bar{A}_1 \times \bar{A}_2' \times \bar{A}_3 \times \ldots \times \bar{A}_{n+1}) \cup \ldots \cup (\bar{A}_1 \times \bar{A}_2 \times \ldots \times \bar{A}_n \times \bar{A}_{n+1}')$.
By the Principle of Mathematical Induction, conclusion (b) holds for each $n \in N$.

Exercises:

1. (a) Define two topologies for R^1 which make the two resulting spaces non-homeomorphic.
   (b) Let $X = \{a, b, c\}$, $\mathscr{T}_X = \{\phi, X, \{a\}, \{b, c\}\}$ and $Y = \{x, y, z\}$. Give a topology for Y which makes X and Y homeomorphic. Give a topology for Y which makes X and Y non-homeomorphic.

2. (a) Give an example of a bijective function $f: X \to Y$ such that f is continuous, but $f^{-1}: Y \to X$ is not continuous.
   (b) Prove that $f: X \to Y$ is a homeomorphism iff $f^{-1}: Y \to X$ is a homeomorphism.

3. Give an example of a closed subset of R^2 whose image under the projection function p_X is not closed. (Thus, projection functions are not always closed. This also shows that a function may be open and not closed.)

4. Prove that the product topology is the smallest topology on $\prod_{\alpha \in \Delta} X_\alpha$, which makes $p_\alpha: \prod_{\alpha \in \Delta} X_\alpha \to X_\alpha$ continuous for each $\alpha \in \Delta$.

5. Prove that the property of a space having an isolated point is a topological property.

6. For any two spaces X and Y, prove $X \times Y \cong Y \times X$.

7. For Theorem 2.6, prove that
   (a) Condition (b) holds iff condition (a) holds.
   (b) Condition (c) holds iff condition (b) holds.
   (c) Condition (d) holds iff condition (b) holds.
   (Now all conditions in Theorem 2.6 are equivalent.)

8. Prove Theorem 2.8.

9. Let $f: X \to Y$ be a function and let $G: X \to f \subset X \times Y$ be given by $G(x) = (x, f(x))$. Prove G is a homeomorphism iff f is continuous.

10. Let $f: X \to Y$ be an open function. Prove that the function $G: X \to f \subset X \times Y$ given in Exercise 9 is an open function.

11. Let $f: X \to Y$ and $g: Y \to Z$ be given functions. Prove

(a) If gf is open (closed) and if f is continuous and surjective, then g is open (closed).

(b) If gf is open (closed) and if g is continuous and injective, then f is open (closed).

12. Prove that $f: X \to Y$ is closed iff for each closed $A \subset X$, then $\{y \in Y : f^{-1}(y) \cap A \neq \phi\}$ is closed in Y.

13. (a) Prove Theorem 2.10(a).
    (b) Prove Theorem 2.10(c).

14. Prove Theorem 2.14(a).

15. Prove Theorem 2.14(c).

16. Let f_0 be a point in the topological product $\prod_{\alpha \in \Delta} X_\alpha$. Let $\beta \in \Delta$ be fixed and define a *slice* through f_0 parallel to X_β as $S(f_0, \beta) = \{f : f(\alpha) = f_0(\alpha)$ for all $\alpha \in \Delta$ where $\alpha \neq \beta\} \subset \prod_{\alpha \in \Delta} X_\alpha$. Prove $S(f_0, \beta) \cong X_\beta$ for each $\beta \in \Delta$.

3. The Identification Topology

The concept of a quotient set has already been discussed in Chapter 2. There, we considered a set X, an equivalence relation R on X, and then formed the set X/R, whose elements were the equivalence classes of X. The set X/R is called the quotient set of X relative to R, and in a sense, we are identifying certain subsets (equivalence classes) of X with points in X/R. Our next main objective is to study quotient sets when considered as a topological space. This will be done specifically in Section 4. For the present section, the manner in which the quotient set is topologized will be studied in more general form. In general, this topology is called an *identification topology*, and we will immediately recognize the method by which it is obtained as one already introduced.

3.1. Definition: Let $(X, \mathcal{T})$ be a space, Y an arbitrary set, and $f: X \to Y$ a surjection. The topology induced on Y by f and $(X, \mathcal{T})$ is called the *identification topology* on Y. If X and Y are both spaces, and $f: X \to Y$ is a continuous surjection, then f is called an *identification function* iff the topology on Y is precisely the identification topology.

An immediate consequence of this definition is that whenever Y has the identification topology, $f: X \to Y$ is continuous. Not every continuous surjection is an identification as is shown by example in Exercise 1 of this section. Every homeomorphism is an identification function as seen from either the definition or the following theorem.

3.2. THEOREM: *Let $(X, \mathcal{T}_X)$ and $(Y, \mathcal{T}_Y)$ be spaces and $f: X \to Y$ a continuous open (closed) surjection. Then f is an identification function.*

Proof: It must be shown that $\mathcal{T}_Y$ is the same topology as that induced by f and $(X, \mathcal{T}_X)$. First, let U be an open set in Y belonging to the topology induced on Y by f and $(X, \mathcal{T}_X)$, and we will show that $U \in \mathcal{T}_Y$. It follows that $f^{-1}(U)$ is open in X. Since f is surjective, then $f(f^{-1}(U)) = U$. Furthermore, since f is open, U is open in Y and hence $U \in \mathcal{T}_Y$.

Next, let $U \in \mathcal{T}_Y$. Then $f^{-1}(U)$ is open in X because f is continuous, which means that U belongs to the induced topology on Y by f and $(X, \mathcal{T}_X)$. Therefore, the two topologies in question are the same. The "closed" part of the theorem is proved in a similar fashion.

Theorem 3.2, along with Theorem 2.3, shows that each projection function $p_\alpha : \prod_{\alpha \in \Delta} X_\alpha \to X_\alpha$ is an identification function.

3.3. THEOREM: *Let $f: X \to Y$ be continuous. If there exists a continuous function $g : Y \to X$ such that $fg = 1_Y$, then f is an identification function.*

Proof: The proof is an exercise.

The next three theorems exhibit fundamental properties of identification functions and topologies.

3.4. THEOREM: *Let $(X, \mathcal{T}_X), (Y, \mathcal{T}_Y)$ and $(Z, \mathcal{T}_Z)$ be spaces, and $f: X \to Y$ and $g : Y \to Z$ both be identification functions. Then $gf: X \to Z$ is an identification function.*

Proof: The proof, a straightforward application of Definition 3.1, is left as an exercise.

3.5. THEOREM: *Let $(X, \mathcal{T}_x)$ and $(Y, \mathcal{T}_Y)$ be spaces and $f: X \to Y$ an identification function. Then the identification topology is the largest topology on Y which makes f continuous.*

Proof: The proof is again left as an easy exercise.

3.6. THEOREM: *Let $f: X \to Y$ be an identification function. Then f is open (closed) iff for each open (closed) set $U \subset X$, then $f^{-1}(f(U))$ is open (closed) in X.*

Proof: Our proof will be for the open part of the theorem. The proof for the closed part is similar. First, suppose f is an open function, and let U be open in X. Then $f(U)$ is open in Y, and since f is an identification function, $f^{-1}(f(U))$ is open in X. Conversely, suppose that for each open set U in X, $f^{-1}(f(U))$ is open in X. Then for each open U in X, $f(U)$ is open in Y. The reason is that for the identification topology on Y, $f(U) \subset Y$ is open iff $f^{-1}(f(U))$ is open in X.

Our final theorem has several applications in future considerations of identification functions.

3.7. THEOREM: *If $(X, \mathcal{T}_x), (Y, \mathcal{T}_Y)$ and $(Z, \mathcal{T}_z)$ are spaces, $f: X \to Y$ an identification and $g: Y \to Z$ any function, then g is continuous iff gf is continuous.*

Proof: If g is continuous, then gf is continuous according to Theorem 1.7.

For the converse, if $gf: X \to Z$ is continuous, it is to be shown that $g: Y \to Z$ is continuous. To do this, let U be any open set in Z and prove that $g^{-1}(U)$ is open in Y. Since gf is continuous, $(gf)^{-1}(U) = f^{-1}(g^{-1}(U))$ is open in X. Thus, $g^{-1}(U) \subset Y$ is open because f is an identification, and by definition, a set $V = g^{-1}(U)$ is open in Y iff $f^{-1}(V)$ is open in X. Consequently, g is continuous.

Exercises:

1. Give an example of a continuous surjection that is not an identification.

2. Prove the "closed" part of Theorem 3.2.

3. Give an example of an identification function that is not open.

4. Prove Theorem 3.3.

5. Prove Theorem 3.4.

6. Prove Theorem 3.5.

7. Let $(X, \mathcal{T})$ be a space and $f : X \to Y$ a surjection. If Y has the identification topology, prove that $A \subset Y$ is closed iff $f^{-1}(A)$ is closed in X.

8. Let $f : X \to Y$ be an identification function. If Z is any set and $g : Y \to Z$ is a surjection, prove that gf is an identification function iff g is an identification function.

9. Let X be a space and $A \subset X$ a subspace. If there exists a continuous function $f : X \to A$ such that $f|A = 1_A$, then f is called a *retraction* of X onto A and A is called a *retract* of X under f. Prove that every retract is an identification function.

10. Let $f : X \to Y$ be a continuous surjection. Then f is an identification function iff the following condition holds: For each space Z and each function $g : Y \to Z$, the continuity of gf implies the continuity of g.

4. Quotient Spaces

Suppose that X is a topological space and R is an equivalence relation on X. As we have seen in Chapter 2, there is a surjective projection function $p : X \to X/R$ given by $p(x) = R(x)$ for each $x \in X$. The projection function is the device by which we topologize X/R.

4.1. Definition: Let $(X, \mathcal{T})$ be a space, R an equivalence relation on X, and $p : X \to X/R$ the projection function. The *quotient topology* on X/R is the identification topology induced by p and $(X, \mathcal{T})$. The set X/R, together with the quotient topology, is called a *quotient space*.

Of course, with only the quotient topology on X/R, p is an identification function so that we have the results of Section 3 at our disposal when studying quotient spaces. We begin our investigation with an example.

4.2. Example: Let $X = \{1, 2, 3, 4\}$ where $\mathcal{T}_X = \{\phi, X, \{1\}, \{2\}, \{3\}, \{1, 2\}, \{1, 3\}, \{2, 3\}\}$ is a topology for X. For any $a, b \in X$, define aRb

iff $a + b$ is an even integer. Then R is an equivalence relation and $R(1) = \{1, 3\}$ and $R(2) = \{2, 4\}$ are the two equivalence classes. Thus, $X/R = \{R(1), R(2)\}$ consists of exactly two points. Now $p^{-1}(R(1)) = \{1, 3\}$ is open in X. This means by Definition 3.1 that the point $R(1)$ belongs to the quotient space topology. However, $p^{-1}(R(2)) = \{2, 4\}$ is not open in X and, therefore, $R(2)$ is not open in X/R. Therefore, the quotient space topology on X/R is $\{\phi, X/R, R(1)\}$.

There are many ways, of course, to describe an equivalence relation on a space $(X, \mathcal{T})$. One way of determining such a relation is to consider any subset $A \subset X$ and define each point in $X \backslash A$ to be related to itself and to no other point in X, and define each point of A to be related to every other point in A. Following this description, the set A determines the equivalence relation, with it being customary to write X/A for the quotient set. Thus, each point in $X \backslash A$ retains its identity in X/A while the entire set $A \subset X$ is *collapsed to a single point* in X/A. The reason is that A itself is a single equivalence class, while each point of $X \backslash A$ represents an equivalence class by itself. An example will illustrate this concept.

4.3. Example: Let $A \subset R^1$ be described as $A = \{x : 0 < x < 1\}$ and consider the quotient space R^1/A obtained by collapsing A to a point $a \in A$.

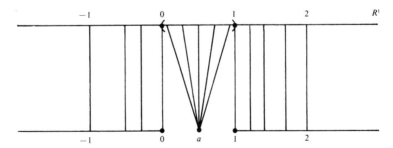

Figure 17

In discussing the quotient space topology, we see the set $V = \{x : 2 < x < 3\} \subset R^1/A$, for example, is open because $p^{-1}(V) = V$, which is open in R^1. Examination shows that open sets in R^1/A are the usual open intervals or unions of open intervals, except when they contain

0, a, or 1. The set $\{a\}$ is itself open in R^1/A, since $p^{-1}(a) = A$ is open in R^1. However, any open set in R^1/A containing either 0 or 1 must also contain a.

4.4. THEOREM: *Let $(X, \mathcal{T})$ be a space, R an equivalence relation on X, and X/R the quotient space. Then the projection $p : X \to X/R$ is open (closed) iff for each open (closed) $U \subset X$, $R(U) = \bigcup_{x \in U} R(x)$ is open (closed) in X.*

Proof: Since $p: X \to X/R$ is an identification function and $p^{-1}(p(U)) = R(U)$, Theorem 3.6 gives the desired conclusion.

Now suppose $f : X \to Y$ is a surjective function and define a relation on X as follows: For any two points $x_1, x_2 \in X$, $x_1 R x_2$ iff $f(x_1) = f(x_2)$. It has been shown in Exercise 6, Section 3 of Chapter 2, that R is an equivalence relation, and because it is induced by the function f we will henceforth denote it by $R(f)$. Having the equivalence relation $R(f)$, we can then form the quotient set $X/R(f)$ and, since f is a surjection, a bijective function $h: Y \to X/R(f)$ may be defined as follows: For each $y \in Y, h(y) = R(a)$ iff $f(a) = y$. This brings us to our main theorem about quotient spaces.

4.5. THEOREM: *Let $f: X \to Y$ be a continuous surjection from the space X to the space Y. Then $h : Y \to X/R(f)$ is a homeomorphism iff f is an identification.*

Proof: Consider the following diagram which the theorem describes.

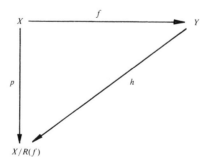

From the remarks preceding the theorem, we have $p = hf$. To prove the theorem, first let f be an identification and we will show that the bijection h is a homeomorphism. By Theorem 3.7, h is continuous because p is continuous and $p = hf$. Also by Theorem 3.7, h^{-1} is continuous because f is continuous by hypothesis, p is an identification, and $f = h^{-1}p$. Therefore, h is a homeomorphism.

Conversely, suppose h is a homeomorphism and we shall show that f is an identification. To do this, note that $f = h^{-1}p$ and that both p and h^{-1} are identification functions. Then by Theorem 3.4, $h^{-1}p = f$ is an identification.

Any partition of X serves to identify certain subsets of X, namely the elements of the partition, as single elements in $\mathscr{P}(X)$. Once we have such an identification, we automatically have an equivalence relation on X, and from there, the quotient space. We now consider some interesting examples of such identifications.

4.6. Example: (The Möbius band.) Let $X = \{(x, y): -1 \leq x \leq 1, -1 \leq y \leq 1\} \subset R^2$ have the subspace topology. Identify each point in X of the form $(x, -1)$ with the point $(-x, 1)$, and identify all other points of X with themselves only. The resulting quotient space is called a Möbius band. Figure 18 gives some visual aid.

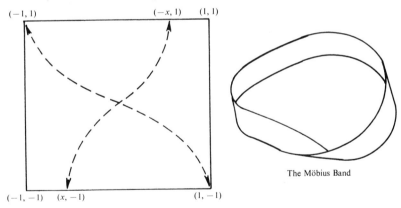

The Möbius Band

Figure 18

The reader will benefit by actually constructing a Möbius band by using a long rectangular piece of paper (a square will be difficult to twist, and besides, squares and rectangles are homeomorphic,) and pasting the identifying edges together to form the band. To do this, you must twist the rectangle once before pasting. If two men started on "opposite sides" of the band, one painting the band blue, the other red, what would they soon discover? Cut the constructed Möbius band lengthwise. Then cut it lengthwise again. What are the results?

4.7. Example: (The Klein Bottle.) Use the same subspace of R^2 as in Example 4.6. Identify each point of X of the form $(x, -1)$ with the point $(-x, 1)$, identify each point of the form $(-1, y)$ with $(1, y)$, and identify all other points of X with themselves. The resulting quotient space is called a *Klein bottle*. Figure 19 gives a visual description of the situation.

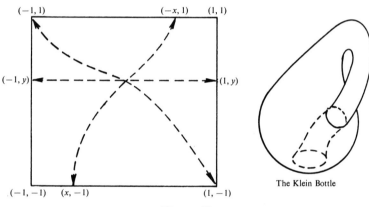

The Klein Bottle

Figure 19

Exercises:

1. Prove the statement $p^{-1}(p(U)) = R(U)$ made in the proof of Theorem 4.4.

2. Prove Theorem 4.4, in which "closed" is used rather than "open."

3. Let $A \subset R^1$ where $A = \{x : x \leq 0\} \cup \{x : x \geq 1\}$. Describe the quotient space topology on R^1/A obtained by collapsing A to a point.

4. Let R^1 have the left ray topology and let $A = \{x : x \leq 0\} \subset R^1$. Describe the quotient space topology on R^1/A.

5. Define an equivalence relation R on R^1 as follows: For each $x, y \in R^1$, xRy iff $x - y$ is an integer. What is the quotient space topology on R^1/R?

6. Identify points of the square $\{(x, y) : -1 \leq x \leq 1, -1 \leq y \leq 1\} \subset R^2$ so that the resulting quotient space will be a torus.

7. Define an equivalence relation R on R^2 as follows: For each (x, y) and (z, w) in R^2, $(x, y)R(z, w)$ iff $x = z$. Describe the quotient space topology on R^2/R, and prove that the quotient space R^2/R is homeomorphic to R^1.

8. Let $f: X \to A$ be a retraction of the space X onto the subset $A \subset X$. (See Exercise 9, Section 3 of this Chapter.) Prove that $A \cong X/R(f)$.

9. Let R_1 and R_2 be equivalence relations on X with the property that if $x_1, x_2 \in X$ and $x_1 R_1 x_2$, then $x_1 R_2 x_2$. Prove that X/R_2 is a quotient space of X/R_1.

10. Prove that if X/R_1 is a quotient space of X and $(X/R_1)/R_2$ is a quotient space of X/R_1, then $(X/R_1)/R_2$ is homeomorphic to a quotient space of X.

6

The Separation and Countability Axioms

1. The Separation Axioms

The examples we have considered up to this point show how widely the behavior of topological spaces may vary. We want to now consider the result of imposing a certain class of restrictions on spaces that will lessen this variation somewhat. Restrictions of other types for spaces will be seen in succeeding chapters. The restrictions we have in mind at the moment are called the *separation axioms*. They give a way of classifying spaces according to the topological distinguishability of points and subsets in the space. Accordingly, if for each two distinct points x and y of a space there exist open disjoint sets containing x and y, respectively, then the greatest possible degree of topological distinguishability for points is obtained. Other degrees of distinguishability for points and sets are given in the separation axioms knows as the "T" axioms which are listed next and studied extensively in the first three sections of this chapter. The T_0, T_1, and T_2 axioms give successively greater distinguishability between points, while the T_3 and T_4 axioms do the same for certain subsets of the space.

1.1. Definition: Let $(X, \mathcal{T})$ be a topological space. Then

(a) The space $(X, \mathcal{T})$ is called a T_0-*space* iff for each pair of distinct points $x, y \in X$, there is either an open set containing x but not y or an open set containing y but not x.

(b) The space $(X, \mathcal{T})$ is called a T_1-*space* iff for each pair of distinct points $x, y \in X$, there is an open set containing x but not y. (Another labeling of the points x and y allows us to say that $(X, \mathcal{T})$ is a T_1-space iff there exists an open set in X containing x but not y, and an open set in X containing y but not x.)

(c) The space $(X, \mathcal{T})$ is called a T_2-*space* or a *Hausdorff space* iff for each pair of distinct points $x, y \in X$, there exist open sets U and V such that $x \in U$, $y \in V$, and $U \cap V = \phi$.

(d) The space $(X, \mathcal{T})$ is called a *regular space* iff for each closed subset $F \subset X$ and each point $x \notin F$, there exist open sets U and V such that $x \in U$, $F \subset V$, and $U \cap V = \phi$. A regular T_1-space is called a T_3-*space*.

(e) The space $(X, \mathcal{T})$ is called a *normal space* iff for each pair of closed disjoint subsets F_1 and F_2 of X there exist open sets U and V such that $F_1 \subset U$, $F_2 \subset V$, and $U \cap V = \phi$. A normal T_1-space is called a T_4-*space*.

It is easily seen that the T_2-condition implies the T_1-condition, which in turn implies the T_0-condition. As suggested by the next example, it is easy to construct examples showing that the reverse implications do not hold. We also shall see later that the T_4-condition implies the T_3-condition, and the T_3-condition implies the T_2-condition, so that the chain of implications is complete in one direction. Again, examples may be given to show that none of the implications can, in general, be reversed. Perhaps it should be pointed out that examples will show that the regular and normal conditions alone do not imply each other, nor will the regular condition imply, or is implied by, the T_2-condition. Also, it might be worthwhile to note here that the definitions for T_3

and T_4-spaces are not standardized in the literature. Some authors will require regularity, with the T_2-axiom for a T_3-space and normality with the T_2-axiom for a T_4-space. At any rate, a space without at least the T_1-axiom is not as interesting in that it lacks many desirable properties.

1.2. Example: (A T_0-space that is not a T_1-space.) Let R^1 have the left ray topology. For any two distinct points x, y where $x < y$, there is an open set (an open left ray) of the form $L_{(x+y)/2}$ containing x but not y, which means that the space is a T_0-space. However, there is no way of getting an open set containing y to exclude x so that our space is not a T_1-space.

1.3. Example: (A regular space that is not Hausdorff.) Let $X = \{a, b, c\}$ with the topology $\mathscr{T} = \{\phi, X, \{a\}, \{b, c\}\}$. Since there are no open disjoint sets in X containing b and c, respectively, the space is not a Hausdorff space. To see that the space is regular, consider all closed sets in X, namely $\phi, X, \{a\}$, and $\{b, c\}$. For each one, except X, consider any point not in the closed set in question. It is easy to exhibit open disjoint sets containing the closed set and the point, respectively.

Example 1.2 shows that a point in a T_0-space, and hence in even more general spaces, need not be closed. As a consequence, T_0-spaces are much more general than the R^n spaces. We furthermore note that if a is any point in the space of Example 1.2, then $\{a\}' = \{x : x > a\}$ and therefore $\overline{\{a\}} = \{x : x \geq a\}$. Thus, if $b \neq a$, then $\overline{\{a\}} \neq \overline{\{b\}}$. This observation leads to an interesting characterization of T_0-spaces.

1.4. THEOREM: *A space $(X, \mathscr{T})$ is a T_0-space iff for each pair of distinct points $x, y \in X$, $\overline{\{x\}} \neq \overline{\{y\}}$.*

Proof: The details are left as an exercise using only the definitions of a T_0-space and the closure of a set.

It might be useful to point out that a space X with the indiscrete topology is always normal because there do not exist two closed disjoint subsets of X other than ϕ and X itself. Thus, the normality condition is satisfied vacuously. Such a space is also regular. If X is nondegenerate, then X is not a T_0-space. One of the properties that makes T_1-spaces so natural and useful is that each point in such a space is a closed subset of that space. In view of this, we shall devote the remainder of this section to properties of T_1-spaces. In fact, closed points characterize T_1-spaces as is shown next.

1.5. THEOREM: *The space $(X, \mathcal{T})$ is a T_1-space iff each point $x \in X$ is a closed subset of X.*

Proof: Let $(X, \mathcal{T})$ be a T_1-space and consider any point $x \in X$. We show that $\{x\}$ is closed by showing $X\backslash\{x\}$ is open. To this end, let $y \neq x$ be any point in X. If no such y exists, then $X = \{x\}$ is closed, and our theorem is proved. Otherwise, for each $y \in X$ where $y \neq x$, there exists an open set $U(y)$ containing y but not x, according to the T_1-axiom. Therefore, Theorem 5.4 of Chapter 3 gives $X\backslash\{x\}$ open and, consequently, $\{x\}$ is a closed subset of X.

For the converse, let each point of X be a closed subset of X and show that $(X, \mathcal{T})$ is a T_1-space. Thus, let x and y be distinct points of X. Then since $\{y\}$ is closed, $X\backslash\{y\}$ is open and contains x but not y. It follows that $(X, \mathcal{T})$ is a T_1-space.

1.6. Corollary: The space R^1 is a T_1-space.

Proof: Corollary 3.5 of Chapter 3, along with Theorem 1.5.

1.7. Corollary: In a T_1-space all finite subsets are closed.

An interesting connection between the confinite topology and T_1-spaces is given next.

1.8. THEOREM: *Of all topologies that may be put on a set X, the cofinite topology is the smallest possible topology which will make X a T_1-space.*

Proof: The set X with the cofinite topology is a T_1-space. To see this, let $x \in X$ be any given point and consider $X\backslash\{x\} = U$. Then U is open in X because the complement of U is the finite set $\{x\}$. Thus, $\{x\}$ is closed since its complement U is open and hence X is a T_1-space according to Theorem 1.5.

Now let $\mathcal{T}$ represent the cofinite topology on X and consider any other topology $\mathcal{T}_1$ on X which is T_1. According to Corollary 1.7, all finite subsets of X are closed, so their complements are open with respect to the $\mathcal{T}_1$ topology. Therefore, if $U \in \mathcal{T}$, then $X\backslash U$ is finite and, thus, closed with respect to the $\mathcal{T}_1$ topology on X. This means $X\backslash(X\backslash U)$

$= U$ is open with respect to $\mathscr{T}_1$ so that $U \in \mathscr{T}_1$. Consequently, $\mathscr{T} \subset \mathscr{T}_1$ and $\mathscr{T}$ is the smallest topology which makes X a T_1-space.

We show next that each subspace of a T_1-space inherits the T_1-property.

1.9. THEOREM: *Any subspace of a T_1-space is also a T_1-space.*

Proof: Let $(A, \mathscr{T}_A)$ be a subspace of the T_1-space $(X, \mathscr{T})$. Consider any two distinct points x and y in A. Then x and y belong to X and, because of the T_1-axiom for X, there exists an open set $U(x)$ containing x but not y. Thus, $U(x) \cap A$ is open in A and contains x but not y. Therefore, $(A, \mathscr{T}_A)$ is a T_1-space.

Each of the T-axiom properties is a topological property. In particular, the proof that the T_1-property is a topological property follows.

1.10. THEOREM: *The property of being a T_1-space (T_0-space) is a topological property.*

Proof: Let $(X, \mathscr{T}_X)$ be a T_1-space and $(Y, \mathscr{T}_Y)$ any space homeomorphic to $(X, \mathscr{T}_X)$. It is required to show that $(Y, \mathscr{T}_Y)$ is also a T_1-space. Therefore, let x and y be any two distinct points of Y and call $h: X \to Y$ the homeomorphism. Then $h^{-1}(x)$ and $h^{-1}(y)$ are single points of X and since X is a T_1-space, there exists an open set $U \in \mathscr{T}_X$ containing $h^{-1}(x)$ but not $h^{-1}(y)$. This means the open set $h(U)$ in Y contains x but not y, and shows that $(Y, \mathscr{T}_Y)$ is a T_1-space. The proof for $(Y, \mathscr{T}_Y)$ being a T_0-space if $(X, \mathscr{T}_X)$ is a T_0-space is similar.

For the product of T_1-spaces we have the following result.

1.11. THEOREM: *Let $\{(X_\alpha, \mathscr{T}_\alpha) : \alpha \in \Delta\}$ be a family of spaces indexed by Δ. Then the product space $\prod_{\alpha \in \Delta} X_\alpha$ is a T_1-space iff each $\alpha \in \Delta$, $(X_\alpha, \mathscr{T}_\alpha)$ is a T_1-space.*

Proof: First, suppose $\prod_{\alpha \in \Delta} X_\alpha$ is a T_1-space and consider the factor space X_β where $\beta \in \Delta$ as fixed for our discussion. Let x and y be distinct points in X_β. According to the definition of the product, there exist two points f and g in $\prod_{\alpha \in \Delta} X_\alpha$ such that $f(\beta) = x$ and $g(\beta) = y$ and

$f(\alpha) = g(\alpha)$ for all $\alpha \in \Delta$ where $\alpha \neq \beta$. Thus, $f \neq g$ and there exists an open set $U \subset \prod_{\alpha \in \Delta} X_\alpha$ containing f but not g, because the product space is a T_1-space. By Theorem 1.4 of Chapter 4, there is a basic open set $V = U_{\alpha_1} \times U_{\alpha_2} \times \ldots \times U_{\alpha_n} \times \prod_{\alpha \in \Delta} \{X_\alpha : \alpha \neq \alpha_1, \alpha_2, \ldots, \alpha_n\}$, where U_{α_k} is open in X_{α_k}, containing f and which is a subset of U. Thus, $f(\alpha_k) \in U_{\alpha_k}$, $k = 1, 2, \ldots, n$ and $f(\alpha) \in X_\alpha$, $\alpha \neq \alpha_1, \alpha_2, \ldots, \alpha_n$. Since f and g differ only in the βth coordinate and $g \notin V$, $g(\beta) \notin U_{\alpha_k}$, $k = 1, 2, \ldots, n$ and $g(\beta) \notin X_\alpha$, $\alpha \neq \alpha_1, \alpha_2, \ldots, \alpha_n$. Consequently, $g(\beta) \in X_{\alpha_j} \backslash U_{\alpha_j}$ for some $1 \leqslant j \leqslant n$, implying $\beta = \alpha_j$. Therefore, U_β is an open set in X_β such that $x = f(\beta) \in U_\beta$, while $y = g(\beta) \in U_\beta$. This shows X_β is a T_1-space.

Conversely, suppose each X_α, $\alpha \in \Delta$, is a T_1-space, and let $f \neq g$ be two points in $\prod_{\alpha \in \Delta} X_\alpha$. Then for at least one $\beta \in \Delta$, $f(\beta) \neq g(\beta)$. Since X_β is a T_1-space, there is an open set U containing $f(\beta)$ but not $g(\beta)$, and hence the basic open set $U \times \prod_{\alpha \neq \beta} X_\alpha$ contains f but not g. It follows that the product space is a T_1-space.

1.12. Corollary: For each $n \in N$, R^n is a T_1-space.

Proof: Corollary 1.6 together with Theorem 1.11.

By its definition, a function f from X to Y is a subset of $X \times Y$. If both X and Y have topologies, then $X \times Y$ has the product topology, and therefore, it is proper to ask about topological properties of f as a subset of the space $X \times Y$. We conclude this section with two such considerations.

1.13. THEOREM: *Let $f : X \to Y$ be any surjection from the space X to the space Y. If f is a closed subset of $X \times Y$, then Y is a T_1-space.*

Proof: Let y and w be two distinct points of Y. Then there is a point $x \in X$ such that $f(x) = w$, i.e., $(x, w) \in f$. Thus, $(x, y) \notin f$. Since f is closed, there is an open set, and hence a basic open set $U \times V$ where U is open in X and V is open in Y, containing (x, y) such that $(U \times V) \cap f = \phi$. Now $x \in U$ and $y \in V$ but $w \notin V$. The reason for this is that if $(x, w) \in U \times V$ and $(x, w) \in f$, then $(U \times V) \cap f \neq \phi$, which is a contradiction. Thus, there is an open set V containing y but not w, and therefore Y is a T_1-space.

1.14. THEOREM: *Let $f : X \to Y$ be any injective function from the space X to the space Y. If f is a closed subset of the space $X \times Y$, then X is a T_1-space.*

Proof.   The proof is left to the reader.

Exercises:

1. (a) Give an example of a T_1-space that is not a T_2-space.

   (b) Give an example of a T_2-space that is not a regular space.

2. (a) Give an example of a normal space that is not a regular space.
   (b) Give an example of a regular space that is not a normal space.

3. Prove that $(X, \mathcal{T})$ is a T_1-space iff for any two distinct points x and y in X, then $\overline{\{x\}} \cap \overline{\{y\}} = \phi$.

4. Prove Theorem 1.4.

5. Prove Theorem 1.14.

6. Prove the property of being a T_0-space is a topological property.

7. (a) Let $f : X \to Y$ be continuous where Y is a T_1-space and X is an arbitrary topological space. For any $y \in Y$, prove that $f^{-1}(y)$ is a closed subset of X.
   (b) Give an example to show part (a) need not hold if Y is a T_0-space.

8. Let A be a subset of the T_1-space X. Prove that x is a cluster point of A iff every open set containing x contains infinitely many distinct points of A.

9. Prove that no finite subset of a T_1-space has a cluster point. (This would also prove all finite subsets of a T_1-space are closed.)

10. Let $(X, \mathcal{T})$ be a T_1-space and $A \subset X$. Prove that the derived set A' is a closed subset of X.

2. Hausdorff Spaces

The Hausdorff or T_2-property is a more stringent restriction on a space than the T_1-property and yet is not so restrictive as to exclude most of the familiar and useful topological spaces in mathematical studies. For this reason, it is a frequently used condition on topological spaces or at least a desirable one for a space to have. In this section we shall deal exclusively with the properties of Hausdorff spaces. Our first observation is that since each T_2-space is also a T_1-space, single points are closed subsets of T_2-spaces because of Theorem 1.5. Next, some properties equivalent to the Hausdorff property are given.

2.1. THEOREM: *The following properties are equivalent:*
   (a) *The space X is Hausdorff.*
   (b) *Let a be a point in the space X. For each $x \in X$, $x \neq a$, there is an open set U in X containing a such that $x \notin \bar{U}$.*
   (c) *The diagonal $D = \{(x, x) : x \in X\}$ is a closed subset of the product $X \times X$.*

Proof: (a) implies (b). Let $a \in X$ be given and consider $x \neq a$. Since X is a T_2-space, there exist open disjoint sets U and V containing a and x, respectively. Thus, x is not a cluster point of the set U and hence $x \notin \bar{U}$.

(b) implies (c). To show D is closed, we shall show the complement of D is open in the product space $X \times X$. Thus, let $(x, y) \in X \times X$ where $(x, y) \notin D$. This implies that $x \neq y$. By (b) there is an open set U containing x such that $y \notin \bar{U}$. But y does belong to the open set $X \backslash \bar{U}$, which means (x, y) belongs to the basic open set $U \times (X \backslash \bar{U})$ in the space $X \times X$. Since $U \cap (X \backslash \bar{U}) = \phi$, no point of the form $(z, z) \in D$ can belong to the open set $U \times (X \backslash \bar{U})$. Therefore, containing each point $(x, y) \notin D$ there is an open set which does not intersect D. It follows that $(X \times X) \backslash D$ is open and that D is closed.

(c) implies (a). Let $x \neq y$ be points in X. Then $(x, y) \notin D$ and there exists an open set W in $X \times X$ containing (x, y) such that $D \cap W = \phi$, since D is closed by (c). Now there is a basic open set of the form $U \times V$, where U and V are open in X, such that $(x, y) \in U \times V \subset W$, which implies $x \in U$, $y \in V$, and $U \cap V = \phi$, because $(U \times V) \cap D = \phi$. The conclusion is that X is Hausdorff.

Three properties analogous to those for T_1-spaces are listed now.

2.2. THEOREM: (a) *Any subspace of a T_2-space is also a T_2-space.*
   (b) *The property of being a T_2-space is a topological property.*
   (c) *The product space $\prod_{\alpha \in \Delta} X_\alpha$ is a T_2-space iff each X_α is a T_2-space.*

Proof: The proof of each part is easily constructed and is left as an instructive exercise.

2.3. Corollary: For each $n \in N$, R^n is a T_2-space.

Proof: It is easy to establish that R^1 is a T_2-space. (See Exercise 6, Section 3, of Chapter 3.) This fact, together with Theorem 2.2(c), gives each R^n space as a T_2-space.

Continuous functions into Hausdorff spaces give the following interesting results.

2.4. THEOREM: *Let X be any topological space, Y a Hausdorff space, and both $f : X \to Y$ and $g : X \to Y$ continuous functions. Then*
   (a) *The set $A = \{x \in X : f(x) = g(x)\}$ is a closed subset of X.*
   (b) *If D is a dense subset of X and $f|D = g|D$, then $f = g$.*

Proof: (a) Assume $a \in X$ is a cluster point of A such that $a \notin A$. Then by definition of A, $f(a) \neq g(a)$. Since Y is a Hausdorff space, there exist open disjoint sets U and V of Y containing $f(a)$ and $g(a)$, respectively. The continuity of both f and g give the existence of open sets U_1 and V_1 of X, each containing a, such that $f(U_1) \subset U$ and $g(V_1) \subset V$. Thus, the open set $W = U_1 \cap V_1$ containing a has the property that $f(W) \subset U$ and $g(W) \subset V$. However, $W \cap A$ contains a point $x \in A$, because a is a cluster point of A, and hence $f(x) = g(x)$. Since $U \cap V = \phi$, it follows that either $f(W)$ is not a subset of U, or $g(W)$ is not a subset of V. This contradicts the earlier fact that $f(W) \subset U$ and $g(W) \subset V$, which implies our initial assumption is false. Consequently, A contains all of its cluster points and, therefore, is closed.

   (b) The set $A = \{x \in X : f(x) = g(x)\}$ is a closed set by (a) and contains the dense set D because $f|D = g|D$. Now, according to Theorem 6.11 (b) of Chapter 3, $A = X$ so that $f = g$.

The next theorem continues the study of properties of a function f as a subset of the space $X \times Y$. It gives conditions under which f is a closed subset of $X \times Y$. The exercises ask for an example to show that the continuity of f is not enough to ensure f closed.

2.5. THEOREM: *Let $f : X \to Y$ be a continuous function from the space X to the Hausdorff space Y. Then f is a closed subset of $X \times Y$.*

Proof: Let (x, w) be any point of $X \times Y$ not in f. Then $f(x) \neq w$ and since Y is Hausdorff, there exist open disjoint sets W and V containing w and $f(x)$, respectively. By continuity of f, there exists an open set $U \subset X$ containing x such that $f(U) \subset V$. This means $U \times W$

can contain no point of the form $(z, f(z))$. However, $U \times W$ does contain (x, w). Therefore, $U \times W$ is an open set in $X \times Y$ containing (x, w) such that $(U \times W) \cap f = \phi$. This means the complement of f is open and, thus, f is a closed subset of the space $X \times Y$.

Along this same line of thought we have the following theorem.

2.6. THEOREM: *Let $f : X \to Y$ be an open surjection where f is a closed subset of the space $X \times Y$. Then Y is Hausdorff.*

Proof: The proof is an exercise.

The idea of a retract has been mentioned earlier in the exercises. However, its connection with Hausdorff spaces prompts us to restate the definition before proving the result we want to obtain.

2.7. Definition: A subset A of a space X is called a *retract* of X iff there exists a continuous function $f : X \to A$ such that $f(x) = x$ for every $x \in A$. The function f is called a *retraction* of X onto A.

2.8. Example: Let $A = \{x : 0 \leqslant x \leqslant 1\}$ be a subset of R^1. Define $f(x) = 0$ if $x < 0$, $f(x) = x$ if $0 \leqslant x \leqslant 1$, and $f(x) = 1$ if $x > 1$. Then f is continuous and A is a retract of R^1 under the retraction f as given.

2.9. THEOREM: *If A is a retract of a Hausdorff space X, then A is closed.*

Proof: Let f be the required retraction and consider any point $x \notin A$. Then $f(x) \in A$ and, therefore, $x \neq f(x)$. Since X is a T_2-space, there exist open disjoint sets U and V containing x and $f(x)$, respectively. Now since f is continuous, there exists an open set $W \subset U$ containing x such that $f(W) \subset V$. The open set W can contain no point of A because these points all remain fixed under f and hence could not have their images in V under f. Thus, containing each $x \notin A$ there is an open set W such that $W \cap A = \phi$. This means that $X \backslash A$ is open so that A is closed.

The final theorem of this section concerns quotient spaces. If $f : X \to Y$ is a given function and the quotient space $X/R(f)$ is formed, there is a function $g : X/R(f) \to Y$ defined as $g(R(a)) = f(a)$ where $R(a)$ is the equivalence class determined by f and a. (Note that in this instance we are not requiring f to be surjective.) As we have seen before, the relationship between f, g and p is given by $g = fp^{-1}$ or $gp = f$ where p is the projection from X onto $X/R(f)$.

2.10. THEOREM: *Let $f : X \to Y$ be continuous where Y is a T_2-space. Then $X/R(f)$ is a T_2-space.*

Proof: First, the projection function $p : X \to X/R(f)$ is an identification and gp equals the continuous function f so that by Theorem 3.7 of Chapter 5, g is continuous. Now let $R(a)$ and $R(b)$ be distinct points in the space $X/R(f)$. This means $f(a) \neq f(b)$, and since Y is Hausdorff, there exist open disjoint sets U and V containing $f(a)$ and $f(b)$, respectively. By continuity of g, $g^{-1}(U)$ and $g^{-1}(V)$ are both open sets in $X/R(f)$ and contain $R(a)$ and $R(b)$, respectively. Furthermore, $g^{-1}(U) \cap g^{-1}(V) = \phi$, giving the desired open disjoint sets about the distinct points $R(a)$ and $R(b)$. Thus, $X/R(f)$ is Hausdorff.

Exercises

1. Let $(X, \mathcal{T})$ be a Hausdorff space. If $\mathcal{T}_1$ is a topology for X such that $\mathcal{T} \subset \mathcal{T}_1$, then prove that $(X, \mathcal{T}_1)$ is also Hausdorff.

2. Prove Theorem 2.2(a) and 2.2(b).

3. Prove Theorem 2.2(c).

4. Give an example of a function $f : X \to Y$ which is continuous but such that f is not a closed subset of $X \times Y$.

5. Give an example of a function $f : X \to Y$ such that f is closed in $X \times Y$, but f is not continuous.

6. Prove Theorem 2.6.

7. Let $f : X \to Y$ be continuous and injective where Y is Hausdorff. Prove that X is Hausdorff.

8. Let X be a finite set. Prove that the only Hausdorff topology on X is the discrete topology.

9. Let $\{x_1, x_2, \ldots, x_n\}$ be a finite subset of a Hausdorff space. Prove there exist n open pairwise disjoint sets $U_1, U_2, \ldots, U_n$ such that $x_1 \in U_1, x_2 \in U_2, \ldots, x_n \in U_n$.

10. Prove the space X is Hausdorff iff for each pair of distinct points $a, b \in X$ there exist closed sets A and B of X such that $A \cup B = X, a \in A$ but $b \notin A$ and $b \in B$ but $a \notin B$.

11. Prove that the space $(X, \mathcal{T})$ is Hausdorff iff for each $x \in X$, $x = \bigcap_{\alpha \in \Delta} \{\bar{U}_\alpha : U_\alpha \text{ is open in } X \text{ and contains } x\}$.

12. Let X be a T_2-space.
   (a) Prove that for each $x \in X$, $x = \bigcap_{\alpha \in \Delta} \{F_\alpha : F_\alpha$ is closed in X and contains $x\}$.
   (b) Prove that for each $x \in X$, $x = \bigcap_{\alpha \in \Delta} \{U_\alpha : U_\alpha$ is open in X and contains $x\}$.
   (c) Give examples to show that neither of the conditions (a) or (b) imply the space is a T_2-space.

13. Let X be an arbitrary space, Y a T_2-space, $f : X \to Y$ and $g : Y \to X$ both continuous and $gf = 1_X$. Prove
   (a) X is a T_2-space.
   (b) $f(X)$ is a closed subset of Y.

14. Prove that every infinite T_2-space contains a countable subset A such that the subspace topology $\mathcal{T}_A$ is discrete.

3. Regular and Normal Spaces

Continuing our study of the hierarchy of topological spaces, we establish the following two results which further show the importance of the T_1-axiom.

3.1. THEOREM: *Every T_3-space is also a T_2-space.*   not ?

Proof: Let $x \neq y$ be two points in the T_3-space X. Since points in X are closed, consider $\{x\}$ as a closed set and y a point in $\{x\}$. Then by regularity, there exist open disjoint sets U and V containing $\{x\}$ and y, respectively. Therefore, X is a T_2-space.

3.2. THEOREM: *Every T_4-space is also a T_3-space.*

Proof: The proof is left to the reader.

3.3. Example: Consider R^1 and the collection of subsets $\mathcal{S} = \{U : U$ is an open interval or U consists of all rational numbers in $R^1\}$. Let $\mathcal{T}(\mathcal{B}[\mathcal{S}])$ be the topology on R^1 having $\mathcal{S}$ as a subbase. It is easy to see $(R^1, \mathcal{T}(\mathcal{B}[\mathcal{S}]))$ is a T_2-space. However, $(R^1, \mathcal{T}(\mathcal{B}[\mathcal{S}]))$ is not a T_3-space because the point $a = 1$ and the closed set consisting of all irrational numbers cannot be separated with open disjoint sets belonging to $\mathcal{T}(\mathcal{B}[\mathcal{S}])$.

Examples of T_3-spaces that are not T_4-spaces are non-trival when considering the aims of this book and will not be given here. Such examples may be found in references 1, 2, and 4 of the Bibliography.

Concerning the closure of points in regular spaces we have the following theorem.

3.4. THEOREM: *Let $(X, \mathcal{T})$ be a regular space and $x \neq y$ two points of X. Then either $\overline{\{x\}} = \overline{\{y\}}$ or $\overline{\{x\}} \cap \overline{\{y\}} = \phi$.*

Proof: The proof is left as an exercise.

This leads us to a relationship between T_0-spaces and T_3-spaces.

3.5. THEOREM: *Every regular T_0-space is a T_3-space.*

Proof: All that needs to be shown is that a regular T_0-space is a T_1-space. The details are an easy exercise.

Important and useful characterizations of regularity and normality are given in the following three theorems.

3.6. THEOREM: *The space X is regular iff for each $x \in X$ and each open set U containing x, there exists an open set V such that $x \in V \subset \bar{V} \subset U$.*

Proof: First, suppose the condition holds and show that X is regular. Let $A \subset X$ be a closed set and $x \notin A$. Then $X \backslash A$ is open and contains x. Thus, by hypothesis there exists an open set V such that $x \in V \subset \bar{V} \subset X \backslash A$. Now $X \backslash \bar{V}$ is open, $x \in V$, $A \subset X \backslash \bar{V}$ and $V \cap (X \backslash \bar{V}) = \phi$ so that x and A are, respectively, contained in open disjoint sets showing X to be a regular space.

Now suppose X is regular and show that the condition holds. Let $x \in X$ and U an open set containing x. Then $X \backslash U$ is closed and does not contain x, so by regularity, there exist open disjoint sets V and W such that $x \in V$ and $X \backslash U \subset W$ and $V \cap W = \phi$. Thus, $V \subset X \backslash W \subset X \backslash (X \backslash U) = U$. But $X \backslash W$ is closed so that $x \in V \subset \bar{V} \subset X \backslash W \subset U$ and V is the desired open set. Therefore, if X is regular, then the condition in the theorem holds.

3.7. THEOREM: *The space X is normal iff for each closed subset $A \subset X$ and open set U containing A, there exists an open set V such that $A \subset V \subset \bar{V} \subset U$.*

Proof: The proof is left as an exercise.

3.8. THEOREM: (*Uryshon's characterization of normality.*) *Let X be Hausdorff and* $I = \{x \in R^1 : 0 \leqslant x \leqslant 1\}$. *Then X is normal iff for each pair of closed disjoint sets A and B in X, there exists a continuous* $f : X \to I$ *such that*
  (*a*) $f(x) = 0$ *for all* $x \in A$ *and*
  (*b*) $f(x) = 1$ *for all* $x \in B$.

We shall not attempt a proof of this theorem here as it is considered more difficult than the level of this book dictates. In spite of this, we should be aware of the existence of this theorem, for it is certainly an important one. A proof for this theorem may be found in any of the references in the Bibliography.

As in the T_1 and T_2-spaces, the property of being a T_3-space is a topological property, with subspaces of T_3-spaces also being T_3-spaces and the product is a T_3-space iff each of the factor spaces is a T_3-space. The property of being a T_4-space is a topological property, but we cannot prove the other two assertions for T_4-spaces. Although we shall not attempt them here, examples exist which show subspaces of normal spaces need not be normal spaces and products of normal spaces which need not be normal. For such examples, consult reference 2 in the Bibliography. It is true, however, that if a product of spaces is normal, then each of the spaces must be normal. We leave the verification of this fact as an exercise.

3.9. THEOREM: (*a*) *A subspace of a T_3-space is also a T_3-space.*
      (*b*) *The property of being a T_3-space (T_4-space) is a topological property.*
      (*c*) *The product space* $\prod_{\alpha \in \Delta} X_\alpha$ *is a T_3-space iff each X_α is a T_3-space.*

Proof: The proofs to parts (a) and (b) are left to the reader. We now consider the proof to part (c). According to Theorem 1.11, the product space is a T_1-space iff each factor space is a T_1-space. Therefore, we need only show that the product is regular iff each space X_α is regular. Figure 20 gives visual aid for the first part of the proof. Let $\prod_{\alpha \in \Delta} X_\alpha$ be regular and consider the space X_β where $\beta \in \Delta$. We wish to show that X_β is regular by using Theorem 3.6. To this end, let $x \in X_\beta$ and V_β be any open set containing x. By the continuity of the βth projection function, p_β, $p_\beta^{-1}(V_\beta) = \{f \in \prod_{\alpha \in \Delta} X_\alpha : f(\alpha) \in V_\beta\}$ is open in the product space and contains a point f_0 such that $f_0(\beta) = x$. Since

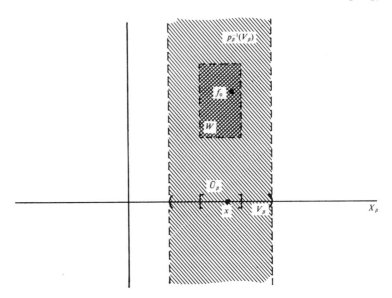

Figure 20

$\prod_{\alpha \in \Delta} X_\alpha$ is regular, there exists an open set $V \subset \prod_{\alpha \in \Delta} X_\alpha$ such that $f_0 \in V$
$\subset \bar{V} \subset p_\beta^{-1}(V_\beta)$ by Theorem 3.6. Thus, there exists a basic open set
$W = U_{\alpha_1} \times U_{\alpha_2} \times \ldots \times U_{\alpha_n} \times \prod_{\alpha \in \Delta} \{X_\alpha : \alpha \neq \alpha_1, \alpha_2, \ldots, \alpha_n\}$, where
U_{α_k} is open in X_{α_k} for $k = 1, 2, \ldots, n$, such that $f_0 \in W \subset V$. From
this is follows that $f_0 \in W \subset \bar{W} \subset \bar{V} \in p_\beta^{-1}(V_\beta)$ by use of Theorem
2.9(d) of Chapter 3. Notice that if $V_\beta = X_\beta$, then X_β could be one of
the factor spaces in the product W. Otherwise, $\alpha_k = \beta$ for some $1 \leqslant$
$k \leqslant n$ is the only possibility. Now applying Theorem 4.9 of Chapter
4, we have

$$\bar{W} = \overline{U_{\alpha_1} \times U_{\alpha_2} \times \ldots \times U_{\alpha_n} \times \prod_{\alpha \in \Delta} \{X_\alpha : \alpha \neq \alpha_1, \alpha_2, \ldots, \alpha_n\}}$$

$$= \bar{U}_{\alpha_1} \times \bar{U}_{\alpha_2} \times \ldots \times \bar{U}_{\alpha_n} \times \prod_{\alpha \in \Delta} \{X_\alpha : \alpha \neq \alpha_1, \alpha_2, \ldots, \alpha_n\}.$$

Making use of the fact that $f_0 \in W \subset \bar{W} \subset \bar{V} \subset p_\beta^{-1}(V_\beta), \bar{U}_\beta \subset V_\beta$
must hold. Since $f_0 \in W, f_0(\beta) = x \in U_\beta$, so that we have found an
open set $U_\beta \subset X_\beta$ such that $x \in U_\beta \subset \bar{U}_\beta \subset V_\beta$, and this makes X_β
regular according to Theorem 3.6.

Conversely, suppose for each $\alpha \in \Delta$, X_α is regular. Let $f \in \prod_{\alpha \in \Delta} X_\alpha$ and U any open set in the product space containing f. Then there exists a basic open set $W = U_{\alpha_1} \times U_{\alpha_2} \times \ldots \times U_{\alpha_n} \times \prod_{\alpha \in \Delta} \{X_\alpha : \alpha \neq \alpha_1, \alpha_2, \ldots, \alpha_n\}$, where U_{α_k} is open in X_{α_k}, such that $f \in W \subset U$. Each X_α regular implies that there exist open sets V_{α_k} of X_{α_k} such that $f(\alpha_k) \in V_{\alpha_k} \subset \bar{V}_{\alpha_k} \subset U_{\alpha_k}$, $k = 1, 2, \ldots, n$. Now $\bar{V}_{\alpha_1} \times \bar{V}_{\alpha_2} \times \ldots \times \bar{V}_{\alpha_n} \times \prod_{\alpha \in \Delta} \{X_\alpha, \alpha \neq \alpha_1, \alpha_2, \ldots, \alpha_n\}$ is closed, again by Theorem 4.9 of Chapter 4, and is a subset of W, hence of U. Thus, we have found an open set $S = V_{\alpha_1} \times V_{\alpha_2} \times \ldots \times V_{\alpha_n} \times \prod_{\alpha \in \Delta} \{X_\alpha : \alpha \neq \alpha_1, \alpha_2, \ldots, \alpha_n\}$ such that $f \in S \subset \bar{S} \subset U$, implying $\prod_{\alpha \in \Delta} X_\alpha$ is regular by Theorem 3.6.

There is one important product of spaces that is normal, namely the R^n spaces. We show this is the case next.

3.10. THEOREM: *For each $n \in N$, R^n is a normal space.*

Proof: Let A and B be disjoint closed subsets of R^n. Then $R^n \backslash A$ is open and contains B. Therefore, for each point $p = (p_1, p_2, \ldots, p_n) \in B$ there is a basic open set $U_1 \times U_2 \times \ldots \times U_n$ lying in $R^n \backslash A$ and containing p where the open sets U_k, $k = 1, 2, \ldots, n$, may be taken as open intervals having the form $U_k = \{x : p_k - \epsilon_{p_k} < x < p_k + \epsilon_{p_k}, \epsilon_{p_k}$ a positive real number$\} = (p_k - \epsilon_{p_k}, p_k + \epsilon_{p_k})$. Define $\epsilon_p = $ minimum $\{\epsilon_{p_k} : k = 1, 2, \ldots, n\}$ and $U(p) = (p_1 - (\epsilon_p)/2, p_1 + (\epsilon_p)/2) \times (p_2 - (\epsilon_p)/2, p_2 + (\epsilon_p)/2) \times \ldots \times (p_n - (\epsilon_p)/2, p_n + (\epsilon_p)/2)$. Now $U(p)$ is a basic open set containing p and is a subset of $R^n \backslash A$. Thus $U = \bigcup_{p \in B} U(p)$ is an open set in $R^n \backslash A$ which contains B. Proceeding in the same manner for each point $q \in A \subset R^n \backslash B$, we have $V = \bigcup_{q \in A} V(q)$ where $V(q) = (q_1 - (\epsilon_q)/2, q_1 + (\epsilon_q)/2) \times (q_2 - (\epsilon_q)/2, q_2 + (\epsilon_q)/2) \times \ldots \times (q_n - (\epsilon_q)/2, q_n + (\epsilon_q)/2)$ and $A \subset V$.

We now show that $U \cap V = \phi$. To do this, suppose there exists a point $r = (r_1, r_2, \ldots, r_n) \in U \cap V$. Then there exist points $a = (a_1, a_2, \ldots, a_n) \in A$ and $b = b_1, b_2, \ldots, b_n) \in B$ such that $r \in V(a)$ and $r \in U(b)$. Thus $a_k - (\epsilon_a)/2 < r_k < a_k + (\epsilon_a)/2$ and $b_k - (\epsilon_b)/2 < r_k < b_k + (\epsilon_b)/2$ for each $k = 1, 2, \ldots, n$. Consequently, $|a_k - r_k| < (\epsilon_a)/2$ and $|b_k - r_k| < (\epsilon_b)/2$ for $k = 1, 2, \ldots, n$. From this we see $|a_k - b_k| = |(a_k - r_k) + (r_k - b_k)| \leqslant |a_k - r_k| + |r_k - b_k| = |a_k - r_k| + |b_k - r_k| < 1/2(\epsilon_a + \epsilon_b)$ by use of the fact that for any two real numbers x and y, $|x + y| \leqslant |x| + |y|$. At this point consider the case where $\epsilon_a \leqslant \epsilon_b$. Then $|a_k - b_k| < 1/2(\epsilon_a + \epsilon_b) \leqslant 1/2(\epsilon_b + \epsilon_b) = \epsilon_b$ implies $b_k - \epsilon_b < a_k < b_k + \epsilon_b$ for each $k = 1, 2, \ldots, n$, which in turn implies $a \in U(b)$. This contradicts the construction of $U(b)$. Now considering the other alternative, $\epsilon_b \leqslant \epsilon_a$, we find $|a_k - b_k| = |b_k - a_k| < \epsilon_a$, which implies $b \in V(a)$. However, this contradicts the construction of $V(a)$. Therefore, there can be no point $r \in U \cap V$ so that $U \cap V = \phi$. We have now found open sets U and V containing B and A, respectively, such that $U \cap V = \phi$, and this shows R^n is normal.

3.11. Corollary: *For each $n \in N$, R^n is a T_4-space.*

Proof: Theorem 3.10 together with Corollary 1.12.

3.12. Corollary: *For each $n \in N$, R^n is a T_3-space.*

Proof: Theorem 3.2.

Concluding this section we shall mention two further separation axioms, but not give extensive consequences of them. Primarily, we want to show that other separation axioms exist besides the ones already mentioned.

3.13. Definition: Let $I = \{x \in R^1 : 0 \leqslant x \leqslant 1\}$. A space X is called *completely regular* iff for each closed set $A \subset X$ and each $p \notin A$, there exists a continuous function $f : X \to I$ such that $f(p) = 1$ and $f(A) = 0$. A completely regular T_1-space is called a $T_{3\frac{1}{2}}$-space.

3.14. Definition: A space is called *completely normal* iff given any two disjoint subsets A and B of X, there exist open disjoint sets U and V of X such that $A \subset U$ and $B \subset V$. A completely normal T_1-space is called a T_5-space.

3.15. THEOREM: *Every T_5-space is a T_4-space, every T_4-space is a $T_{3\frac{1}{2}}$-space, and every $T_{3\frac{1}{2}}$-space is a T_3-space.*

Proof: That the T_5-condition implies the T_4-condition follows readily from the definitions. Uryshon's characterization of normality shows that the T_4-condition implies $T_{3\frac{1}{2}}$-condition. To show the $T_{3\frac{1}{2}}$-condition implies the T_3-condition, let $A \subset X$ be closed and $p \notin A$. Then there exists a continuous $f : X \to I$ such that $f(p) = 0$ and $f(A) = 1$. The two open disjoint subsets of I, $U = \{x : x < 1/4\}$ and $V = \{x : x > 3/4\}$, yield $f^{-1}(U)$ and $f^{-1}(V)$ as open disjoint subsets of X

containing p and A, respectively. Therefore, X is regular. Since X is also a T_1-space, X is a T_3-space.

Again, examples exist which show that the implications of Theorem 3.15 may not be reversed.

Exercises

1. Prove Theorem 3.2.

2. Prove Theorem 3.4.

3. Prove Theorem 3.5.

4. Prove Theorem 3.7.

5. Prove parts (a) and (b) of Theorem 3.9.

6. Let $f: X \rightarrow Y$ be an open surjection from the normal space X into the space Y. Disregarding the fact that normality is a topological property, prove that Y is a normal space. (This is a stronger result than Theorem 3.9(b).)

7. Let R^1 have the left ray topology.
   (a) Prove that this space is normal.
   (b) Prove that this space is not regular.
   (c) Prove that this space is not a T_4-space.

8. Prove that if X is regular and Hausdorff and $A \subset X$ is closed, then the quotient space X/A is a Hausdorff space.

9. Let $(X, \mathcal{T})$ be a regular space and A a closed subset of X. Prove that A is the intersection of all open sets in X that contain A.

10. Let $\{(X_\alpha, \mathcal{T}_\alpha) : \alpha \in \Delta\}$ be an indexed family of spaces. Prove that if $\prod_{\alpha \in \Delta} X_\alpha$ is a T_4-space, then $(X_\alpha, \mathcal{T}_\alpha)$ is a T_4-space for each $\alpha \in \Delta$.

4. The First Axiom of Countability

We now turn to another type of restriction on a topology in the form of countability of certain sets belonging to the topology. This restriction is called the *First Axiom of Countability* and will be used extensively in the study of sequences in the next chapter. For the present section, we shall investigate some of the more general aspects of first countable spaces. Before defining first countability, we need a preliminary definition.

4.1. Definition: Let $(X, \mathcal{T})$ be a topological space and $p \in X$. A family $\mathcal{B}_p$ of open sets, each of which contains p, is called a *local base* at p iff given any open set U of X containing p, there exists a $V \in \mathcal{B}_p$ such that $p \in V \subset U$.

A local base at each point p of a space X exists because the family $\mathcal{B}_p$ of all possible open sets in the space containing p would surely satisfy the definition. A useful classification of spaces in terms of whether a countable local base at each point exists or not will be given next. It is the first of our two countability axioms.

4.2. Definition: (The First Axiom of Countability.) A space X is called *first countable* iff for each point $x \in X$ there exists a countable local base at x.

Every finite space is first countable. Also, any space with the discrete topology is first countable because for any open set U containing the point x, $\{x\}$ is itself an open set and $x \in \{x\} \subset U$. Therefore, $\{x\}$ is a local base at x.

4.3. THEOREM: *The space R^1 is first countable.*

Proof: Let p be any given point in R^1 and consider the family $\mathcal{B}_p = \{B_n : n \in N\}$ where $B_n = \{x : p - 1/n < p < p + 1/n$ where $n \in N\}$. This family $\mathcal{B}_p$ of open intervals is countable and if U is any open set in R^1 containing p, there exists an open interval $(a, b) = \{x : a < x < b\} \subset U$ containing p by definition of an open set in R^1. Now let $k \in N$ be such that $1/k < \text{minimum } \{p - a, b - p\}$. Then the open interval $B_k = \{x : p - 1/k < x < p + 1/k\} \subset (a, b) \subset U$. Definition 4.1 is now satisfied for the family $\mathcal{B}_p$ giving R^1 as a first countable space.

One of the very useful facts about first countable spaces is given next.

4.4. THEOREM: *Let X be a first countable space and let $p \in X$. Then there exists a countable local base $\{U_n : n \in N\}$ at p such that*

(a) $U_{n+1} \subset U_n$ for all $n \in N$.
(b) $\bigcap_{n \in N} U_n = p$ provided X is also a T_1-space.

Proof: The first countability of X implies there is a countable local base $\{V_n : n \in N\}$ at the point p. We proceed to define a countable local base satisfying (a). To this end, let $U_1 = V_1, U_2 = V_1 \cap V_2$, and in general, $U_n = V_1 \cap V_2 \cap \ldots \cap V_n$ for each $n \in N$. Certainly $U_{n+1} \subset U_n$ for all $n \in N$, and because $U_n \subset V_n$, the family $\{U_n : n \in N\}$ is a local base at the point p.

Finally we show the family $\{U_n : n \in N\}$ satisfies (b) provided X is a T_1-space. By construction of the U_n, the point $p \in U_n$ for each $n \in N$, and hence $p \in \bigcap_{n \in N} U_n$. Now let $x \neq p$ be any point in X. Since X is a T_1-space, there exists an open set $W \subset X$ containing p but not x. Therefore there is some element U_k from the local base $\{U_n : n \in N\}$ such that $p \in U_k \subset W$ and hence $x \notin U_k$. This implies $x \notin \bigcap_{n \in N} U_n$ and consequently $p = \bigcap_{n \in N} U_n$.

Subspaces inherit the property of first countability.

4.5. THEOREM: *If X is a first countable space, then every subspace of X is first countable.*

Proof: The proof is a direct application of the definitions of first countability and the subspace topology. The details are left to the reader.

4.6. THEOREM: *First countability is a topological property.*

Proof: Again, the proof is left as an exercise.

We now turn to a consideration of the product of first countable spaces. If Δ is an arbitrary indexing set and $\{X_\alpha : \alpha \in \Delta\}$ is a family of first countable spaces, then $\prod_{\alpha \in \Delta} X_\alpha$ may not be a first countable space. It may be proved, however, that if Δ is a countable indexing set and $\{X_\alpha : \alpha \in \Delta\}$ is a family of first countable spaces, then $\prod_{\alpha \in \Delta} X_\alpha$ is first countable. For a general discussion of the product of first countable spaces, the reader is referred to reference 1 in the Bibliography. It is always true, however, that if $\prod_{\alpha \in \Delta} X_\alpha$ is first countable, then each X_α is first countable. This fact is stated next in theorem form and the proof left as a straightforward exercise involving product spaces.

4.7. THEOREM: Let $\{(X_\alpha, \mathcal{T}_\alpha) : \alpha \in \Delta\}$ be a family of space. If the product space $\prod_{\alpha \in \Delta} X_\alpha$ is first countable, then $(X_\alpha, \mathcal{T}_\alpha)$ is first countable for each $\alpha \in \Delta$.

Proof: The proof is an exercise.

Our goal now will be to prove that a finite product of spaces is first countable iff each of the factor spaces is first countable. From this we can infer that for each $n \in N$, R^n is a first countable space. The next three results will establish our assertions.

4.8. THEOREM: *The product space $X \times Y$ is first countable iff X and Y are both first countable.*

Proof: First, suppose $X \times Y$ is first countable and let $(x, y) \in X \times Y$. Then the subspace $X \times \{y\}$ of $X \times Y$ is first countable by Theorem 4.5. According to Theorem 2.12 of Chapter 5, X is homeomorphic to $X \times \{y\}$ so that it follows from Theorem 4.6 that X is first countable. Similarly, Y is first countable.

Conversely, let both X and Y be first countable and $(x, y) \in X \times Y$. We must show that there is a countable local base at the point (x, y). Since X is first countable, there is a countable local base $\mathscr{B}_x = \{U_n : n \in N\}$ at the point $x \in X$. Likewise, there is a countable local base $\mathscr{B}_y = \{V_n : n \in N\}$ at the point $y \in Y$. Define $\mathscr{B}_x \times \mathscr{B}_y = \{U_n \times V_m : n, m \in N\}$. If we think of $\mathscr{B}_x \times \mathscr{B}_y = \bigcup_{n \in N} A_n$, where $A_n = \{U_n \times V_m : m \in N\}$, then $\mathscr{B}_x \times \mathscr{B}_y$ is countable by Theorem 5.9 of Chapter 2. The proof is completed by showing $\mathscr{B}_x \times \mathscr{B}_y$ is a local base at the point (x, y). Certainly $(x, y) \in U_n \times V_m$, for all $n, m \in N$. Now let U be any open set in $X \times Y$ containing (x, y). Then there exists a basic open set $W \times T$ such that $(x, y) \in W \times T \subset U$ where W is open in X and contains x, and T is open in Y and contains y. Thus, there are local base elements $U_n \in \mathscr{B}_x$ and $V_m \in \mathscr{B}_y$ such that $x \in U_n \subset W$ and $y \in V_m \subset T$. Consequently, $(x, y) \in U_n \times V_m \subset W \times T \subset U$. This shows $\mathscr{B}_x \times \mathscr{B}_y$ is a local base at (x, y) and thus, $X \times Y$ is first countable.

4.9. THEOREM: *Let $\{(X_k, \mathscr{T}_k) : k = 1, 2, \ldots, n\}$ be a finite family of spaces. Then the product space $\prod_{k=1}^{n} X_k$ is first countable iff X_k is first countable for each $k = 1, 2, \ldots, n$.*

Proof: The proof is by mathematical induction. Theorem 4.8 establishes the result for the case when $n = 2$. Now let $n \in N$ be a natural number such that $\prod_{k=1}^{n} X_k$ is first countable iff $X_k, k = 1, 2, \ldots, n$ is first countable. We need to show that $\prod_{k=1}^{n+1} X_k$ is first countable iff $X_k, k = 1, 2, \ldots, n + 1$, is first countable. To do this, we use Theorem 2.13 of Chapter 5 to write $\prod_{k=1}^{n+1} X_k \cong \left(\prod_{k=1}^{n} X_k \right) \times X_{n+1}$. If each $X_k, k = 1, 2, \ldots, n + 1$, is first countable, then $\prod_{k=1}^{n} X_k$ is first countable by the inductive hypothesis. Therefore, $\left(\prod_{k=1}^{n} X_k \right) \times X_{n+1}$ is first countable by Theorem 4.8, and this in turn implies that $\prod_{k=1}^{n+1} X_k$ is first countable according to Theorem 4.6. On the other hand, if $\prod_{k=1}^{n+1} X_k$ is first countable, then both $\prod_{k=1}^{n} X_k$ and X_{n+1} are first countable by the fact that $\prod_{k=1}^{n+1} X_k \cong \left(\prod_{k=1}^{n} X_k \right) \times X_{n+1}$ and the results of Theorem 4.8. Now the inductive hypothesis gives $X_k, k = 1, 2, \ldots, n$, first countable so that $X_k, k = 1, 2, \ldots, n + 1$, is first countable. In summary, we have used the inductive hypothesis to prove that $\prod_{k=1}^{n+1} X_k$ is first countable iff each of the spaces $X_k, k = 1, 2, \ldots, n + 1$, is first countable. By the Principle of Mathematical Induction, our theorem is now proved.

4.10. Corollary: *For each $n \in N$, R^n is a first countable space.*

Proof: Theorem 4.3 gives R^1 first countable and, since R^n is the product of R^1 with itself n times, Theorem 4.9 gives R^n first countable.

Exercises

1. Give a countable local base for the point $p = (x, y) \in R^2$.

2. If $(X, \mathcal{T})$ is a T_1-space, $p \in X$ and $\mathcal{B}_p$ is a local base at the point p, prove $\cap \{B : B \in \mathcal{B}_p\} = p$.

3. Prove Theorem 4.5.

4. Prove Theorem 4.6.

5. Prove Theorem 4.7.

6. Prove that R^1 with the left ray topology is first countable.

7. Show that Theorem 4.6 may be strengthened by proving the following: If $f : X \to Y$ is a continuous open surjection and X is a first countable space, then Y is a first countable space.

8. Prove or disprove that R^1 with the cofinite topology is first countable.

9. (a) If $(X, \mathscr{T})$ is first countable and $\mathscr{T}_1$ is a topology for X such that $\mathscr{T}_1 \subset \mathscr{T}$, then prove $(X, \mathscr{T}_1)$ is first countable.
   (b) Give an example of a space that is first countable and a topology $\mathscr{T}_1$ for X such that $\mathscr{T} \subset \mathscr{T}_1$ and $(X, \mathscr{T}_1)$ is not first countable.

10. Prove that if $\{X_\alpha : \alpha \in \Delta\}$ is a family of first countable spaces indexed by a countable indexing set, then $\prod_{\alpha \in \Delta} X_\alpha$ is first countable.

5. The Second Axiom of Countability

The second of the countability axioms will now be stated.

5.1. Definition: (The second Axiom of Countability.) A space $(X, \mathscr{T})$ is called *second countable* iff there is a countable base for $\mathscr{T}$.

The relationship between first and second countable spaces is given first.

5.2. THEOREM: *Every second countable space $(X, \mathscr{T})$ is first countable.*

Proof: Let $\mathscr{B}$ be a countable base for $\mathscr{T}$. Then for each $x \in X$ the family $\mathscr{B}_x$ of all elements of $\mathscr{B}$ countaining x forms a local base at x. Since there are only a countable number of elements in $\mathscr{B}$, then $\mathscr{B}_x$ consists of a countable number of elements of $\mathscr{B}$. Hence, $(X, \mathscr{T})$ is first countable.

5.3. Example: (Not every first countable space is second countable.) Consider R^1 with the discrete topology. This space is first countable, but no countable base exists for the discrete topology. Therefore, the space is not second countable.

We now prove R^1 is a second countable space. This fact will be used to prove R^n, $n \in N$, is second countable.

5.4. THEOREM: *The space R^1 is second countable.*

Proof: To prove our assertion, we must find a countable base for the standard topology on R^1. To do this, let Q be the set of all rationals and consider the countable local base of open intervals $\mathscr{B}_q = \{(q - 1/m, q + 1/m) : m \in N\}$ at each point $q \in Q$. The collection $\mathscr{B} = \{\mathscr{B}_q : q \in Q\}$ is a countable family of countable sets and hence is countable by Theorem 5.9 of Chapter 2. We shall show $\mathscr{B}$ is a base for the standard topology. If U is an open set in R^1 where $U \neq \phi$, then U may be expressed as the union of a countable collection of open disjoint intervals according to Theorem 3.4 of Chapter 3. That is, $U = \bigcup_{n \in M \subset N} V_n$, where each V_n is an open interval. Therefore, the problem at hand reduces to showing each V_n, $n \in M$, is the union of elements from $\mathscr{B}$. Fix n and note that V_n contains a rational number by Theorem 3.1 of Chapter 3. For each rational $q \in V_n$ consider all of those elements of $\mathscr{B}_q$ which are subsets of V_n. Call the union of these elements A_q. Then certainly $\bigcup_{q \in V_n} A_q \subset V_n$. Furthermore, if $x \in V_n$, there is an open interval $(a, b) \subset V_n$ containing x and, consequently, there is a natural number k such that $1/k < 1/2$ (minimum $\{x - a, b - x\}$). Now, within the open interval $(x - 1/k, x + 1/k)$ lies a rational number q and hence $x - 1/k < q < x + 1/k$ or $-(1/k) < q - x < 1/k$, which implies $q - 1/k < x < q + 1/k$. Also $a < q - 1/k$ and $q + 1/k < b$ so that $x \in (q - 1/k, q + 1/k) \subset \bigcup_{q \in V_n} A_q$ or $V_n \subset \bigcup_{q \in V_n} A_q$. Thus, $V_n = \bigcup_{q \in V_n} A_q$ and $U = \bigcup_{n \in M \subset N} V_n$ is the union of elements from the countable collection $\mathscr{B}$. This completes the proof that $\mathscr{B}$ is a countable base for the standard topology on R^1.

The second countable property is inherited by subspaces.

5.5. THEOREM: *Any subspace of a second countable space is second countable.*

Proof: Let X be second countable and $\mathscr{B} = \{U_n : n \in N\}$ a countable base for the topology on X. For any nonempty subset $A \subset X$, the collection $\mathscr{B}_A = \{U_n \cap A : n \in N \text{ and } U_n \in \mathscr{B}\}$ is a base for the subspace topology $\mathscr{T}_A$ by Theorem 1.6 of Chapter 4. Since $\mathscr{B}_A$ is countable, $(A, \mathscr{T}_A)$ is second countable.

Closely related to the condition of second countability is a condition known as *separability*. We will now give its definition and investigate some of its properties.

5.6. Definition: A space X is called *separable* iff there exists a countable subset of X which is dense in X.

The space R^1 provides a non-trivial example of a separable space since the rationals form a countable dense subset of R^1. Any countable space is also separable.

5.7. THEOREM: *Every second countable space is separable.*

Proof: A set consisting of a point from each element of a countable base will be countable and dense in the space. The details are left as an exercise.

5.8.   Example: (A separable space that is not first countable and hence not second countable.) Let R^1 have the cofinite topology. The set of all rational numbers Q is a countable dense subset of this space. On the other hand, Exercise 11(a), Section 1, of Chapter 4, tells us that this space is not first countable. Finally, the contrapositive of Theorem 5.2 implies that the space is not second countable. Details are left as an exercise.

5.9. THEOREM: *If $f : X \to Y$ is a continuous surjection from the separable space X to the space Y, then Y is separable.*

Proof: Let D be a countable dense subset of X. Then $\bar{D} = X$, and by continuity of f, it follows that $f(X) = Y = f(\bar{D}) \subset \overline{f(D)}$. Since $\overline{f(D)} \subset Y$, we have $Y = \overline{f(D)}$ and Exercise 8, Section 5, of Chapter 2, gives $f(D)$ countable. As a consequence, $f(D)$ is a countable dense subset of Y, implying that Y is separable.

Actually, Theorem 5.9 is a stronger result than we proved in the section on homeomorphisms in Chapter 5. We can infer from Theorem 5.9 that separability is a topological property. We show next that second countability is also a topological property.

5.10.   THEOREM: *The property of being second countable is a topological property.*

Proof: Let X be second countable and $h : X \to Y$ a homeomorphism. Denote by $\mathscr{B} = \{U_n : n \in N\}$ a countable base for the topology on X and consider the family of sets $\mathscr{B}_Y = \{h(U_n) : n \in N\}$. Certainly $\mathscr{B}_Y$ is countable. We need to show $\mathscr{B}_Y$ is a base for the topology on Y. To do this, let V be any open subset of Y. Then $h^{-1}(V)$ is open in X and for some subset M of N we have $h^{-1}(V) = \bigcup_{j \in M} U_{n_j}$ where $U_{n_j} \in \mathscr{B}$ because $\mathscr{B}$ is a base for the topology on X. Thus, $h(h^{-1}(V)) = V =$

$h(\bigcup_{j \in M} U_{n_j}) = \bigcup_{j \in M} h(U_{n_j})$ showing V to be expressed as the union of members of $\mathscr{B}_Y$. Therefore, $\mathscr{B}_Y$ is a base for the topology on Y and Y is second countable.

It is not true that, if Δ is an arbitrary indexing set and $\{(X_\alpha, \mathscr{T}_\alpha) : \alpha \in \Delta\}$ is a family of second countable (or separable) spaces, then the product space $\prod_{\alpha \in \Delta} X_\alpha$ is second countable (separable). This statement may be proved, however, if Δ is a countable indexing set, although we shall not attempt to do so here. For a proof of this fact and a statement of a more general condition, see reference 1 of the Bibliography, for instance. On the other hand, we have the following theorem.

5.11. THEOREM: *If $\{(X_\alpha, \mathscr{T}_\alpha) : \alpha \in \Delta\}$ is a family of spaces and the product space $\prod_{\alpha \in \Delta} X_\alpha$ is second countable (separable), then X_α is second countable (separable) for each $\alpha \in \Delta$.*

Proof: The proof for separable spaces only will be given, while that for second countable spaces is left as an exercise. Let D be a countable dense subset of $\prod_{\alpha \in \Delta} X_\alpha$, and consider the space X_β for $\beta \in \Delta$. We are going to show that the set $p_\beta(D) \subset X_\beta$ is a countable dense subset of X_β. Since D is countable, certainly $p_\beta(D)$ is a countable subset of X_β. Now let $x \in X_\beta$ and U_β be any open subset of X_β containing x. Then $p_\beta^{-1}(V_\beta)$ is open in $\prod_{\alpha \in \Delta} X_\alpha$ and contains a point f_0 such that $f_0(\beta) = x$. According to Theorem 6.11(c) of Chapter 3, there exists a point $f \in D$ such that $f \in p_\beta^{-1}(V_\beta)$. Therefore, $p_\beta(f) \in p_\beta(p_\beta^{-1}(V_\beta)) = V_\beta$, the equality holding because p_β is surjective. Thus, there is a point $p_\beta(f)$ of $p_\beta(D)$ belonging to U_β which makes $p_\beta(D)$ dense in X by Theorem 6.11(c) of Chapter 3.

We are now going to give a theorem which shows that a finite product is second countable (separable) iff each of the factor spaces is second countable (separable). This will allow us to conclude that $R^n, n \in N$, is second countable (separable).

5.12. THEOREM: *The product space $X \times Y$ is second countable (separable) iff both X and Y are second countable (separable) spaces.*

Proof: Only the proof for second countability will be given here. If $X \times Y$ is second countable, we may use Theorem 5.11 to obtain the second countability of both X and Y. We might also use the fact that if $(x, y) \in X \times Y$, then $X \times \{y\}$ is a second countable subspace of $X \times Y$ that is homeomorphic to X. The fact that second countability is a topological property gives X as a second countable space. The space Y could be shown second countable in the same manner. Perhaps we should point out that this technique is not valid in showing the separable part of the proof since it is not true that subspaces of separable spaces are always separable.

For the converse, let X and Y be second countable. Then there exist countable bases $\mathscr{B}_X = \{U_n : n \in N\}$ and $\mathscr{B}_Y = \{V_m : m \in N\}$ for the spaces X and Y, respectively. The family $\mathscr{B}_X \times \mathscr{B}_Y = \{U_n \times V_m : n, m \in N\}$ is a countable collection which is a base for the product topology on $X \times Y$. Verification that this family actually is a base for the product topology is an easy instructive matter and is left as an exercise.

5.13. THEOREM: *A finite product space $\prod_{k=1}^{n} X_k$ is second countable (separable) iff each space X_k is second countable (separable).*

Proof: (For second countable spaces.) If $n = 2$, Theorem 5.12 gives $X_1 \times X_2$ second countable iff X_1 and X_2 are both second countable spaces. Assuming $\prod_{k=1}^{n} X_k$ second countable iff each X_k, $k = 1, 2, \ldots, n$, is second countable, $\prod_{k=1}^{n+1} X_k \cong \left(\prod_{k=1}^{n} X_k \right) \times X_{n+1}$ gives $\prod_{k=1}^{n+1} X_k$ second countable iff each space X_k, $k = 1, 2, \ldots, n + 1$, is second countable, by Theorem 5.11, the inductive hypothesis, and the fact that second countability is a topological property. This completes the inductive proof.

5.14. Corollary: *For each $n \in N$, R^n is second countable (separable).*

Proof: Theorem 5.4, along with Theorem 5.13.

This chapter concludes with an example of a separable space which has a non-separable subspace.

5.15. Example: In R^1, consider the set $\mathscr{B}$ of all intervals having the form $[a, b)$ where a and b are real numbers. For each $x \in R^1$, there exists a real number b such that $b > x$ and, consequently, $x \in [x, b) \in \mathscr{B}$. If $[a, b)$ and $[c, d)$ are any two members of $\mathscr{B}$ such that $x_0 \in$

$[a, b) \cap [c, d)$, then both a and c are less than or equal to x_0, while both b and d are greater than x_0. From this, it follows that $[a, b) \cap [c, d)$ is one of the following intervals: $[a, b)$, $[c, d)$, $[a, d)$ or $[c, b)$, all of which belong to $\mathscr{B}$. By Theorem 1.7 and 1.8 of Chapter 4, there is a unique topology $\mathscr{T}(\mathscr{B})$ generated by $\mathscr{B}$ and having $\mathscr{B}$ as a base. Since every interval of the form (a, b) contains a rational, every interval of the form $[a, b)$ contains a rational number. This means that the set of all rationals Q is a countable dense subset of our space

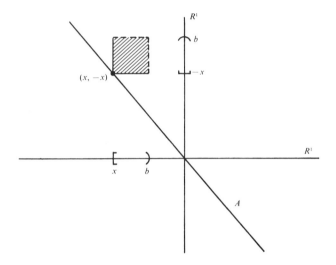

Figure 21

$(R^1, \mathscr{T}(\mathscr{B}))$. Thus, since this space is separable, the product of this space with itself is separable according to Theorem 5.13. However, the subset $A = \{(x, -x) : x \in R^1\}$ of $R^1 \times R^1$ has a discrete subspace topology $\mathscr{T}_A$. The reason is that for each $(x, -x) \in A$, the basic open set $[x, b) \times [-x, b)$ intersects A in only the point $(x, -x)$. Figure 21 gives some visual aid.

Exercises

1. In the proof to Theorem 5.12, give details of the proof that $\mathscr{B}_X \times \mathscr{B}_Y$ is a base for the product topology.

2. Prove Theorem 5.7.

3. (a) Prove that $X \times Y$ is separable iff both X and Y are separable spaces.

   (b) Use part (a) to prove that the finite product of spaces is separable iff each of the factor spaces is separable.

4. Consider R^1 with the left ray topology. Is this space second countable? Separable?

5. If $(X, \mathscr{T})$ is separable and $\mathscr{T}_1 \subset \mathscr{T}$, prove that $(X, \mathscr{T}_1)$ is separable.

6. If X is a second countable space and $A \subset X$ is uncountable, prove that there exists some point $a \in X$ which is a cluster point of A.

7. Prove that the set of isolated points in a second countable space is countable.

8. Prove the "second countable" part of Theorem 5.11.

9. Let $(X, \mathscr{T})$ be a second countable space, and let $\mathscr{B}$ be a base for $\mathscr{T}$. Prove that there exists a countable subset of $\mathscr{B}$ which is a base for $\mathscr{T}$.

10. Verify the details of Example 5.8.

7

Convergence

1. Sequences

Calculus offers many uses of sequences in R^n with which we are already familiar. Our study of sequences at the present time will be directed toward more general topological spaces with limited reference to the R^n spaces. Sequences in the R^n spaces will be more thoroughly discussed in Chapter 9. For general topological spaces, sequences have the distinct disadvantage of not being able to describe all of the topological properties of the space. We are going to specifically point out three such deficiencies of sequences in this chapter, see what conditions on topological spaces correct them, and then see how they may be generally corrected by the notions of filters and nets. We begin with the definition of a sequence.

1.1. Definition: A *sequence* in the set X is a function $f: N \rightarrow X$ from the set of natural numbers N to the set X. If $f(N) = \{a\}$ where $a \in X$, then f is called a *constant sequence*. If f is injective, then f is called a sequence of distinct points.

Whenever a sequence $f: N \rightarrow X$ in X is described, we are many times interested in the value of the images $f(n) \in X$ for each $n \in N$. Thus, if the notation $f(n) = x_n$ is used, it is sometimes convenient to denote the sequence as merely (x_n) since these functional values completely describe the function $f: N \rightarrow X$, as we have seen before. In so doing, we obscure somewhat the functional aspect of a sequence and tend to think of a sequence as a subset of X indexed by the natural numbers. Even though this may be the case, if the notation (x_n) or $(f(n))$ is used for a sequence, keep in mind that it actually describes a function $f: N \rightarrow X$ such that $f(n) = x_n$ for each $n \in N$. In no instance is a sequence $f: N \rightarrow X$ a subset of X according to our definition.

1.2. Example: Each of the following functions is a sequence in R^1: (a) $f(n) = n^2$ for each $n \in N$, (b) $f(n) = 0$ for each $n \in N$, (c) $f(n) = (-1)^n(1/n)$ for each $n \in N$, and (d) $f(n) = 1$ if n is odd and $f(n) = n$ if n is even, $n \in N$.

Whether or not a sequence converges is of prime importance to our studies. Next we give the definition of convergence for general topological spaces. Notice how strongly dependent convergence is upon the topology of the space X and the behavior of the set of images $\{f(n)\}$ in X. Of course, the fact that N is linearly ordered plays an important role, too.

1.3. Definition: Let $(X, \mathcal{T})$ be a topological space and $f: N \rightarrow X$ a sequence in X. Then f *converges* to the point $a \in X$ iff given any open set $U(a)$ containing a there exists a natural number n_0 such that if $n > n_0$, $f(n) \in U(a)$. If $f: N \rightarrow X$ converges to $a \in X$, we may write the notation as $f \rightarrow a$ or $(x_n) \rightarrow a$ or $\lim x_n = a$. If $f \rightarrow a$, then the point a is called a *sequential limit point* of f.

If X has the indiscrete topology and $a \in X$, then every sequence in X converges to the point a. If X has the discrete topology, then the only sequences in X which converge are the ones for which there exists a natural number n_0 such that f restricted to the set $\{n \in N : n > n_0\}$ is a constant function.

1.4. Example: Consider the space R^1. The sequence $f(n) = 1/n$ for each $n \in N$ in R^1 converges to $a = 0$. To see this, let U be any open set containing $a = 0$. Then there exists a natural number n_0 such that $V = \{x \in R^1 : -(1/n_0) < x < 1/n_0\} \subset U$. Thus, for $n > n_0$, $f(n) = 1/n \in V \subset U$ and hence $f \to 0$. The sequence $f(n) = n$ from N to R^1 does not converge to any point in R^1 because for each $a \in R^1$ the open set $\{x \in R^1 : a - 1 < a < a + 1\}$ contains no more than two points of the set $f(N) = N$. Hence, f cannot converge to a.

This example brings us to a condition telling when a sequence does not converge.

1.5. THEOREM: *(A non-convergence criterion.) Let $(X, \mathcal{T})$ be a space and $f : N \to X$ a sequence in X. Then f does not converge to the point $a \in X$ iff there exists some open set $U(a) \in \mathcal{T}$ containing a such that for every natural number n_0 there is some $n > n_0$ such that $f(n) \notin U(a)$.*

Proof: The proof is left as an exercise.

1.6 Example: Let $X = \{a, b, c\}$ with topology $\{\phi, X, \{a\}, \{a, b\}\}$. For $n \in N$, define $f : N \to X$ as $f(n) = a$ if n is even and $f(n) = b$ if n is odd. It is easy to verify that $f \to b$ and also $f \to c$, but f does not converge to the point $a \in X$.

Example 1.6 is an example of how eratic sequences can behave in general topological spaces. That is, sequences may converge to more than one point. The following theorem gives a condition which overcomes such an undesirable event.

1.7. THEOREM: *Let $(X, \mathcal{T})$ be Hausdorff space and $f : N \to X$ a sequence in X. If the sequence f converges, it converges to only one point in X.*

Proof: Let $f \to a$. Assume $f \to b$ where $b \neq a$. Since X is a Hausdorff space, there exist open disjoint sets $U(a)$ and $V(b)$ containing a and b, respectively. Since $f \to a$, there exists a natural number n_1 such that if $n > n_1$, then $f(n) \in U(a)$. Similarly, $f \to b$ implies that there exists a natural number n_2 such that if $n > n_2$, then $f(n) \in V(b)$. Now, if $n > \text{maximum}\{n_1, n_2\}$, then $f(n) \in U(a) \cap V(b) = \phi$, which is impossible. Therefore f can converge only to a.

Since R^n is Hausdorff for each $n \in N$, the following corollary holds.

1.8. Corollary: For each $n \in N$, each convergent sequence $f: N \to R^n$ converges to only one point in R^n.

For the space R^1, convergence of sequences may be described in the familiar notation of calculus.

1.9. THEOREM: *Let $f: N \to R^1$ be a sequence in R^1. Then $f \to a \in R^1$ iff for each real number $\epsilon > 0$ there exists a natural number n_0 such that for all $n > n_0$, $|f(n) - a| < \epsilon$.*

Proof: The straightforward proof is left as an exercise.

For sequences $f: N \to R^1$ and $g: N \to R^1$ we may form the sequences $(f \pm g): N \to R^1$ defined by $(f \pm g)(n) = f(n) \pm g(n)$, $(cf): N \to R^1$ defined by $(cf)(n) = c \cdot f(n)$ for the real number c, $(f \cdot g): N \to R^1$ defined by $(f \cdot g)(n) = f(n) \cdot g(n)$ and $(f/g): N \to R^1$ defined by $(f/g)(n) = f(n)/g(n)$ provided $g(n) \neq 0$. Once this is done we may repeat a familiar theorem from calculus.

1.10. THEOREM: *Let $f. N \to R^1$ and $g: N \to R^1$ be sequences in the space R^1. If $f \to a$ and $g \to b$, then*
   (a) *The sequence $f \pm g$ converges to $a \pm b$.*
   (b) *The sequence cf converges to $c \cdot a$.*
   (c) *The sequence $f \cdot g$ converges to $a \cdot b$.*
   (d) *If $g(n) \neq 0$ for all $n \in N$ and $b \neq 0$, then the sequence f/g converges to a/b.*

Proof: (a) We prove (a) for the sequence $f + g$. Let $\epsilon > 0$ be any real number. Since $f \to a$, there exists an $n_1 \in N$ such that if $n > n_1$, then $|f(n) - a| < \epsilon/2$. Similarly, $g \to b$ implies the existence of an $n_2 \in N$ such that if $n > n_2$, then $|g(n) - b| < \epsilon/2$. Now if we let $n_0 = $ maximum $\{n_1, n_2\}$ and consider any natural number $n > n_0$, then $|f(n) - a| < \epsilon/2$ and $|g(n) - b| < \epsilon/2$ are both true. Therefore, by use of the triangle inequality for real numbers ($|x + y| \leq |x| + |y|$ for all $x, y \in R^1$), we have for each $n > n_0$

$$|(f(n) + g(n)) - (a + b)| = |((f(n) - a) + (g(n) - b)|$$
$$\leq |f(n) - a| + |g(n) - b| < \epsilon/2 + \epsilon/2 = \epsilon.$$

The fact that $(f + g) \to (a + b)$ now follows from Theorem 1.9

because for each real number $\epsilon > 0$ we have shown the existence of a natural number n_0 such that if $n > n_0$, then $|(f(n) + g(n)) - (a + b)| < \epsilon$.

(b) The proof of (b) is left as an exercise.

(c) It must be shown that if $\epsilon > 0$, there exists an $n_0 \in N$ such that if $n > n_0$, then $|f(n) \cdot g(n) - a \cdot b| < \epsilon$. We first write

$$
\begin{aligned}
|f(n) \cdot g(n) - a \cdot b| &= |(f(n) \cdot g(n) - f(n) \cdot b) + (f(n) \cdot b - a \cdot b)| \\
&\leq |f(n) \cdot g(n) - f(n) \cdot b| + |f(n) \cdot b - a \cdot b| \\
&= |f(n) \cdot (g(n) - b)| + |b \cdot (f(n) - a)|.
\end{aligned}
$$

Then, using the fact that the absolute value of the product is the same as the product of the absolute values, we have

$$
|f(n) \cdot g(n) - a \cdot b| \leq |f(n)| \cdot |g(n) - b| + |b| \cdot |f(n) - a|.
$$

Our task now is to find an $n_0 \in N$ such that the quantity on the right of the last inequality is less than ϵ. Since $f \to a$, it is a straightforward exercise to show that there exists a $k \in N$ such that $|f(n)| < k$ for all $n \in N$. Thus, if we let $M = \text{maximum}\{k, b\}$, we have $|f(n) \cdot g(n) - a \cdot b| \leq M \cdot |g(n) - b| + M \cdot |f(n) - a|$ for all $n > M$. From $f \to a$ and $g \to b$ we have the existence of $n_1, n_2 \in N$ such that if $n > n_1$, then $|f(n) - a| < \epsilon/2M$ and if $n > n_2$, then $|g(n) - b| < \epsilon/2M$. Therefore, if we let $n_0 = \text{maximum}\{n_1, n_2, k\}$, it follows that if $n > n_0$, then $|f(n) \cdot g(n) - a \cdot b| < M\epsilon/2M + M\epsilon/2M = \epsilon$, which concludes the proof of (c).

(d) Since $g \to b$, where $b \neq 0$, there exists a natural number $k \in N$ such that if $n > k$, $|g(n) - b| < |b|/2$. Thus, $|b| = |(b - g(n)) + g(n)| \leq |g(n) - b| + |g(n)| < |b|/2 + |g(n)|$. This means $|b|/2 < |g(n)|$ for all $n > k$. Therefore,

$$
\begin{aligned}
\left| \frac{f(n)}{g(n)} - \frac{a}{b} \right| &= \left| \frac{b \cdot f(n) - a \cdot g(n)}{g(n) \cdot b} \right| \\
&= \left| \frac{b \cdot (f(n) - a) + a \cdot (b - g(n))}{g(n) \cdot b} \right| \\
&\leq \frac{|b| \cdot |f(n) - a| + |a| \cdot |b - g(n)|}{|g(n)||b|} \\
&< \frac{|b| \cdot |f(n) - a| + |a| \cdot |b - g(n)|}{1/2|b| \cdot |b|}
\end{aligned}
$$

for all $n > k$. (In the last expression we have replaced $|g(n)|$ with the smaller quantity $|b|/2$, thereby making the quotient larger.) Now let $\epsilon > 0$. Since $f \to a$ and $g \to b$, there exist natural numbers n_1 and n_2 such that if $n > n_1$, then $|f(n) - a| < \epsilon |b|/4$ and if $n > n_2$, then $|g(n) - b| < \epsilon |b|^2/4|a|$. At this point let $n_0 = $ maximum $\{n_1, n_2, k\}$. Then for all $n > n_0$ we have

$$\left| \frac{f(n)}{g(n)} - \frac{a}{b} \right| \leq \frac{|b| \cdot |f(n) - a| + |a| \cdot |b - g(n)|}{|b|^2/2}$$

$$< \frac{|b|\epsilon|b|/4 + |a|\epsilon|b|^2/4|a|}{|b|^2/2}$$

$$= \frac{\epsilon|b|^2/4 + \epsilon|b|^2/4}{|b|^2/2} = \epsilon.$$

This proves part (d).

Exercises:

1. Prove Theorem 1.5.

2. (a) Let $(X, \mathcal{T})$ be a space and $f: n \to X$ a sequence. If $f \to a \in X$ and $U(a)$ is any open set in X containing a, prove that $X \backslash U(a)$ contains only a finite number of points of $f(N)$.
   (b) Give an example to show that the converse of part (a) is not true.

3. Let R^1 have the left ray topology.
   (a) Does the sequence $f: N \to R^1$ given by $f(n) = 1/n$ converge? If so, to what point or points?
   (b) Does the sequence $f: N \to R^1$ given by $f(n) = -n$ converge? If so, to what points?

4. (a) Does every constant sequence converge?
   (b) Which sequences in R^1 with the cofinite topology converge?
   (c) Which sequences in R^1 with the co-countable topology converge?

5. Prove Theorem 1.9.

6. Prove Theorem 1.10(b).

7. Let $f: N \to X$ be a sequence in the space X. Prove that $f \to a \in X$ iff for each basic (subbasic) open set W in X there exists a natural number $n_0 \in N$ such that if $n > n_0$, then $f(n) \in W$.

8. Give an example of two sequences $f: N \to R^1$ and $g: N \to R^1$

such that the sequence $f + g$ converges but neither f nor g converges. Do the same for the sequence f/g. (This shows that the converses of Theorem 1.10(a) and (d) do not hold.)

9. Prove that if $c \neq 0$ is a real number and the sequence (cf): $N \to R^1$ converges to the point $b \in R^1$, then $f: N \to R^1$ converges to b/c.

10. Prove that if the sequence $f: N \to R^1$ converges to the point $a \in R^1$, then there exists a natural number k such that $|f(n)| < k$ for all $n \in N$.

11. Prove that if $f: N \to R^1$ is a sequence in R^1 given by $f(n) = (2n + 1)/(n + 3)$ for each $n \in N$, then $f \to 2$.

12. Prove that if a is a real number such that $0 < a < 1$ and $f: N \to R^1$ is given by $f(n) = a^n$, then the sequence f converges to 0.

13. Give an example to show the converse of Theorem 1.7 does not hold. That is, give an example of a space X where each sequence in X that converges, converges to exactly one point in X, but X is not Hausdorff.

2. Convergence in First Countable Spaces

In this section we are going to consider two deficiencies that sequences in general topological spaces have and give conditions that will overcome them.

2.1. THEOREM: *Let A be a subset of the space X. If $a \in X$ and there exists a sequence $f: N \to A\backslash\{a\}$ in $X\backslash\{a\}$ that converges to the point a, then a is a cluster point of A.*

Proof: Let $U(a)$ be any open set in X containing the point a. Then there exists an $n_0 \in N$ such that for all $n > n_0$, $f(n) \in U(a)$ because $f \to a$. Since each $f(n) \in A\backslash\{a\}$, it follows that $f(n) \in A\backslash\{a\} \cap U(a)$ for all $n > n_0$. Therefore, a is a cluster point of A.

The converse to Theorem 2.1 is false, however. The next example shows why.

2.2. Example: Let R^1 have the co-countable topology. That is, a subset U of R^1 is open iff $R^1\backslash U$ is a countable set. Consider the set

$A = \{x \in R^1 : 0 \le x \le 1\}$. The point $a = 5$, for instance, is a cluster point of A because every open set U belonging to the co-countable topology containing $a = 5$ is such that $R^1 \backslash U$ is countable and hence $U \cap A \ne \phi$. In fact, $\bar{A} = R^1$. However, there is no sequence $f : N \to A$ in A converging to $a = 5$. The reason is that for any sequence $f : N \to A$ in A, $R^1 \backslash f(N)$ is an open set containing $a = 5$, but no point of $f(N)$ and, therefore, f cannot converge to $a = 5$.

In view of Example 2.2, sequences are not adequate for characterizing the cluster points of subsets of general topological spaces. This is the first of the three deficiencies we spoke of earlier. There are conditions which may be imposed on spaces which will overcome this somewhat undesirable situation, however. The next two theorems will clarify the situation.

2.3. THEOREM: *Let $(X, \mathscr{T})$ be a first countable T_1-space. Then $a \in X$ is a cluster point of $A \subset X$ iff there exists a sequence of distinct points $f : N \to A$ in A which converges to a.*

Proof: First, let a be a cluster point of $A \subset X$. We shall now construct a sequence of distinct points in A which converges to a. Note that by Exercise 8, Section 1, of Chapter 6, A must be an infinite set. By Theorem 4.4 of Chapter 6, there exists a countable local base $\{U_n : n \in N\}$ at a such that $U_{n+1} \subset U_n$ and $\bigcap_{n \in N} U_n = a$. Thus, the open set U_1 contains a point $x_1 \in A$ such that $x_1 \ne a$ because a is a cluster point of A. The point x_1 is a closed subset in the T_1-space X so that $X \backslash \{x_1\}$ is open, contains a and, therefore, has a nonempty intersection with U_2 because $a \in U_2$. Since a is a cluster point of A, there exists a point $x_2 \in A$, $x_2 \ne x_1$, belonging to the open set $(X \backslash \{x_1\}) \cap U_2$. Inductively, there exists a point $x_n \in A$ belonging to the open set $(X \backslash \bigcup_{k=1}^{n-1} x_k) \cap U_n$ such that $x_n \ne x_k$ for $k = 1, 2, \ldots, n - 1$. The reasons are that $\bigcup_{k=1}^{n-1} x_k$ is closed in the T_1-space X implying $X \backslash \bigcup_{k=1}^{n-1} x_k$ is open, $a \in (X \backslash \bigcup_{k=1}^{n-1} x_k) \cap U_n \ne \phi$ and a is a cluster point of A. Thus, if we let $f(n) = x_n$ for each $n \in N$, then we have defined a sequence $f : N \to A$ which converges to a. The reason is that $f \to a$ comes from the properties of $\{U_n : n \in N\}$ and the fact that $f(n) \in U_n$ for each $n \in N$.

For the converse, suppose $a \in X$ is a point such that there exists a sequence of distinct points $f: N \to A$ in $A \subset X$ for which $f \to a$. We need to show a is a cluster point of A. To do this, let $U(a)$ be any open set in X containing a. Since $f \to a$, there exists an $n_0 \in N$ such that if $n > n_0$, then $f(n) \in U(a)$. This, along with the hypothesis that f is injective, gives $f(N) \cap U(a)$ as an infinite set which implies a is a cluster point of A.

2.4. Corollary: The point $a \in R^n$ is a cluster point of $A \subset R^n$ iff there exists a sequence of distinct points $f: N \to A$ in A converging to a.

Proof: For each $n \in N$, R^n is T_1 and first countable by Theorem 1.12 and Corollary 4.10, respectively, of Chapter 6.

Notice in the last half of the proof to Theorem 2.3 that no use was made of either the T_1-axiom or the First Axiom of Countability. Therefore, this part of Theorem 2.3 is true in any topological space whatever. That is, in any space X, if $f: N \to A$ is a sequence of distinct points in $A \subset X$ converging to $a \in X$, then a is a cluster point of A. Recall also that in the first part of the proof to Theorem 2.3, the T_1-axiom was used only to ensure construction of a sequence of *distinct* points $f: N \to A$ in A. Therefore, with first countability we may now prove the converse of Theorem 1.1.

2.5. THEOREM: *Let X be a first countable space and $A \subset X$. Then $a \in X$ is a cluster point of A iff there exists a sequence $f: n \to A \backslash \{a\}$ in $A \backslash \{a\}$ which converges to the point a.*

Proof: If there exists a sequence in $A \backslash \{a\}$ converging to a, then a is a cluster point of A by Theorem 1.1.

To prove the converse we follow the pattern of the first part of the proof to Theorem 2.3. Thus, let a be a cluster point of A. Due to the first countable hypothesis, there exists a countable local base $\{U_n : n \in N\}$ at the point a such that $U_{n+1} \subset U_n$. Now, for each $n \in N$, there exists a point $x_n \in U_n \cap A$ such that $x_n \neq a$. The reason is that a is a cluster point of A and U_n is an open set in X containing a. It follows that if we define $f: N \to A \backslash \{a\}$ as $f(n) = x_n$ for each $n \in N$, then f is a sequence in $A \backslash \{a\}$ converging to the point a.

In view of Theorem 2.5, cluster points in first countable spaces are completely characterized by sequences. As a consequence of this fact, we show next that closed sets in first countable spaces are characterized by sequences. As we have observed before, if we can determine which sets are to be closed in a space, then we know the open sets and thus the topology of the space. Therefore, sequences in first countable spaces completely determine the topology of those spaces.

2.6. THEOREM: *Let X be a first countable space and $A \subset X$. Then $a \in \bar{A}$ iff there exists a sequence in A converging to the point a.*

Proof: Suppose first that $a \in \bar{A}$. If $a \in A$, then the sequence f: $N \to A$ given by $f(n) = a$ for all $n \in N$ is of the desired type. If $a \notin A$, then $a \in A'$ and, by Theorem 2.5, there exists a sequence in A converging to a.

Conversely, suppose there exists a sequence in A converging to the point a. If $a \in A$, then $a \in \bar{A}$. If $a \notin A$, then Theorem 2.5 tells us that a is a cluster point of A and this, in turn, tells us that $a \in \bar{A}$.

We now turn our attention to the second of the deficiencies of sequences in topological spaces. This one is the fact that sequences do not characterize continuous functions from one topological space to another. For general spaces we do have the following result, however.

2.7. THEOREM: *Let $(X, \mathcal{T}_X)$ and $(Y, \mathcal{T}_Y)$ be spaces and $g: X \to Y$ a continuous function at the point $a \in X$. For each sequence $f: N \to X$ in X converging to the point a, the sequence $gf: N \to Y$ converges to the point $g(a) \in Y$.*

Proof: Let $a \in X$ and $f: N \to X$ such that $f \to a$. For any open set V containing $g(a)$, there exists an open set $U \subset X$ containing a such that $g(U) \subset V$. Thus there exists an $n_0 \in N$ such that if $n > n_0$, then $f(n) \in U$, due to the fact that $f \to a$. Consequently, $g(f(n)) \in V$ for all $n > n_0$. The existence of n_0 is precisely what is needed to ensure that $gf: N \to Y$ converges to $g(a)$ according to Definition 1.3.

The converse to this theorem is false, as Example 2.8 shows.

2.8. Example: Consider $X = R^1$ with the co-countable topology as the domain of the function $g : X \to Y$ given by $g(x) = x$ where $Y = R^1$ with the standard topology. Suppose a is any point belonging to the domain X, and let $f : N \to X$ be any sequence in X converging to a. At this point, we need to observe that for the co-countable topology, the only sequences converging to a are those where $f(n) = a$ for all n greater than some fixed natural number n_0. Thus, the sequence $gf : N \to Y = R^1$ converges to $g(a)$ in Y because for $n > n_0$, $g(f(n)) = g(a)$. However, g is not continuous at $a \in X$. The reason is that for the open set $\{y \in Y : g(a) - 1 < y < g(a) + 1\}$ no open set $U \subset X$ belonging to the cofinite topology and containing a has the property that $g(U) \subset V$.

This situation may be corrected by imposing the condition of first countability on the domain space X as set forth in the next theorem. We include the results of Theorem 2.7 in this theorem to emphasize that a first countable domain space enables us to characterize continuity of a function with sequences.

2.9. THEOREM: *Let $g : X \to Y$ be a function from the first countable space X to the arbitrary space Y. Then g is continuous at $a \in X$ iff given any sequence $f : N \to X$ in X converging to a, the sequence $gf : N \to Y$ converges to $g(a)$.*

Proof: If g is continuous and $f : N \to X$ converges to $a \in X$, then $gf : N \to Y$ converges to $g(a) \in Y$ by Theorem 2.7.

For the converse, let $a \in X$, and suppose for every sequence $f : N \to X$ in X converging to a, the sequence $gf : N \to Y$ in Y converges to $g(a)$. We shall prove that g is continuous at the point a by contradiction. Assume g is not continuous at a. Then there exists an open set $V \subset Y$ containing $g(a)$ such that no open set $U \subset X$ containing a has the property that $g(U) \subset V$. Consider the countable local base $\{U_n : n \in N\}$ such that $U_{n+1} \subset U_n$, as given in Theorem 4.4(a) of Chapter 6. Such a local base at a exists since X is first countable by hypothesis. Then there is at least one point $x_1 \in U_1$ such that $g(x_1) \notin V$ and, inductively, for each $n \in N$ there is a point $x_n \in U_n$ such that $g(x_n) \notin V$. The sequence $f : N \to X$ in X given by $f(n) = x_n$ converges to the point a, but the sequence $gf : N \to Y$ in Y cannot converge to $g(a)$ because $g(a) \in V$ and $g(f(n)) \notin V$ for each $n \in N$. This contradiction to the hypothesis means that the assumption made earlier is false and, therefore, g is continuous at the point $a \in X$.

2.10. Corollary: The function $g : R^n \to R^m$ is continuous at $a \in R^n$ iff given any sequence $f : N \to R^n$ in R^n converging to $a \in R^n$, the sequence $gf : N \to R^m$ in R^m converges to the point $g(a) \in R^m$.

2.11. THEOREM: *Let X be first countable space and $h : X \to R^1$ and $g : X \to R^1$ be continuous functions. Then*
  (a) *The function $(h \pm g) : X \to R^1$ is continuous.*
  (b) *The function $(cf) : X \to R^1$ is continuous for each $c \in R^1$.*
  (c) *The function $(h \cdot g) : X \to R^1$ is continuous.*
  (d) *The function $(h/g) : X \to R^1$ is continuous provided $g(x) \neq 0$ for all $x \in X$.*

Proof: The proof is a straightforward application of Corollary 2.10 and Theorem 1.10.

Exercises:

1. Let X be a first countable space. Prove that $U \subset X$ is open iff there is no sequence $f : N \to X \backslash U$ in $X \backslash U$ which converges to a point of U.

2. Prove that the sequence constructed in the first part of the proof to Theorem 2.3 converges to the point a as asserted.

3. Prove that covergence of sequences is a topological property.

4. Prove Theorem 2.11.

5. Let $g : X \to Y$ be a continuous function and let $f : N \to X$ be a sequence in X such that the sequence $gf : N \to Y$ converges to the point $g(a)$. Prove or disprove that $f : N \to X$ converges to the point $a \in X$.

6. (The next two examples delineate between cluster points and sequential limit points.)
   (a) Give an example of a sequence $f : N \to X$ in a space X such that $a \in X$ is a cluster point of $f(N)$ but the sequence f does not converge to the point a.
   (b) Give an example of a sequence $f : N \to X$ that converges to the point $a \in X$ but a is not a cluster point of $f(N)$.

7. Let X be a space having the property that if a sequence in X converges, it converges to exactly one point in X. Prove X is a T_1-space.

8. Let $f : N \to X$ and $g : N \to Y$ be sequences in the spaces X and Y, respectively. Prove that the sequence $h : N \to X \times Y$ given by $h(n) = (f(n), g(n))$ converges to the point $(a, b) \in X \times Y$ iff $f \to a$ and $g \to b$.

9. (a) Prove that the function $g : R^1 \to R^1$ given by $g(x) = x^2$, for each $x \in R^1$, is continuous at each point $a \in R^1$.

   (b) Give an example of a function $g : R^2 \to R^2$ such that every sequence f in R^2 which converges to $(0, 0)$, except one, has the property that the sequence gf converges to $g(0, 0)$.

10. Prove that every polynomial equation $a_n x^n + a_{n-1} x^{n-1} + \cdots + a_1 x + a_0 = 0$, with real coefficients and of odd degree has a real root.

11. Let X be a space and $f : N \to \mathscr{P}(X)$ a *sequence of sets* in X. Define $\overline{\mathrm{Lim}}\, f = \{x \in X : \text{for each open set } U(x) \subset X \text{ containing } x, U(x) \cap f(n) \neq \phi \text{ for infinitely many } n \in N\}$ and $\underline{\mathrm{Lim}}\, f = \{x \in X : \text{for each open set } U(x) \subset X \text{ containing } x \text{ there exists an } n_0 \in N \text{ such that if } n > n_0, U(x) \cap f(n) \neq \phi\}$. ($\overline{\mathrm{Lim}}\, f$ and $\underline{\mathrm{Lim}}\, f$ are called the *limit superior* and *limit inferior* of the sequence of sets f, respectively. The sequence f converges iff $\overline{\mathrm{Lim}}\, f = \underline{\mathrm{Lim}}\, f$ and converges to $\overline{\mathrm{Lim}}\, f = \underline{\mathrm{Lim}}\, f$.)

   (a) Give an example of a sequence of sets f in a space X such that $\overline{\mathrm{Lim}}\, f \neq \underline{\mathrm{Lim}}\, f$.

   (b) Prove that the sets $\overline{\mathrm{Lim}}\, f$ and $\underline{\mathrm{Lim}}\, f$ are closed subsets of X.

3. Subsequences

3.1. Definition: Let $f : N \to X$ be a sequence in X and $g : N \to N$ an injective function which is strictly increasing. That is, if $m > n$, then $g(m) > g(n)$. The function $fg : N \to X$ is called a *subsequence* of f in X.

In other words, a subsequence of f in X is itself a sequence in X that has a special relationship with the original sequence f. This special relationship is that the function g selects natural numbers $g(1) = n_1$, $g(2) = n_2$, $g(3) = n_3, \ldots$ from N such that $n_1 < n_2 < n_3 < \cdots$ and f in turn selects elements $f(n_1)$, $f(n_2)$, $f(n_3), \ldots$ from $f(N)$ to form the subsequence $fg = \{(n, f(g(n))) : n \in N\}$.

3.2. Example: Consider the sequence $f: N \to R^1$ given by $f(n) = (-1)^n(1 - 1/n)$. If $g: N \to N$ is given by $g(n) = 2n$, then the subsequence $fg: N \to R^1$ given by $(fg)(n) = f(g(n)) = f(2n) = 1 - 1/(2n)$ converges to the point $a = 1$. For $g: N \to N$ given by $g(n) = 2n - 1$, the subsequence $fg: N \to R^1$ given by $(fg)(n) = -(1 - 1/(2n - 1))$ converges to the point $b = -1$ in R^1. The original sequence f does not converge to any point of R^1.

Example 3.2 points out that a sequence may have convergent subsequences whether or not the original sequence converges. However, when the original sequence converges, each subsequence must also converge to the same point as is shown by the next theorem.

3.3. THEOREM: *If f is a sequence in the space X such that $f \to a \in X$, then each subsequence of f in X converges to the point a.*

Proof: Let $U(a)$ be any open set in X containing the point a. Then $f \to a$ implies that there exists an $n_0 \in N$ such that if $n > n_0$, then $f(n) \in U(a)$. Now let $g: N \to N$ be any strictly increasing function and consider the natural number $g(n_0)$. It is an easy inductive proof to establish that $g(n) \geq n$ for all $n \in N$ so that $g(n_0) \geq n_0$. Therefore, if n is any natural number such that $n > n_0$, then $g(n) > g(n_0) \geq n_0$ and hence $f(g(n)) \in U(a)$. Thus, we have demonstrated an $n_0 \in N$ such that if $n > n_0$, then $f(g(n)) \in U(a)$. This shows the subsequence $fg: N \to X$ of f in X converges to the point a.

The following definition will lead us to a third and final consideration of why sequences lack certain desirable properties in topological spaces.

3.4. Definition: Let X be a space and $f: N \to X$ a sequence in X. Then f *accumulates* at the point $a \in X$ iff for every open set $U(a)$ containing a and for every natural number $n_0 \in N$, there exists an $n \in N$ such that $n > n_0$ and $f(n) \in U(a)$.

3.5. Example: The sequence $f: N \to R^1$ given by $f(n) = (-1)^n$ accumulates to the points 1 and -1. The sequence f does not converge, however.

Example 3.5 shows that a sequence may accumulate to a point which is not a sequential limit point of the sequence. It also shows that a sequence may accumulate to a point which is not a cluster point of $f(N)$. In all T_1-spaces, a cluster point of $f(N)$ is a point of accumulation of the sequence f. A proof of this fact is left as an exercise, as is the fact that this statement is false if the T_1 hypothesis is removed. However, for any space it is always true that a sequential limit point of a sequence is a point of accumulation for the sequence. Our next theorem relates convergent subsequences to points of accumulation.

3.6. THEOREM: *Let X be a topological space and f a sequence in X. If fg is any subsequence of f in X that converges to the point $a \in X$, then f accumulates to the point a.*

Proof: The proof is left to the reader.

By examining sequences in some of our familiar spaces, it may appear that the converse of Theorem 3.6 is also true. Example 3.7 shows this is not the case, however. That is, if the sequence f in X accumulates to the point $a \in X$, there may be no subsequence of f which converges to a. This is the third of the reasons that we are considering which point out why sequences are unsatisfactory in general topological spaces.

3.7. Example: Let $X = (N \times N) \cup \{(0, 0\} \subset R^2$ with the following topology: Each point of $N \times N$ is open and open sets containing $(0, 0)$ are all of those subsets U of X which contain $(0, 0)$ and for which there exists a natural number n_0 such that for all $n \geq n_0$, U contains all of $\{n\} \times N$ except for a finite number of points. Now consider the sequence $f : N \to X$ given by $f(1) = (1, 1)$, $f(2) = (1, 2)$, $f(3) = (2, 1)$, $f(4) = (3, 1)$, $f(5) = (2, 2)$, etc., sometimes called the diagonal enumeration of $N \times N$. Some of the points of $f(N)$ are shown in Figure 22.

The sequence f has $(0, 0)$ as a cluster point due to the definition of the open sets containing $(0, 0)$. However, no subsequence fg of f can converge to $(0, 0)$. The reason is that for each natural number n_0, there is an $n \in N$ such that $g(n) > n_0$ and hence $f(g(n)) = (p, q)$ is not contained in the open set $\{(0, 0)\} \cup (\{p\} \times N) \backslash \{(p, q)\} \bigcup_{m > p} (\{m\} \times N)$.

Thus, the following statement is not true in general topological spaces: The sequence f in X accumulates to the point $a \in X$ iff there

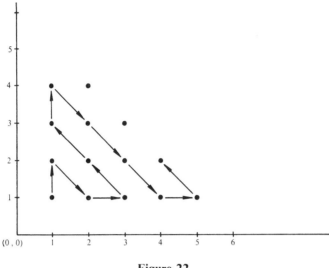

Figure 22

is some subsequence fg of f in X converging to a. We again call on the First Axiom of Countability to remedy our problem.

3.8. THEOREM: *Let X be a first countable space. Then the sequence f in X accumulates to the point $a \in X$ iff there exists a subsequence fg of f in X converging to a.*

Proof: Theorem 3.6 proves half of our present theorem. For the other half, let f be a sequence in X which accumulates to the point $a \in X$. Then let $\{U_n : n \in N\}$ be a countable local base of open sets at the point a such that $U_{n+1} \subset U_n$ for each $n \in N$ as given by Theorem 4.4(a) of Chapter 6. For $n = 1$ there exists a natural number $n_1 > 1$ such that $f(n_1) \in U_1$, due to the fact that f accumulates at the point a. Inductively, for each natural number k there exists a natural number $n_k > n_{k-1}$ such that $f(n_k) \in U_k$. If we now define g: $N \to N$ as $g(k) = n_k$, then $fg : N \to X$ is a subsequence of f in X which converges to a.

3.9. Corollary: The sequence f in R^n accumulates to the point $a \in R^n$ iff there exists a subsequence fg of f in R^n converging to a.

To summarize for the R^n spaces, we have shown in this chapter that sequences give complete information about the standard topology on these spaces.

Exercises:

1. (a) Would the sequence $g(n) = 1/2$, $n \in N$, be a subsequence of the sequence given in Example 3.2? Explain.
    (b) Give an example of a subsequence of the sequence in Example 3.2 which does not converge.

2. Prove Theorem 3.6.

3. Prove, or give a counterexample to the following statement: A sequence in a space is convergent iff it accumulates to exactly one point in X.

4. State the condition under which the sequence f in X does not accumulate to the point $a \in X$.

5. Prove that the sequence f in X converges to $a \in X$ iff every subsequence of f in X converges to a.

6. If f is a sequence of distinct points in a space X and f accumulates to the point $a \in X$, prove a is a cluster point of $f(N)$.

7. (a) Let X be a T_1-space and $f: N \to X$ a sequence in X. Prove that each cluster point of $f(N)$ is a point of accumulation of f.
    (b) Give an example to show part (a) is false if the T_1-hypothesis is removed.

8. Let $g : N \to N$ be a strictly increasing function. Prove that $g(n) \geq n$ for all $n \in N$.

9. (a) Show that the asserted topology on $X = (N \times N) \cup \{(0, 0)\}$ described in Example 3.7 is actually a topology on X.
    (b) Is this topology T_1? T_2? Regular?

10. Let $f: N \to \mathscr{P}(X)$ be a sequence of sets in the space X (see Exercise 11, Section 2) and fg any subsequence of f. Prove that $\underline{\text{Lim}}\ f \subset \underline{\text{Lim}}\ fg \subset \overline{\text{Lim}}\ fg \subset \overline{\text{Lim}}\ f$.

11. Let f be a sequence of sets in the space X, and let $a \in X$ be a point of $\overline{\text{Lim}}\ f$ but not of $\underline{\text{Lim}}\ f$. Prove that there exists an open set $U(a)$ containing a and a subsequence of sets fg in X such that $f(g(n)) \cap U(a) = \phi$ for each $n \in N$.

12. Let f be a sequence of sets in the space X and $A \subset X$. Prove that f converges (see Exercise 12, Section 2) to A iff every subsequence of f converges to A.

4. A Glimpse of Filters

We have seen some of the faults of sequences, topologically speaking, and what conditions imposed on the space will correct time. It would be desirable to find a substitute for sequences which could overcome the difficulties we have encountered. Two widely used substitutes are *filters* and *nets* on a space. We shall give a brief account of each to illustrate why they are sometimes more desirable than sequences in a space.

4.1. Definition: Let $X \neq \phi$ be a set and $\mathscr{F} = \{A_\alpha : \alpha \in \Delta\}$ a family of subsets of X indexed by Δ. Then $\mathscr{F}$ is a *filter* in X iff the following three conditions hold:
   (a) $A_\alpha \neq \phi$ for every $\alpha \in \Delta$.
   (b) For any $A_\alpha \in \mathscr{F}$, $A_\beta \in \mathscr{F}$, then $A_\alpha \cap A_\beta \in \mathscr{F}$.
   (c) If $A_\alpha \in \mathscr{F}$ and $A_\alpha \subset A \subset X$, then $A \in \mathscr{F}$.

As in the case of a topology on X, $\mathscr{F}$ defines a structure in X and (a), (b), and (c) of Definition 4.1 are the axioms of the structure. Notice that no two elements of $\mathscr{F}$ are disjoint. Some examples of filters might be helpful at this point.

4.2. Example: Let $X = \{a\}$ be a single point. Then the only filter in X is $\mathscr{F} = \{a\}$. If $X = \{a, b\}$ consists of exactly two points, then each of the following are filters in $X : \mathscr{F}_1 = \{\{a\}, \{a, b\}\}$, $\mathscr{F}_2 = \{\{b\}, \{a, b\}\}$ and $\mathscr{F}_3 = \{a, b\}$.

From Definition 4.1, a filter in any set $X \neq \phi$ can be generated by any nonempty subset $A \subset X$ by defining the filter to consist of all possible subsets of X which contain A. If X is a topological space, it is also easily verified that the set of all subsets M of X containing a fixed point $a \in X$ and such that there exists an open set U where $a \in U \subset M$, forms a filter in X. Such a filter is called the *neighborhood filter of a*.

Analogous to the idea of a base for a topology, we have the idea of a filterbase.

4.3. Definition: Let $X \neq \phi$ be a set and $\mathscr{B} = \{B_\alpha : \alpha \in \Delta\}$ a family of subsets of X. Then $\mathscr{B}$ is a *filterbase* in X iff the following two conditions hold :

    (a) $B_\alpha \neq \phi$ for every $\alpha \in \Delta$.

    (b) If $B_\alpha \in \mathscr{B}$ and $B_\beta \in \mathscr{B}$, then there exists a $B_\gamma \in \mathscr{B}$ such that $B_\gamma \subset B_\alpha \cap B_\beta$.

A simple example of a filterbase in X would be a single nonempty subset $A \subset X$.

4.4. THEOREM: *Let $\mathscr{B} = \{B_\alpha : \alpha \in \Delta\}$ be a filterbase in X. Then the family $\mathscr{F}(\mathscr{B}) = \{A \subset X : A$ contains an element B_α for some $\alpha \in \Delta\}$ forms a filter in X. The family $\mathscr{F}(\mathscr{B})$ is called the filter generated by $\mathscr{B}$.*

Proof: The conditions (a), (b), and (c) of Definition 4.1 are to be established. The definition of a filterbase implies that (a) of this definition is satisfied. Now let A_1 and A_2 be any elements of $\mathscr{F}(\mathscr{B})$. Then there is some B_α such that $B_\alpha \subset A_1$ and a B_β such that $B_\beta \subset A_2$ by definition of the family $\mathscr{F}(\mathscr{B})$. By (b) of Definition 4.3, there exists a $B_\gamma \in \mathscr{B}$ such that $B_\gamma \subset B_\beta \cap B_\beta$, and since $B_\alpha \cap B_\beta \subset A_1 \cap A_2$, condition (b) of Definition 4.1 holds for the family $\mathscr{F}(\mathscr{B})$. Finally, if $A \in \mathscr{F}(\mathscr{B})$, then there is some $B_\alpha \in \mathscr{B}$ such that $B_\alpha \subset A$ so that if $M \subset X$ is any set such that $A \subset M$, then $B_\alpha \subset A \subset M$, which means $M \in \mathscr{F}(\mathscr{B})$ by definition of $\mathscr{F}(\mathscr{B})$. Therefore, (c) of Definition 4.1 is satisfied, and we have completed the verification that $\mathscr{F}(\mathscr{B})$ is a filter in X.

We are now ready to consider concepts with filters that parallel those of convergence for sequences. The next three definitions accomplish this task.

4.5. Definition: Let $\mathscr{F} = \{A_\alpha = \alpha \in \Delta\}$ be a filter in X add $A \subset X$. Then $\mathscr{F}$ is *eventually in A* iff $A \in \mathscr{F}$. The filter $\mathscr{F}$ is *frequently in A* iff $A \cap A_\alpha \neq \phi$ for every $\alpha \in \Delta$.

4.6. Definition: For a family of filters $\{\mathscr{F}_\alpha : \alpha \in \Delta\}$ in X, $\mathscr{F}_\alpha$ is a subfilter of $\mathscr{F}_\beta$ iff $\mathscr{F}_\beta \subset \mathscr{F}_\alpha$.

Definition 4.6 will enable us to draw an analogy between subsequences and subfilters.

4.7. Definition: Let X be a topology space, $\mathscr{F}$ a filter in X and $a \in X$. Then $\mathscr{F}$ *converges* to a iff for each open set U in X containing a, $\mathscr{F}$ is eventually in U. The filter $\mathscr{F}$ *accumulates* to the point a iff for each open set U of X containing a, $\mathscr{F}$ is frequently in U.

We are now in a position to see how the use of filters overcomes some of the objections of sequences in general topological spaces.

4.8. THEOREM: Let $(X, \mathscr{T})$ be a space and $A \subset X$. Then $a \in X$ is a cluster point of A iff there exists a filterbase $\mathscr{B}$ in $A\backslash\{a\}$ such that the filter $\mathscr{F}(\mathscr{B})$ generated by $\mathscr{B}$ converges to a.

Proof: Let a be a cluster point of A. Then each open set U containing a contains a point of A different from a. Therefore, the family $\mathscr{B} = \{U_\alpha \cap (A\backslash\{a\}) : U_\alpha$ is an open set in X containing $a\}$ is a filterbase in $A\backslash\{a\}$. The reason is that every element of $\mathscr{B}$ is nonempty and if $U_\alpha \cap (A\backslash\{a\})$ and $U_\beta \cap (A\backslash\{a\})$ belong to $\mathscr{B}$, then $(U_\alpha \cap (A\backslash\{a\})) \cap (U_\beta \cap (A\backslash\{a\})) = (U_\alpha \cap U_\beta) \cap (A\backslash\{a\}) \in \mathscr{B}$, since $U_\alpha \cap U_\beta$ is an open set containing a. Consequently, the filter $\mathscr{F}(\mathscr{B})$ generated by $\mathscr{B}$ has among its elements all open sets containing a, and this implies $\mathscr{F}(\mathscr{B})$ converges to a by Definition 4.7.

Conversely, suppose there is a filterbase $\mathscr{B}$ in $A\backslash\{a\}$ such that the filter $\mathscr{F}(\mathscr{B})$ converges to the point a. Let U be any open set containing a. Then the convergence of $\mathscr{F}(\mathscr{B})$ implies that $\mathscr{F}(\mathscr{B})$ is eventually in U. Thus, $U \in \mathscr{F}(\mathscr{B})$ by Definition 4.5. Therefore, there is an element B of $\mathscr{B}$ such that $B \subset U$. Since $\mathscr{B}$ is defined on $A\backslash\{a\}$, $B \subset A\backslash\{a\}$. The fact that $B \neq \phi$ implies that there is a point of A different from a in the open set U. It follows that a is a cluster point of A.

4.9. THEOREM: Let $f : X \to Y$ be a given function and $\mathscr{F} = \{A_\alpha : \alpha \in \Delta\}$ a filter in X. Then the collection $\{f(A_\alpha) : A_x \in \mathscr{F}\}$ is a filterbase in Y. The filter generated by this filterbase will henceforth be denoted by $f(\mathscr{F})$.

Proof: The easy verification is left to the reader.

4.10. THEOREM: Let $f : X \to Y$ be a function from one topological space into another. Then f is continuous at $a \in X$ iff for every filter $\mathscr{F}$ in X which converges to a, the filter $f(\mathscr{F})$ converges to $f(a)$.

Proof: The proof is left as an exercise.

4.11.   THEOREM: *Let $(X, \mathcal{T})$ be a topological space and $\mathcal{F} = \{A_\alpha : \alpha \in \Delta\}$ a filter in X. Then $\mathcal{F}$ accumulates to the point $a \in X$ iff there exists a subfilter of $\mathcal{F}$ which converges to a.*

Proof: First, let $\mathcal{F}$ accumulate to the point a. Then, for any open set U containing a, $\mathcal{F}$ is frequently in U, i.e., $U \cap A_\alpha \neq \phi$ for all $\alpha \in \Delta$. Thus, considering the collection $\{U_\gamma : \gamma \in \Omega\}$ of all open sets containing a, the family $\{A_\alpha \cap U_\gamma : \alpha \in \Delta, \gamma \in \Omega\}$ forms a filterbase $\mathcal{B}$ in X, as is easily verified. Then, by contruction, the filter $\mathcal{F}(\mathcal{B})$ generated by $\mathcal{B}$ has the property that $\mathcal{F} \subset \mathcal{F}(\mathcal{B})$ or that $\mathcal{F}(\mathcal{B})$ is a subfilter of $\mathcal{F}$. Furthermore, $\mathcal{F}(\mathcal{B})$ converges to a.

Now let $\mathcal{F}$ have a subfilter $\mathcal{F}_1$ that converges to the point a, and let U be any open set containing a. Then $U \in \mathcal{F}_1$, by definition of convergence. Let $A_\alpha \in \mathcal{F}$. Then $A_\alpha \in \mathcal{F}_1$, since $\mathcal{F} \subset \mathcal{F}_1$ and, therefore, $U \cap A_\alpha \neq \phi$ for every $A_\alpha \in \mathcal{F}$, showing $\mathcal{F}$ accumulates to the point a.

Exercises:

1. List all possible filters for the set $X = \{a, b, c\}$. How many filters are there for a set X having n distinct elements?

2. Let X be any infinite set. Prove that the set of complements of finite subsets of X forms a filter in X. That is, $\mathcal{F} = \{A_\alpha : X \backslash A_\alpha$ is finite$\}$ is a filter in X.

3. Give an example of a filter in R^1 which does not converge. Then give an example of a filter which has no accumulation points in R^1 with the left ray topology.

4. Prove Theorem 4.9.

5. Prove Theorem 4.10.

6. In the proof of Theorem 4.11, prove that the collection $\{A_\alpha \cap U_\gamma : \alpha \in \Delta, \gamma \in \Omega\}$ is a filterbase. Also prove that $\mathcal{F}(\mathcal{B})$ converges to a.

7. Prove that the space X is Hausdorff iff each convergent filter in X converges to exactly one point.

8. Prove or give a counterexample to the following statement: If $\mathcal{F}_1$ and $\mathcal{F}_2$ are filters in X, then $\mathcal{F}_1 \cap \mathcal{F}_2$ is a filter in X.

9. Let X be a space and $A \subset X$. Prove that $a \in \bar{A}$ iff there is a filterbase in A which converges to a.

10. Let $f : X \to Y$ be a function and $\mathscr{F}$ a filterbase in Y. Prove that $\{f^{-1}(B) \in \mathscr{P}(X) : B \in \mathscr{F}\}$ is a filterbase in X iff $f^{-1}(B) \neq \phi$ for each $B \in \mathscr{F}$.

5. A Glimpse of Nets

The other substitute for sequences that overcomes some of their undesirable features is a structure called a *net*. We shall give some of the elementary definitions and point out why nets are an improvement over sequences in that they overcome the three deficiencies of sequences that we mentioned earlier. In so doing, we will make use of relations on sets which, in the past, have always been denoted by R. In this section, however, we use the symbol $\geq$ to represent a relation. Aside from being a rather standard symbol for a relation when discussing nets in the literature, special cases of nets will be more meaningful with this notation.

5.1. Definition: Let D be a nonempty set. The relation $\geq$ *directs* D iff the following three conditions hold:
    (a) For every $a \in D$, $a \geq a$. That is, $\geq$ is reflexive.
    (b) If $a \geq b$ and $b \geq c$, then $a \geq c$. That is, $\geq$ is transitive.
    (c) For each $a, b \in D$, there exists an element $c \in D$ such that $c \geq a$ and $c \geq b$. This is called the *directive* property of $\geq$.
    A *directed set* is a set D together with a relation $\geq$ that directs D and is denoted by $(D, \geq)$.

5.2. Example: Let R^1 be the real numbers where $\geq$ means, as usual, "greater than or equal to." Then $(R^1, \geq)$ is a directed set. The natural numbers N are directed by the same relation and, therefore, $(N, \geq)$ is a directed set.

5.3. Definition: Let $(D, \geq)$ be a directed set and $E \subset D$. Then E is called a *residual subset* of D iff there exists an element $a \in D$ such that if $d \in D$, and $d \geq a$, then $d \in E$. The subset $E \subset D$ is called a *cofinal subset* of D iff for each $d \in D$ there exists an element $e \in E$ such that $e \geq d$.

5.4. Example: Consider the directed set $(R^1, \geq)$ of Example 5.2. The set $E = \{x : x \geq 0\}$ is a residual subset of R^1 as well as a cofinal subset of R^1. The set $E_1 = \{x : x \text{ is an even integer}\}$ is a cofinal subset of R^1 but not a residual subset of R^1.

5.5. Definition: A *net* in the set X is a function $F : D \to X$ where $(D, \geq)$ is a directed set. In particular, if the directed set is $(N, \geq)$ as given in Example 5.2, then the net F is a sequence.

5.6. Definition: Let $F : D \to X$ be a net in the space X and $A \subset X$.
    (a) The net F is in A iff $F(D) \subset A$.
    (b) The net F is *eventually in* A iff there exists a residual $E \subset D$ such that $F(E) \subset A$.
    (c) The net F is *frequently in* A iff there exists a cofinal set $E \subset D$ such that $F(E) \subset A$.

5.7. Example: Consider the directed set $(N, \geq)$ as in Example 5.2 and define $F : N \to R^1$ as $f(n) = (-1)^n(1 - 1/n)$ for each $n \in N$. Then F is a net in R^1 and, in particular, is a sequence. The reader should convince himself that if $A = \{x : x \geq 0\} \subset R^1$, then F is not in A, F is not eventually in A, but F is frequently in A. Examine other subsets of R^1 and ask which of the conditions (a), (b), and (c) of Definition 5.6 hold.

We are now ready to consider the concepts of *convergence* and *accumulation points* of nets in a topological space.

5.8. Definition: Let $F : D \to X$ be a net in the space $(X, \mathcal{T})$. The net F *converges* to $a \in X$ iff for each open set U containing a, the net F is eventually in U. The point $a \in X$ is called an *accumulation point* of the net F iff for each open set U containing a the net F is frequently in U.

It might be well to restate Definition 5.8 using Definitions 5.3 and 5.6 to point out the similarity between the convergence of sequences and the convergence of nets. The same might be helpful for accumulation points of sequences and nets. For instance, the net F converges to $a \in X$ iff for each open set U containing a there exists an $e \in D$ such that if $d \in D$ and $d \geq e$, then $F(d) \in U$. In addition, F accumulates to $a \in X$ iff for each open set U containing a and $d \in D$ there exists an $e \in D$ such that $e \geq d$ and $F(e) \in U$.

The next theorem gives our first reason why nets are more desirable than sequences in a topological space.

5.9. THEOREM: *Let $(X, \mathcal{T})$ be a topological space and $A \subset X$. Then $a \in X$ is a cluster point of A iff there exists a net $F : D \to X$ which is in $A \setminus \{a\}$ and converges to a.*

Proof: Suppose there exists a net $F : D \to X$ which is in $A \setminus \{a\}$ and converges to a. Let U be any open set containing a. Then, by Definition 5.8. F is eventually in U so that $U \cap (A \setminus \{a\}) \neq \phi$. It follows that a is a cluster point of A.

Conversely, suppose a is a cluster point of A. We are required to construct a net which is in $A \setminus \{a\}$ and converges to a. Every open set $U(a)$ in X containing a contains at least one point $x \in A$ such that $x \neq a$. Now define $D = \{(x, U(a)) : U(a)$ is an open set containing a and $x \in A \cap U(a), x \neq a\}$. Then define $(x_1, U_1(a)) \geq (x_2, U_2(a))$ iff $U_1(a) \subset U_2(a)$. It is easy to verify that $(D, \geq)$ is a directed set. Once this is done, we may define a net $F : D \to X$ as $F(x, U(a)) = x$. Then, by its definition, the net F is in $A \setminus \{a\}$, and we shall now show it converges to a. To do this, let $U(a)$ be any open set in X containing a. Then there is an $x \in A \cap U(a), x \neq a$, so that $(x, U(a)) \in D$ and $F((x, U(a)) = x$. Also, for any $(x_1, U_1(a)) \geq (x, U(a))$, then $U_1(a) \subset U(a)$ so that $F((x_1, U_1(a)) = x_1 \in U_1(a) \subset U(a)$. Therefore, F is eventually in $U(a)$ and by Theorem 5.8 converges to a.

5.10. THEOREM: *A function $f : X \to Y$, where X and Y are topological spaces, is continuous at $a \in X$ iff for every net F in X which converges to a, the net fF converges to $f(a)$.*

Proof: The construction of the required net is the same as in the proof of Theorem 5.9. Otherwise, a straightforward application of the definitions gives the proof. The details are left to the reader.

For a final consideration of nets, we define the concept of a *subnet* and give the analog to Theorem 3.8.

5.11. Definition: Let $F : D \to X$ be a net in the space X. Then the net $G : D_1 \to X$ is a *subnet* of F iff there exists a function $f : D_1 \to D$ from the directed set D_1 to the directed set D such that
   (a) $G = Ff$ and
   (b) for each $d \in D$ there exists a $d_1 \in D_1$ such that $f(x) \geq d$ for every $x \geq d_1$.

5.12.   THEOREM: *Let $F: D \to X$ be a net in the space X. Then the net F accumulates to the point $a \in X$ iff there exists a subnet $G: D_1 \to X$ of $F: D \to X$ which converges to a.*

Proof: Suppose first that $G: D_1 \to X$ is a subnet of $F: D \to X$ which converges to $a \in X$. Then Definition 5.11 gives the existence of a function $f: D_1 \to D$ obeying conditions (a) and (b). Select an element $d_0 \in D$. Then according to (b) of Theorem 5.11, there exists an element $d_1 \in D_1$ such that $f(x) \geq d_0$ for every $x \geq d_1$. Since G converges to a, if U is any open set containing a, there exists an element $d_2 \in D_1$ such that $d_2 \geq d_1$ and $G(d_2) \in U$. If we now let $d = f(d_2) \in D$, we have $f(d_2) \geq d_0$ and $F(d) = F(f(d_2)) = G(d_2) \in U$. Consequently, the net $F: D \to X$ accumulates to the point a.

Conversely, suppose $F: D \to X$ accumulates to the point $a \in X$. Define D_1 to be the family $\{(d, U(a)) : U(a)$ is an open set in X containing a, $d \in D$ and $F(d) \in U(a)\}$. Since F accumulates to the point a, ordered pairs of this type exist for every open set $U(a)$. Now order D_1 as follows: $(d_1, U_1(a)) \geq (d_2, U_2(a))$ iff $d_1 \geq d_2$ and $U_1(a) \subset U_2(a)$. It is not a difficult exercise to show that $(D_1, \geq)$ is a directed set. Once this is done, define $f((d, U(a)) = d$ for all $(d, U(a)) \in D_1$. Then our proof will be complete when we show the net $G = Ff$ is a subnet of F that converges to a. To show G is a subnet of F, only part (b) of Definition 5.11 needs to be verified. Therefore, let $d \in D$ and let $U(a)$ be any open set containing a. Then there exists an element $d_0 \in D$ such that $d_0 \geq d$ and $F(d_0) \in U(a)$. If we let $(d_0, U(a)) = d_1 \in D_1$ and choose any $(x, U_1(a)) \geq (d_0, U(a))$, then $x \geq d_0$ so that $f((x, U_1(a)) = x \geq d_0 \geq d$. Thus we have shown that for each $d \in D$, there exists an element $(d_0, U(a)) \in D_1$ such that $f((x, U_1(a)) \geq d$ for every $(x, U_1(a)) \geq (d_0, U(a))$, and condition (b) of Definition 5.11 is satisfied. The fact that the subnet $G = Ff$ converges to a is left as an exercise.

Exercises:

1. In the proof of Theorem 5.9, verify that $(D, \geq)$, as defined, is a directed set.

2. Prove Theorem 5.10.

3. Give an example of a net in a space which converges to more than one point.

4. Prove that in a Hausdorff space, a net can converge to at most one point.

5. Prove that a set U in the space X is open iff no net in $X \backslash U$ can converge to a point of U.

6. Let $(X, \mathcal{T})$ be a topological space and $A \subset X$. Prove that $a \in \bar{A}$ iff there exists a net in A which converges to the point a.

7. Prove that a net F in a topological product $\prod\limits_{\alpha \in \Delta} X_\alpha$ converges to $f \in \prod\limits_{\alpha \in \Delta} X_\alpha$ iff $p_\alpha(F)$ converges to $p_\alpha(f) = f(\alpha)$ for each $\alpha \in \Delta$.

8. Show that the pair $(D_1, \geq)$ in the proof of Theorem 5.12 is a directed set.

9. Prove that the subnet $G = Ff$ in the proof of Theorem 5.12 converges to a.

10. Let X be a space and $F : D \to X$ a net in X. Prove that the set of all points to which F accumulates is a closed subset of X.

8

Connected and Compact Spaces

1. Connected Spaces

No doubt we can all give several uses of the word "connected" as applied to our everyday activities. In mathematics, too, this word has several uses and meanings. The meaning we wish to attach to it here is an attempt to formalize the intuitive concept of a topological space being all in "one piece." The precise definition is given now.

1.1. Definition: The space $(X, \mathscr{T})$ is *disconnected* iff there exist two open disjoint nonempty sets U and V such that $U \cup V = X$. The sets U and V form a *separation* of X. The space $(X, \mathscr{T})$ is *connected* iff it is not disconnected. A subset $A \subset X$ is connected iff the space $(A, \mathscr{T}_A)$ is connected.

We should notice that in a separation of the space X, then open sets U and V are also closed subsets of X. Definition 1.1 gives the definition of connectedness in a negative fashion so that to prove that a space is connected, we must establish that there is *no possible* separation of the space. Generally speaking, this will be difficult to do in a direct manner and, therefore, an indirect proof can be used to great advantage when proving theorems about connectedness. Some examples will provide more insight into our definition.

1.2. Example: The space $X = \{a, b, c\}$ with the topology $\mathscr{T} = \{\phi, X, \{a\}, \{b, c\}\}$ is a disconnected space since the open sets $U = \{a\}$ and $V = \{b, c\}$ form a separation of X. The space $X = \{a, b, c\}$ with topology $\mathscr{T} = \{\phi, X, \{a\}, \{a, b\}\}$ is a connected space since there is no separation possible of the space $(X, \mathscr{T})$.

1.3. Example: The subset $A = \{x : 1 < x \leq 5\} \cup \{x : 7 \leq x \leq 10\}$ is not connected in R^1. To see this, observe $U = \{X : 1 < x \leq 5\}$ and $V = \{x : 7 \leq x \leq 10\}$ form a separation of the space A with the subspace topology inherited from R^1.

The importance of the space R^1 prompts us to investigate as one of our first objectives the concept of connectedness in the R^1 space. The next two theorems lead to Theorem 1.6, which gives us full information about which subsets are and are not connected in R^1. The result is that a subset A of R^1 is connected iff A is an interval.

1.4. THEOREM: *A subset A of R^1 containing more than one point is connected iff given any two points $a, b \in A$ where $a < b$, then each $x \in R^1$ satisfying $a < x < b$ also belongs to A.*

Proof: First, let A be connected and let $a, b \in A$ where $a < b$. If $x \in R^1$ and satisfies $a < x < b$, we are to show $x \in A$. To do this, suppose $x \notin A$. Then $\{z \in A : z < x\} = U$ and $\{z \in A : z > x\} = V$ are open subsets in the subspace A and, furthermore, $U \cup V = A$ and $U \cap V = \phi$. Therefore, A is not connected. This contradiction to the hypothesis that A is connected means our assumption $x \notin A$ is false. It follows that $x \in A$.

Now examine the hypothesis that for $a, b \in A$ where $a < b$, each

$x \in R^1$ satisfying $a < x < b$ also belongs to A. We are to show A is connected. To this end, assume A is not connected. Then, in the subspace $A \subset R^1$, there exist two nonempty disjoint open sets U and V such that $A = U \cup V$. Therefore, we may assume (relabeling points and sets if necessary) that there is a point $a \in U$ and a point $b \in V$ such that $a < b$. Now define $C = \{c \in R^1 : [a, c] \cap V = \phi\}$ and note that since $a \in C, C \neq \phi$. Furthermore, b is an upper bound for C so that C has a least upper bound x_0 and $a \leq x_0 \leq b$. By hypothesis, it follows that $x_0 \in A$, which means $x_0 \in U$ or $x_0 \in V$. We first consider the consequences of $x_0 \in U$. Since U is an open subset of the subspace A, there exists an open interval (y, w) in R^1 such that $x_0 \in (y, w)$ and $(y, w) \cap A \subset U$. Therefore, $(y, w) \cap V = \phi$ so that $([a, x_0) \cup (y, w)) \cap V = \phi = [a, w) \cap V$. Now, choosing any point x_1 such that $x_0 < x_1 < w$, we see $[a, x_1] \cap V = \phi$ which implies $x_1 \in C$. This contradicts the fact that x_0 is the least upper bound of C so that $x_0 \in U$ cannot hold. Now let us explore the possiblity of $x_0 \in V$. If this is the case, there exists an open interval (s, t) in R^1 such that $x_0 \in (s, t)$ and $(s, t) \cap A \subset V$ which means $(s, t) \cap U = \phi$. Now, considering any point $z \in R^1$ such that $z \in (s, x_0)$, we see $[a, z] \cap V = \phi$ because $z < x_0$. Therefore, $z \notin V$. Since $z \in (s, t)$ and $(s, t) \cap U = \phi$, it follows that $z \notin U$. Thus, we have a point $z \in R^1$ such that $a < z < x_0 \leq b$ but $z \notin U \cup V = A$. This contradicts our hypothesis and means that $x_0 \notin V$. Since $x_0 \notin U$ and $x_0 \notin V$, it follows that $x_0 \notin A$, which is contrary to the established fact that $x_0 \in A$. Consequently, the assumption that A is not connected is false and, therefore, gives A connected.

1.5. THEOREM: *Let A be a subset of R^1 which contains more than one point. Then A is an interval iff given $a, b \in A$ where $a < b$, then each $x \in R^1$ satisfying $a < x < b$ also belongs to A.*

Proof: First, suppose $A \subset R^1$ is an interval. Considering each of the possible forms for A (See Definition 1.4 of Chapter 3), it is left to the reader to verify that if $a, b \in A$ where $a < b$, then each $x \in R^1$ satisfying $a < x < b$ also belongs to A.

Now take the hypothesis that given $a, b \in A$ with $a < b$, then each $x \in R^1$ satisfying $a < x < b$ also belongs to A. The fact that A is an interval may be established by considering four cases after observing that if $x \in R^1$ and $x \notin A$, then x is either an upper bound or a lower bound for A. The reason is that if this is not the case, there are two points $a, b \in A$ with $a < b$, such that $a < x < b$, which implies

$x \in A$ by our hypothesis. We now list and examine the four cases, concluding that in each case A is an interval.

Case 1. The set A has no upper bound and no lower bound. Thus, for each $x \in R^1$, there are two points $a, b \in A$ such that $a < x < b$ and, therefore, by hypothesis, $x \in A$. Consequently, A is the interval R^1.

Case 2. The set A has an upper bound but no lower bound. Since $A \neq \phi$, A has a least upper bound which will be denoted by a. If $x \in R^1$ and $x > a$, then $x \notin A$, due to the definition of a. If $x \in R^1$ and $x < a$, we shall show $x \in A$. When $x < a$, there is a point $w \in A$ such that $x < w < a$, again by the definition of a. Since A has no lower bound, there is a point $z \in A$ such that $z < x$. Therefore, we have $z, w \in A$ and $z < x < w$, which implies $x \in A$. Now if $a \in A$, $A = \{x \in R^1 : x \leq a\}$ and if $a \notin A$, $A = \{x \in R^1 : x < a\}$. In either event, A is an interval.

Case 3. The set A has a lower bound but no upper bound. The conclusion that A is an interval follows from an argument similar to the one in Case 2.

Case 4. The set A has both an upper bound and a lower bound. Since $A \neq \phi$, A has a least upper bound b and a greatest lower bound a. Since A has at least two distinct points, it follows that $a < b$. Now let x be any point in R^1 such that $a < x < b$, and we will show $x \in A$. By definition of a and b, there exist points z and w in A such that $a < z < x$ and $x < w \leq b$. Thus, we have $z, w \in A$ and $z < x < w$ which implies $x \in A$. Therefore, $(a, b) = \{x : a < x < b\} \subset A$ and A has one of the forms (a, b), $(a, b]$, $[a, b)$ or $[a, b]$ depending upon which, if either, of the points a or b belong to A. It follows that A is an interval.

1.6. THEOREM: *A subset A of R^1 is connected iff A is an interval. In particular, R^1 is connected.*

Proof: If A contains more than one point, Theorems 1.4 and 1.5 give A connected iff A is an interval. If A consists of a single point, A is a closed interval and is connected by Exercise 1(b) at the end of this section.

Some conditions equivalent to connectedness of general topological spaces will be given next. Some of these may be easier to use in certain situations than others.

1.7. THEOREM: *The following conditions are equivalent:*
   (a) *The space X is connected.*
   (b) *The only subsets of the space X which are both open and closed are X and ϕ.*
   (c) *There is no continuous surjection $f : X \rightarrow \{a, b\}$ from the space X onto the space $\{a, b\}$ having two distinct points and the discrete topology.*

Proof: (a) implies (b). Suppose $A \subset X, A \neq \phi, A \neq X$, and A is both open and closed in X. Then $X \backslash A$ is both open and closed and is nonempty. Therefore, $A \cup (X \backslash A) = X$ is a separation of X, which means X is disconnected. This contradicts the hypothesis (a) that X is connected. Therefore, if (a) is true, then (b) is true.

   (b) implies (c). Suppose a continuous surjection $f : X \rightarrow \{a, b\}$, where $\{a, b\}$ has the discrete topology, exists. Then $f^{-1}(a)$ and $f^{-1}(b)$ are both nonempty, both open, both closed, $f^{-1}(a) \cap f^{-1}(b) = \phi$ and $X = f^{-1}(a) \cup f^{-1}(b)$. Thus $f^{-1}(a)$, as well as $f^{-1}(b)$, is a proper subset of X, which is both open and closed, contradicting (b).

   (c) implies (a). If X is not connected, there is a separation of X into open nonempty disjoint sets U and V. Define $f : X \rightarrow \{a, b\}$ as follows: $f(x) = a$ for all $x \in U$ and $f(x) = b \neq a$ for all $x \in V$. Then f is a continuous surjection contradicting (c).

Another convenient way of telling when a subset of a space (or the space itself) is connected is given after the following definition.

1.8. Definition: Two subsets A and B of the space X are *separated* in X iff $A \neq \phi, B \neq \phi, A \cap \bar{B} = \phi$, and $B \cap \bar{A} = \phi$.

1.9. THEOREM: *A subset C of a space $(X, \mathcal{T})$ is connected iff C cannot be written as the union of two separated sets in X, each of which has a nonempty intersection with C.*

Proof: First, let C be connected and suppose $C = A \cup B$ where A and B are separated sets in $X, C \cap A \neq \phi$, and $C \cap B \neq \phi$. Then $\bar{B}$ is closed in X so that $C \cap \bar{B} = B$ is a closed subset of the subspace $(C, \mathcal{T}_c)$. Also, $\bar{A}$ is closed in X so that $X \backslash \bar{A}$ is open in X and hence $C \cap (X \backslash \bar{A})$ is open in $(C, \mathcal{T}_c)$. By use of Exercise 9, Section 1, of Chapter 1, we may write $C \cap (X \backslash \bar{A}) = (C \cap X) \backslash (C \cap \bar{A}) = C \backslash (C \cap \bar{A})$

and by DeMorgan's Theorem $C\backslash(C \cap \bar{A}) = (C\backslash C) \cup (C\backslash\bar{A}) = C\backslash\bar{A}$. However, $C = A \cup B$ and $\bar{A} \cap B = \phi$, so that $C\backslash\bar{A} = B$. Therefore, we have shown $B = C \cap (X\backslash\bar{A})$ is an open subset of $(C, \mathcal{T}_c)$. Now, since $B \neq \phi$ and $A \neq \phi$, we have a proper subset B of C that is both open and closed in the subspace C, implying that C is disconnected by Theorem 1.7(b). This contradiction to the fact that C is connected means the supposition $C = A \cup B$ where A and B are separated in X is false and, as a consequence, there is no way of expressing a connected set C as the union of two separated sets in X each having a nonempty intersection with C.

For the converse, take the hypothesis that C cannot be written as the union of two separated sets in X and show that C is a connected subset of X. Assume C is not connected. Then by Theorem 1.7(b) there is some proper subset A of C which is both open and closed in the subspace $(C, \mathcal{T}_c)$. For notation, let $B = C\backslash A$. Then it follows that B is also a proper subset of C which is both open and closed in $(C, \mathcal{T}_c)$. Since A is open in $(C, \mathcal{T}_c)$, there exists an open subset $U \in \mathcal{T}$ such that $U \cap C = A$. Therefore, $U \cap B = \phi$ and, since $A \subset U$, no point of A is a cluster point of B so that $\bar{B} \cap A = \phi$. In the same way we can show that $\bar{A} \cap B = \phi$ because B is open in $(C, \mathcal{T}_c)$. We now have shown that $C = A \cup B$ where A and B are separated in X and $A \cap C \neq \phi$ and $B \cap C \neq \phi$, which contradicts our hypothesis. Therefore, our assumption is false, and C is connected.

The next two theorems may be proved using Theorem 1.9.

1.10. THEOREM: *If the connected set C is contained in the union of two separated sets, then C must be contained in one of the sets.*

Proof: Assume that the conclusion is false and apply Theorem 1.9. The details are left to the reader.

1.11. THEOREM: *Let C be a connected subset of the space X. Then every set B such that $C \subset B \subset \bar{C}$ is also connected. In particular, the closure of any connected set is connected.*

Proof: Suppose B is not connected. Then $B = A \cup D$ where A and D are separated in X according to Theorem 1.9. Since $C \subset B$ and C is connected by hypothesis, either $C \subset A$ or $C \subset D$ by Theorem 1.10. Consider $C \subset A$. Because $C \subset B \subset \bar{C}$, every point in $D \neq \phi$ is a cluster point of C and hence of A, which means $D \subset \bar{A}$ so that A and

D are not separated. This contradiction implies B is connected. A similar conclusion results if $C \subset D$.

1.12. THEOREM: *Let* $\{C_\alpha : \alpha \in \Delta\}$ *be a family of connected subsets of the space* X. *If for every* $\alpha, \beta \in \Delta, C_\alpha \cap C_\beta \neq \phi$, *then* $\bigcup_{\alpha \in \Delta} C_\alpha$ *is connected.*

Proof: An indirect proof using Theorems 1.9 and 1.10 is suitable. The details are left as an exercise.

1.13. THEOREM: *Connectedness is a topological property.*

Proof: The proof is an exercise.

1.14. THEOREM: *In a* T_1-*space* X *every connected set containing more than one point is infinite.*

Proof: Let $A = \{x_1, x_2, \ldots, x_n\}, n \geq 2$, be a finite subset of distinct points in X. The subspace A has the discrete topology. To see this, observe first that since points in X are closed, A is closed in the space X. Therefore, for each $x_k \in A, k = 1, 2, \ldots, n, \{x_k\} \cap A = \{x_k\}$ is closed in the subspace A. As a consequence, every subset of the finite subspace A is closed in A, which means every subset of A is also open in the space A. It follows that A has the discrete topology. Now $\{x_1\} \cup A\backslash\{x_1\}$ is a separation of A because $A\backslash\{x_1\} \neq \phi$ by hypothesis. Therefore, since every finite subset of X containing more than one point is disconnected, every connected subset of X having more than one point must be infinite.

Exercises:

1. Let X be a topological space.
   (a) Prove that the empty set is connected.
   (b) Prove that a single point in X is connected.

2. (a) Prove that the set Q of rational numbers in R^1 is not connected.
   (b) Prove that R^1 with the left ray topology is connected.
   (c) Prove that R^1 with the cofinite topology is connected.

3. (a) If C is a connected subset of the space $(X, \mathscr{T})$ and $\mathscr{T}_1$ is a topology for X such that $\mathscr{T}_1 \subset \mathscr{T}$, prove that C is connected in $(X, \mathscr{T}_1)$.
   (b) Give an example to show part (a) may be untrue if $\mathscr{T}$ is replaced with a larger topology.

4. Prove Theorem 1.10.

5. Prove Theorem 1.12.

6. Prove Theorem 1.13.

7. Let X be a connected space. Let C be a connected subset of $(X, \mathcal{T})$. If $X\backslash C = B \cup D$ where B and D are separated sets in X, then prove that $C \cup B$ and $C \cup D$ are connected sets.

8. Prove that the space $(X, \mathcal{T})$ is connected iff each open subset $U \subset X$ distinct from X and ϕ has the property that $\text{Bd}(U) \neq \phi$.

9. Let A be a connected subset of the space X. If $B \subset X$ is both open and closed and $A \cap B \neq \phi$, prove $A \subset B$.

10. Let $\{C_\alpha : \alpha \in \Delta\}$ be a family of connected subsets of a space X such that $\bigcap_{\alpha \in \Delta} C_\alpha \neq \phi$. Prove that $\bigcup_{\alpha \in \Delta} C_\alpha$ is connected.

11. Let X be a space such that each pair of points in X is contained in a connected subset of X. Prove that X is a connected space.

2. More Properties of Connected Spaces

This section will be devoted to a further study of connected spaces. In Section 1 it was noted that connectedness is a topological property. Our next theorem will be a more general result in that any property preserved by a continuous function will also be preserved by a homeomorphism, and therefore that property will be a topological property.

2.1. THEOREM: *If C is a connected subset of the space X and $f : X \to Y$ is continuous, then $f(C)$ is connected in the space Y.*

Proof: If we assume $f(C)$ is not connected, there exists a continuous surjection $g : f(C) \to \{a, b\}$, where $\{a, b\}$ has the discrete topology, according to Theorem 1.7(c). But $f|C$ is continuous by Theorem 1.8 of Chapter 5 so that the composition $gf : C \to \{a, b\}$ is a continuous surjection. Therefore, Theorem 1.7(c) implies C is not connected, and thus contradicts the hypothesis of our theorem. This means our assumption is false and $f(C)$ is connected.

From studying calculus we are all undoubtedly familiar with the *Intermediate Value Theorem*. There, the function under consideration was in all likelihood from R^1 to R^1. We now give a generalization of this theorem by showing that the domain space R^1 may be replaced with any connected space.

2.2. THEOREM: *(The Intermediate Value Theorem.) Let $f: X \to R^1$ be continuous where X is a connected space. If the points a and b belong to $f(X)$ where $a < b$, and c is any real number such that $a < c < b$, then there is at least one point $x_0 \in X$ such that $f(x_0) = c$.*

Proof: Assume that there is no point $x_0 \in X$ such that $f(x_0) = c$. Then $U = \{x \in R^1 : x < c\}$ and $V = \{x \in R^1 : x > c\}$ are open intervals of reals with $a \in U$ and $b \in V$. Hence $f^{-1}(U)$ and $f^{-1}(V)$ are nonempty open sets in X, $f^{-1}(U) \cap f^{-1}(V) = \phi$ and $f^{-1}(U) \cup f^{-1}(V) = X$. Therefore, X is disconnected, which contradicts the hypothesis. The conclusion follows.

Theorem 2.2 is often a useful tool in locating roots of continuous functions $f: R^1 \to R^1$. If it is known that $f(x_1) > 0$ and $f(x_2) < 0$, then this theorem tells us there is some point x_0 between x_1 and x_2 such that $f(x_0) = 0$ and hence x_0 is a root to the equation $f(x) = 0$.

Theorem 2.2 also leads us to another interesting fact concerning the real numbers. That fact is that if $[a, b]$ is a closed interval of reals, then for each continuous $f: [a, b] \to [a, b]$ there exists a point $c \in [a, b]$ (depending upon f) such that $f(c) = c$. In other words, f leaves the point c fixed. In general, when a space X has the property that each continuous $f: X \to X$ leaves a point of X fixed, we say X has the *fixed point property*.

2.3. THEOREM: *Let $[a, b]$ be a closed interval of real numbers. Then $[a, b]$ has the fixed point property.*

Proof: If $a = b$, our theorem is proved. Otherwise, assume there exists a function $f: [a, b] \to [a, b]$ that leaves no point of $[a, b]$ fixed. That is, $f(x) \neq x$ for all $x \in [a, b]$. This means the function $g: [a, b] \to R^1$ defined by $g(x) = 1/(x - f(x))$ is a well-defined function on $[a, b]$ because the denominator cannot be zero. Now g is continuous by Theorem 2.10 of Chapter 7 since the constant function $h(x) = 1$, $1_x(x) = x$ and $f(x)$ are all continuous on $[a, b]$. Furthermore, $g(a) < 0$ while $g(b) > 0$. Applying Theorem 2.2, there exists a point $x_0 \in [a, b]$ such that $g(x_0) = 0$. However, this is impossible by the definition of g. As a consequence, our assumption must be false and hence every continuous function from $[a, b]$ to $[a, b]$ leaves a point fixed.

A natural question to ask about connected spaces is whether $\prod_{\alpha \in \Delta} X_\alpha$ is connected iff each space X_α, $\alpha \in \Delta$, is connected. The answer is yes. The fact that if $\prod_{\alpha \in \Delta} X_\alpha$ is connected, then each space X_α is connected, is easy to prove and is left as an exercise. However, we shall not attempt to prove here that if each space X_α, $\alpha \in \Delta$, is connected, then $\prod_{\alpha \in \Delta} X_\alpha$ is connected. (A proof may be found in any of the references in the Bibliography.) Rather, our objective will be to prove this fact for the case of a finite indexing set. The following theorem and its corollary do this for us.

2.4. THEOREM: *The product space $X \times Y$ is connected iff both of the spaces X and Y are connected.*

Proof: First, let $X \times Y$ be connected. Then the projection function $p_X : X \times Y \to X$ is a continuous surjection so that by Theorem 2.1, X is connected. Similarly $p_Y : X \times Y \to Y$ gives Y connected.

For the converse, suppose X and Y are both connected and we shall show $X \times Y$ is connected. For purposes of obtaining a contradiction, suppose $X \times Y$ is disconnected. Then there exists, by Theorem 1.7(c), a continuous surjection $f : X \times Y \to \{a, b\}$ where the space $\{a, b\}$ has the discrete topology. Therefore, there exist points (x_0, y_0) and (x_1, y_1) in $X \times Y$ such that $f(x_0, y_0) = a$ and $f(x_1, y_1) = b$. Consider now the point $(x_1, y_0) \in X \times Y$ for which there are two possibilities: $f(x_1, y_0) = a$ or $f(x_1, y_0) = b$. If the first possibility prevails, define a function $h : Y \to \{x_1\} \times Y$ given by $h(y) = (x_1, y)$. The function h is a homeomorphism by the proof to Theorem 2.12 of Chapter 5. Thus $fh : Y \to \{a, b\}$ is a continuous surjection so that Y is disconnected and we have a contradiction. For the other possibility, $f(x_1, y_0) = b$, construct the function $g : X \to X \times \{y_0\}$ given by $g(x) = (x, y_0)$. This function is also a homeomorphism so that $fg : X \to \{a, b\}$ is a continuous surjection, implying that X is disconnected. Therefore, a contradiction is reached in either possible case, implying that the assumption $X \times Y$ is disconnected is false.

2.5. Corollary: For a finite family of spaces, $X_1, X_2, \ldots, X_n$, the product space $\prod_{k=1}^{n} X_k$ is connected iff each of the spaces X_k, $k = 1, 2, \ldots, n$ is connected.

Proof: The proof is an easy application of mathematical induction, Theorem 2.4 and Theorem 2.13 of Chapter 5.

2.6. Corollary: For each $n \in N$, the space R^n is connected.

Proof: Theorem 1.6 and Corollary 2.5.

We are now going to show that R^1 and R^2 are not homeomorphic. Actually, it may be shown that R^n and R^m, $n \neq m$, are not homeomorphic, although this will not be done here. First, we can easily verify that for any point $a \in R^1$, $R^1 \backslash \{a\}$ is not connected. On the contrary, we have for R^2 the following theorem.

2.7. THEOREM: *For any point $(a, b) \in R^2$, $R^2 \backslash \{(a, b)\}$ is connected.*

Proof: Let $(a, b) \in R^2$ and $(x_0, y_0) \neq (a, b)$ any other point in R^2. Consider the straight line L joining (x_0, y_0) to (a, b). If a point (x, y) in the plane is not on L, join (x, y) and (x_0, y_0) with a straight line in R^2. This line is homeomorphic to R^1 and, therefore, connected by Theorem 1.13. If $(x, y) \in L$, join (x, y) to any point $(c, d) \notin L$ with a straight line and then join (c, d) to (x_0, y_0) with another straight line. These two lines are connected and have (c, d) as a common point: hence, the union of the two lines is a connected set by Theorem 1.12. Now every point in $R^2 \backslash \{(a, b)\}$ lies in a connected set with the point (x_0, y_0) and, therefore, $R^2 \backslash \{(a, b)\}$ is connected by Theorem 1.12.

2.8. THEOREM: *The spaces R^1 and R^2 are not homeomorphic.*

Proof: Suppose, on the contrary, R^1 is homeomorphic to R^2 where $h : R^1 \to R^2$ is a homeomorphism. Let $x_0 \in R^1$ and form $R^1 \backslash \{x_0\}$. Then by Theorem 2.7 of Chapter 5, $R^1 \backslash \{x_0\}$ is homeomorphic to $R^2 \backslash \{h(x_0)\}$. However, $R^1 \backslash \{x_0\}$ is not connected, while $R^2 \backslash \{h(x_0)\}$ is connected, thus contradicting Theorem 1.13 that connectedness is a topological property. Consequently, our assumption is false and R^1 is not homeomorphic to R^2.

To conclude this section, we want to lead up to the fact that R^1 is homeomorphic to each subspace $\{x : a < x < b\} \subset R^1$. We first have a preliminary result.

2.9. THEOREM: *If $f : R^1 \to R^1$ is a continuous injective function, then f is an open function.*

Proof: Since each open set in R^1 is the union of open intervals, we need only show the image of each open interval is open under f. Thus, let $A = \{x \in R^1 : a < x < b\}$ be an open interval. Now Theorem 1.6 tells us A is connected, Theorem 2.1 gives $f(A)$ connected in R^1, and Theorem 1.6, again, gives $f(A)$ an interval. Since f is injective, $f(a) \notin f(A)$ and $f(b) \notin f(A)$. However, the continuity of f gives $f(a)$ and $f(b)$ as end points of the interval $f(A)$. To see this, assume $f(a)$ is not an end point of $f(A)$. Then there exists an open interval (c, d) containing $f(a)$ such that $(c, d) \cap f(A) = \phi$. Consequently, no open set U in R^1 containing a has the property that $f(U) \subset (c, d)$ because a is a cluster point of A, and this contradicts the continuity of f. The same holds for $f(b)$ and, since $f(a) \neq f(b)$, $f(a)$ and $f(b)$ are the two end points of $f(A)$ so that either $f(A) = x \in R^1 : f(a) < x < f(b)\}$ or $\{f(A) = \{x \in R^1 : f(b) < x < f(a)\}$. In any event, $f(A)$ is an open interval.

2.10. Corollary: If $f : R^1 \to R^1$ is a continuous injective function, then f is a homeomorphism from R^1 to $f(R^1)$.

Proof: The fact that f is injective gives f a bijection from R^1 to $f(R^1)$. The fact that f is an open function gives $f^{-1} : f(R^1) \to R^1$ continuous. Therefore, f is a bijection where both f and f^{-1} are continuous, implying that f is a homeomorphism by Theorem 2.6(b) of Chapter 5.

2.11. THEOREM: *The space R^1 is homeomorphic to the subspace* $A = \{x \in R^1 : a < x < b\}$.

Proof: Let $c = (a + b)/2$ and define $f : R^1 \to A$ as follows:

$$f(x) = \frac{c - ax}{1 - x} \quad \text{if} \quad x \leq 0$$

$$f(x) = \frac{c + bx}{1 + x} \quad \text{if} \quad x \geq 0.$$

Notice that $f(0) = c$ and, if $x \leq 0$, then $a < f(x) \leq c$ and, if $x \geq 0$, then $c \leq f(x) < b$. It is an easy matter to show f is injective. To show f is surjective, let $w \in A$. If $w = c$, then $f(0) = c$. If $a < w < c$, then $(c - ax)/(1 - x) = w$ may be solved to obtain $x = (w - c)/(w - a) \in R^1$ (because $w \neq a$) so that $f(x) = f((w - c)/(w - a)) = w$. Similarly, if $c < w < b$, an $x \in R^1$ may be found such that $f(x) = w$. Therefore, $f(R^1) = A$.

To show f is continuous at the point $x_0 \in R^1$, there are three cases to be considered: (1) $x_0 = 0$, (2) $x_0 < 0$, and (3) $x_0 > 0$. Pursuing case (2), we find that $a < f(x_0) < c$ and for any open set U containing $f(x_0)$ an open interval $(d, e) \subset U$ exists such that $a < d < x_0 < e < c$. Now $f^{-1}(d) = (d - c)/(d - a) < x_0 < f^{-1}(e) = (e - c)/(e - a)$ and, furthermore, for any $f^{-1}(d) < x < f^{-1}(e)$, we have $d < f(x) < e$. Therefore, the open interval $(f^{-1}(d), f^{-1}(e))$ in R^1 contains x_0 and under f has its image in the interval $(d, e) \subset U$. This shows f is continuous at x_0. Continuity of f for the other two cases may be shown in a similar manner.

Corollary 2.10 now tells us f is a homeomorphism.

Exercises:

1. Give a different proof of Theorem 2.2 based on Theorem 2.1.

2. Give an example of several subsets of R^2 which are connected and several which are not connected.

3. By a method similar to the proof of Theorem 2.8, show that the intervals $\{x : a \leq x \leq b\}$ and $\{x : c < x < d\}$ in R^1 are not homeomorphic.

4. Prove that the intervals $A = \{x: a < x < b\}$ and $B = \{x : c < x < d\}$ in R^1 are homeomorphic by first defining a linear function $f(x) = px + q$ (p and q are constants to be determined) from $\bar{A}$ onto $\bar{B}$ such that $f(a) = c$ and $f(b) = d$. Then prove that this function is a homeomorphism.

5. Let $A = \{x : a \leq x \leq b\}$ and $B = \{x : c \leq x \leq d\}$ be closed intervals in R^1. Prove that any homeomorphism $h : A \to B$ must have the property that $h(a)$ and $h(b)$ are end points of B.

6. In the proof to Theorem 2.11.
   (a) Show that the function f is injective.
   (b) Prove that the stated inequalities hold in showing f is continuous at the point $x_0 < 0$ in case (2).

7. In the proof of Theorem 2.11, prove that f is continuous at each point $x_0 > 0$.

8. Prove that if the product space $\prod_{\alpha \in \Delta} X_\alpha$ is connected, then each of the spaces $X_\alpha, \alpha \in \Delta$, is connected.

9. Prove that if X is a connected space and R is an equivalence relation on X, then the quotient space X/R is a connected space.

10. Give an example of a space that does not have the fixed point property.

11. Prove that the set $A = \{(x, y) \in R^2 : \text{either } x \text{ or } y \text{ is irrational}\}$ is a connected subset of R^2.

3. Components and Locally Connected Spaces

If a space X is not connected, then X consists of several "pieces" each of which is connected. These pieces are called components of the space and will be precisely defined by means of an equivalence relation. First notice, however, that since each point $x \in X$ is connected, x lies in at least one connected subset of X containing x. We shall show that the largest or maximal connected subset of X containing x is the component of X containing x.

3.1. THEOREM: *For any x and y belonging to the space X, define xCy iff there exists a connected set $M \subset X$ such that x and y both belong to M. Then C is an equivalence relation on X.*

Proof: From the comments above the statement of Theorem 3.1, xCx for every $x \in X$ and, therefore, C is reflexive. If xCy, then there is some connected set M containing both x and y and hence y and x. Thus, yCx, and this implies C is symmetric. If xCy and yCz, there exists a connected set M_1 containing x and y and also a connected set M_2 containing y and z. Therefore, $y \in M_1 \cap M_2$, and, by Theorem 1.12, $M_1 \cup M_2$ is connected. Since $M_1 \cup M_2$ contains x and z, xCz, which means C is transitive. We have now shown C is an equivalence relation on X.

3.2. Definition: Let $x \in X$ and C be the equivalence relation described in Theorem 3.1. The equivalence class determined by x relative to the equivalence relation C, denoted as usual by $C(x)$, is called the *component* of X determined by the point x. For $A \subset X$, a component of the space $(A, \mathcal{T}_A)$ will be called a component of the set A.

Observe that any connected space X has exactly one component, namely X, because for each $x \in X, C(x) = X$. After two examples we will give two characterizations of components that will shed more light on their structure.

3.3. Example: Consider the subspace of all integers Z in R^1. Then for each $x \in Z, C(x) = x$. That is, each point in Z is itself a component of Z which show Z has a countably infinite number of components. The subspace $A = \{x \in R^1 : 0 < x < 1\} \cup \{x \in R^1 : 5 \leq x < 7\}$ has exactly two components.

Example 3.3 leads us to the fact that in any space with the discrete topology, each point is a component of the space. In any space where each point of the space is a component of that space, we say the space is *totally disconnected*.

3.4. Example: Let $A = \{(x, y) : 0 < x < 1, 0 \leq y \leq 1\}$ and $B = \{(x, y) : x = 2\}$ be subsets of R^2. The subspace $A \cup B$ of R^2 has exactly two components, namely A and B. The subspace A has only one component. The subspace $A \cup B$ where $A = \{(x, y) : 0 < x < 1, 0 \leq y \leq 1\}$ and $B = \{(x, y) : 1 < x < 2, 0 \leq y \leq 1\}$ also has two components.

3.5. THEOREM: *For each $a \in X, C(a)$ is the union of all connected subsets of X which contain a. In particular, $C(a)$ is connected.*

Proof: Let $a \in X$ be fixed and let $\{M_\alpha : \alpha \in \Delta\}$ be the family of all connected subsets of X each of which contains a. It is to be shown that $C(a) = \bigcup_{\alpha \in \Delta} M_\alpha$. The details, involving proving a standard set equality, are left as an exercise.

3.6. THEOREM: *For each $a \in X, C(a)$ is the largest connected subset of X which contains a.*

Proof: By Theorem 3.5, $C(a)$ is connected. It remains to show that there is no larger connected subset containing the point a. To this end, let A be any connected subset of X containing a. Then for each point $x \in A$, xCa and, therefore, $x \in C(a)$. Thus, $A \subset C(a)$ and $C(a)$ is then the largest connected subset of X containing a.

3.7. THEOREM: *For each $a \in X, C(a)$ is a closed subset of X.*

Proof: Suppose $C(a)$ is not closed. Then there is some point $x \in X$ such that x is a cluster point of $C(a)$ but $x \notin C(a)$. Since $C(a)$ is connected according to Theorem 3.5, and $C(a) \subset C(a) \cup \{x\} \subset \overline{C(a)}$, Theorem 1.11 implies $C(a) \cup \{x\}$ is connected. However, this contradicts Theorem 3.6. It follows that $C(a)$ is closed subset of X.

In a moment we shall see that in certain types of spaces each component $C(a)$ of X is also an open subset of X.

Even though a space X may not be connected, it may be possible to find a "small" connected open set about each point of the space X. In such a case, connectedness in a local sense is being described rather than connectedness in a global sense for the space X. We define such a concept now.

3.8. Definition: The space X is *locally connected* at the point $x \in X$ iff given any open set $U \subset X$ containing x there exists an open *connected* subset $V \subset X$ such that $x \in V \subset U$. If X is locally connected at each point $x \in X$, then X is called a *locally connected space*. A subset $A \subset X$ is locally connected iff the space $(A, \mathscr{T}_A)$ is locally connected.

3.9. Example: The space R^1 is both connected and locally connected. To see that R^1 is locally connected, we first observe that for each point $x \in X$ and any open set U containing x there is an open interval $V = (a, b)$ such that $x \in (a, b) = V \subset U$. Then, by Theorem 1.6, the interval (a, b) is connected so that Definition 3.8 is satisfied for the point x. Therefore, R^1 is locally connected. If $Z \subset R^1$ is the set of integers, then the subspace Z is locally connected but not connected.

3.10. Example: (A space that is connected but not locally connected.) For each $n \in N$, define $A_n = \{(1/n, y) : 0 \le y \le 1\}$ to be a

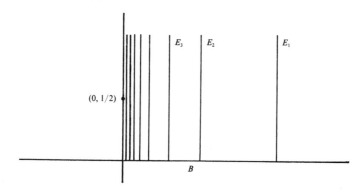

Figure 23

subset of R^2. Let $B = \{(x, 0) : 0 \le x \le 1\}$ and X the subspace of R^2 defined as $X = \bigcup_{n \in N} A_n \cup B \cup \{(0, 1/2)\}$. Figure 23 gives a visual description of the set X. Since the point $(0, 1/2)$ is a cluster point of the connected set $(\bigcup_{n \in N} A_n) \cup B$, Theorem 1.11 implies X is connected. The space X is locally connected at every point except $(0, 1/2)$. To see X is not locally connected at $(0, 1/2)$, consider the open set $U = \{(x, y) : x^2 + (y - 1/2)^2 < 1/4\} \cap X$ and observe that U consists of countably many components and that one of these components is the point $(0, 1/2)$ itself. However, $\{(0, 1/2)\}$ is itself not an open set in the space X. Thus, no *connected* open set V of X exists such that $(0, 1/2) \in V \subset U \subset X$.

Two useful characterizations of local connectedness are given next.

3.11. THEOREM: *The space X is locally connected at the point $x \in X$ iff there is a local base at x consisting of connected open subsets of X.*

Proof: Suppose first that X is locally connected at $x \in X$. Then if U is any open set containing x, there exists a connected open set $V(x)$ such that $x \in V(x) \subset U$. Thus, let $\{V(x)\}$ be the family of all connected open sets in X which contain x. This family then forms a local base at x.

Conversely, suppose there is a local base at x consisting of connected open sets. Then for any open set U containing x there is a connected open set $V(x)$ in the local base at x such that $x \in V(x) \subset U$. Thus, X is locally connected at x.

3.12. THEOREM: *The space X is locally connected iff the components of each open subset of X are themselves open subsets of X.*

Proof: First, suppose X is locally connected and let $U \subset X$ be any open set. We are required to show that for each $a \in U$ the component $C(a)$ of U is an open subset of X. Now each $x \in C(a)$ is also in the open set U. Hence, by local connectedness, there exists a connected open set V such that $x \in V \subset U$. But $C(a)$ is the largest connected set containing x and, therefore, $V \subset C(a)$. Thus, for each $x \in C(a)$ there is an open set V such that $x \in V \subset C(a)$, which implies $C(a)$ is an open subset of X.

Now suppose the components of each open set are themselves open in X. Let $x \in X$ and U be an open set in X containing x. Then the components of U are open by hypothesis, and one of these components $V = C(x)$ contains x. Thus, V is an open set such that $x \in V \subset U$, implying X is locally connected at each point $x \in X$.

3.13. Corollary: If X is a locally connected space and $C(x)$ is a component of X, then $C(x)$ is open in X.

Proof: The space X is itself open, so that by Theorem 3.12, $C(x)$ is open.

Again the question arises as to whether the product space $\prod_{\alpha \in \Delta} X_\alpha$ is locally connected iff each space X_α is locally connected. The answer in this case is no. If $\prod_{\alpha \in \Delta} X_\alpha$ is locally connected, then each space X_α may be shown to be locally connected. However, if $\{X_\alpha : \alpha \in \Delta\}$ is a family of locally connected spaces, it need not follow that $\prod_{\alpha \in \Delta} X_\alpha$ is locally connected. A further insight into this question may be gained by consulting references 1 and 2 of the Bibliography. Our approach will now be to consider finite indexing sets.

3.14. THEOREM: *The space $X \times Y$ is locally connected iff both of the spaces X and Y are locally connected.*

Proof: Suppose first that $X \times Y$ is locally connected. Let $a \in X$ and U be any open set in X containing a. Then $U \times Y$ is open in $X \times Y$ and contains at least one point (a, y) such that $y \in Y$. Consequently, there is an open connected set V containing (a, y) such that $V \subset U \times Y$. The projection function $p_X : X \times Y \to X$ gives $a \in p_X(V) \subset U$ where $p_X(V)$ is open and connected according to Theorem 2.3 of Chapter 5 and Theorem 2.1 of this chapter. Thus, X is locally connected for each $a \in X$ and hence is locally connected. In a similar manner, Y may be shown to be locally connected.

Conversely, let X and Y both be locally connected. Consider any point $(a, b) \in X \times Y$ and any open set $V \subset X \times Y$ containing (a, b). Then there exist open sets U and W in X and Y, respectively, such that $a \in U, b \in W$ and $U \times W \subset V$. Since X and Y are locally connected, there exist open connected sets U_1 and W_1 such that $a \in U_1 \subset U$ and $b \in W_1 \subset W$. Hence $U_1 \times W_1 \subset V$ and contains (a, b), is open in $X \times Y$, and is connected by Theorem 2.4. Thus $X \times Y$ is locally connected at the point (a, b). It follows that $X \times Y$ is locally connected.

3.15. Corollary: The product space $\prod\limits_{k=1}^{n} X_k$ is locally connected iff each space $X_k, k = 1, 2, \ldots, n$, is locally connected.

Proof: The proof follows the familiar pattern.

3.16. Corollary: For each $n \in N$, R^n is a locally connected space.

Since R^2 is locally connected, Example 3.10 shows subspaces of locally connected spaces may not themselves be locally connected. Of course, local connectedness is a topological property. In fact, we can prove a stronger theorem from which immediately follows the topological invariance.

3.17. THEOREM: *Let $f: X \to Y$ be an open continuous surjection from the locally connected space X to the arbitrary space Y. Then Y is locally connected.*

Proof: The proof is an exercise.

Exercises:

1. Show the details of the proof to Theorem 3.5.

2. In R^2 sketch the function $f: R^1 \to R^1$ given by $f(x) = \sin(1/x)$, $x \neq 0$, and $f(0) = 0$.
   (a) If f a connected set in R^2?
   (b) At which points is f locally connected and at which points is f not locally connected?

3. Prove Theorem 3.17.

4. Let X be a space having a finite number of components. Prove that each of these components is both open and closed in X.

5. (a) Prove that the subspace $A = \{1/n : n \in N\} \cup \{0\}$ in R^1 is not locally connected.
   (b) Give an example to show that if X is locally connected and $f: X \to Y$ is a continuous surjection, then Y need not be locally connected.

6. Prove that components of a space are preserved under homeomorphisms.

7. Let A be a connected subset which is both open and closed in the space X. Prove that A is a component of the space X.

8. Prove that if the product space $\prod_{\alpha \in \Delta} X_\alpha$ is locally connected, then each space X_α is locally connected.

9. Let X be a locally connected space and $U \subset X$ an open subset of X. If V is a component of U, prove that $V \cap \text{Bd}(V) = \phi$.

10. Let X be a locally connected space and $A \subset X$. If B is a component of A, prove that $\text{Int}(B) = B \cap \text{Int}(A)$.

11. Let $f: X \to Y$ be an identification function from the locally connected space X to the space Y. Prove that Y is locally connected.

12. If X is a locally connected space and R an equivalence relation on X, prove that the quotient space X/R is locally connected.

4. Compact Spaces

Compact topological spaces are important in that certain families of open sets may be reduced to a finite number of open sets. That this is an obvious advantage over noncompact spaces will be seen as our studies unfold. The concept of a compact space concerns coverings of the space, which will be defined next.

4.1. Definition: Let $\{A_\alpha : \alpha \in \Delta\}$ be a family of subsets of the space X and $B \subset X$. The family $\{A_\alpha : \alpha \in \Delta\}$ *covers* B iff $B \subset \bigcup_{\alpha \in \Delta} A_\alpha$. If Δ is finite and $\{A_\alpha : \alpha \in \Delta\}$ covers B, then $\{A_\alpha : \alpha \in \Delta\}$ is called a *finite cover* of B. If each $A_\alpha, \alpha \in \Delta$, is open (closed) in X and $\{A_\alpha : \alpha \in \Delta\}$ covers B, then $\{A_\alpha : \alpha \in \Delta\}$ is called an *open (closed) cover* of B.

4.2. Example: Let $B = \{x : 0 < x \leq 1\}$ be a subset of R^1. For each $n \in N$, let $A_n = \{x : 1/n < x < 2\}$. Then $\{A_n : n \in N\}$ is an open cover of B but is not a finite cover of B. Furthermore, there is no finite number of sets among $\{A_n : n \in N\}$ that will cover B. The family consisting of the two sets $\{x : -1 < x < 3\}, \{x : 10 < x < 40\}$ is a finite open cover of the set B.

4.3. Definition: Let $\{A_\alpha : \alpha \in \Delta\}$ be a cover of $B \subset X$. Then the family $\{A_\beta : \beta \in \Omega \subset \Delta\}$ is a *subcover* of $\{A_\alpha : \alpha \in \Delta\}$ for B iff $\{A_\beta : \beta \in \Omega \subset \Delta\}$ is a cover of B.

4.4. Example: Consider the cover $\{A_n : n \in N\}$ of the set B as given in Example 4.2. The family $\{A_m : m = 2, 4, 6, \ldots\}$ is a subcover of

$\{A_n : n \in N\}$ for B. The finite cover of B given in Example 4.2 is not a subcover of $\{A_n : n \in N\}$ for B.

We might note that any cover of $B \subset X$ is also a subcover of itself for B. With the concepts discussed thus far in this section, a definition of compact spaces may now be given.

4.5. Definition: A space X is called *compact* iff each open cover of X has a finite subcover for X. A subset A of the space $(X, \mathscr{T})$ is compact iff the space $(A, \mathscr{T}_A)$ is compact.

In order to prove a space X is compact by use of Definition 4.5, it must be shown that every open cover of X has a finite subcover for X. In practice, this is usually very difficult to do and, therefore, we shall later seek some more useful characterizations of compactness, at least for certain types of spaces. On the other hand, to prove a space is not compact, we need only exhibit one open cover which has no finite subcover.

4.6. Example: The space R^1 is not compact. Consider the open cover $\{A_k : k \text{ an integer}\}$ where $A_k = \{x : k - \frac{1}{4} < x < k + \frac{5}{4}\}$. The cover $\{A_k : k \text{ is an integer}\}$ is not finite and, furthermore, if any A_j is deleted from the cover, then the set $\{x : j + \frac{1}{4} < x < j + \frac{3}{4}\}$ is not contained in any $A_k, k \neq j$. Therefore, no subcover, and hence no finite subcover, of $\{A_k : k \text{ is an integer}\}$ for R^1 exists. This shows R^1 is not compact. In a similar fashion, it may be shown that each R^n, $n \in N$, is not compact.

In Example 4.2, the subspace $B = \{x : 0 < x \leq 1\}$ is not compact because the first cover of B given in that example has no finite subcover for B. Notice further that any infinite space with the discrete topology is not compact.

4.7. Example: Let R^1 have the cofinite topology. This space is compact. To see this, let $\{A_\alpha : \alpha \in \Delta\}$ be any open cover of this space, and remember A_α is open iff $R^1 \backslash A_\alpha = \{x_1, x_2, \ldots, x_n\}$ is a finite set. Now select any $A_{\alpha_0} \in \{A_\alpha : \alpha \in \Delta\}$. This set covers all of R^1 except a finite set of points $\{x_1, x_2, \ldots, x_n\}$. Next select any A_{α_1} containing x_1, then some A_{α_2} containing x_2, and in general, select an A_{α_k} containing $x_k, k = 1, 2, \ldots, n$. Such A_{α_k} exist because $\{A_\alpha : \alpha \in \Delta\}$ is a cover for R^1. Thus $\{A_{\alpha_k} : k = 0, 1, 2, \ldots, n\}$ is a finite subcover of $\{A_\alpha : \alpha \in \Delta\}$ for R^1. Hence R^1 with the cofinite topology is compact.

From the definition it is easy to see that any finite space is compact. In applying the definition of compactness to subspaces, examples are easy to find of compact subspaces which are not closed subsets of the original space. Theorem 4.10 will show that compact subsets of Hausdorff spaces are always closed, however. Also, examples easily show that closed subspaces may not be compact. On the other hand, if the original space X is compact, closed subspaces are always compact, as the next theorem shows.

4.8. THEOREM: *Let X be a compact space and $B \subset X$ a closed subset of X. Then B is compact.*

Proof: Let $\{A_\alpha : \alpha \in \Delta\}$ be any open cover of B. To prove the theorem we need to show there is a finite subcover of $\{A_\alpha : \alpha \in \Delta\}$ for B. Since B is closed, $X \backslash B$ is open and hence $\{A_\alpha : \alpha \in \Delta\} \cup (X \backslash B)$ is an open cover of the entire space X. The compactness of X implies there is a finite subcover of this cover for X. This subcover must contain some elements of $\{A_\alpha : \alpha \in \Delta\}$ and perhaps $X \backslash B$. Therefore, the subcover of X can be expressed as $A_{\alpha_1}, A_{\alpha_2}, \ldots, A_{\alpha_n}, X \backslash B$. In any event, $X \backslash B$ covers no points of B so that $B \subset \bigcup_{k=1}^{n} A_{\alpha_k}$ and thus $A_{\alpha_1}, A_{\alpha_2}, \ldots, A_{\alpha_n}$ forms a finite subcover of $\{A_\alpha : \alpha \in \Delta\}$ for B. Consequently, B is compact.

4.9. THEOREM: *Let A be a compact subset of the Hausdorff space X and $x \notin A$. Then there exist disjoint open sets U and V containing x and A, respectively.*

Proof: Since X is Hausdorff, for each $a \in A$ there exist disjoint open sets $V(a)$ and $U_a(x)$ containing a and x, respectively. The family $\{V(a) : a \in A\}$ found in this manner forms an open cover of A. The compactness of A then implies a finite subcover $V(a_1), V(a_2), \ldots, V(a_n)$ of $\{V(a) : a \in A\}$ for A exists. For each $V(a_k), k = 1, 2, \ldots, n$, there is a corresponding $U_{a_k}(x)$ and hence $\bigcap_{k=1}^{n} U_{a_k}(x) = U$ is open and contains x. But U does not intersect any $V(a_k)$, $1 \leq k \leq n$. The reason is that if $U \cap V(a_i) \neq \phi$ for some $1 \leq i \leq n$, then $V(a_i) \cap U_{a_i}(x) \neq \phi$ since $U \subset U_{a_k}(x)$ for $k = 1, 2, \ldots, n$. However, this is contrary to the way $V(a_k)$ and $U_{a_k}(x)$ were chosen. Thus, if we define $V = \bigcup_{k=1}^{n} V(a_k)$, $U \cap V = \phi$, $x \in U$ and $A \subset V$. This proves the theorem.

4.10. THEOREM: *A compact subset of a Hausdorff space X is closed.*

Proof: Let $A \subset X$ be compact and $x \notin A$. By Theorem 4.9 there exist open sets U and V containing x and A, respectively, such that $U \cap V = \phi$. Thus, $U \cap A = \phi$. Therefore, containing each $x \notin A$ there is an open set U such that $U \cap A = \phi$, which implies $X \backslash A$ is open so that A is closed.

4.11. THEOREM: *Let X be a compact Hausdorff space. Then $A \subset X$ is compact iff A is closed.*

Proof: Theorems 4.8 and 4.10.

4.12. THEOREM: *Every compact Hausdorff space is normal.*

Proof: Let A and B be any two disjoint closed subsets of the compact Hausdorff space X. By Theorem 4.8, both A and B are compact. Now use Theorem 4.9 for each $a \in A$ and the set B. Repeat the technique in the proof of Theorem 4.9. Details are left as an exercise.

4.13. THEOREM: *Compactness is a topological property.*

Proof: The proof is left to the reader.

This section is concluded with a characterization of compactness for general topological spaces. This characterization uses a property known as the finite intersection property.

4.14. Definition: (The Finite Intersection Property.) The family $\{A_\alpha : \alpha \in \Delta\}$ has the *finite intersection property* iff every finite non-empty subcollection $A_{\alpha_1}, A_{\alpha_2}, \ldots, A_{\alpha_n}$ of $\{A_\alpha : \alpha \in \Delta\}$ has the property that $\bigcap_{k=1}^{n} A_{\alpha_k} \neq \phi$.

4.15. THEOREM: *The space X is compact iff every family $\{A_\alpha : \alpha \in \Delta\}$ of closed sets having the finite intersection property has a nonempty intersection.*

Proof: First, let X be compact and $\{A_\alpha : \alpha \in \Delta\}$ be a family of closed sets having the finite intersection property. We are to show

$\bigcap_{\alpha \in \Delta} A_\alpha \neq \phi$. Suppose $\bigcap_{\alpha \in \Delta} A_\alpha = \phi$. Then $\{X \backslash A_\alpha : \alpha \in \Delta\}$ is a collection of open sets and $\bigcup_{\alpha \in \Delta} X \backslash A_\alpha = X \backslash \bigcap_{\alpha \in \Delta} A_\alpha = X \backslash \phi = X$. By compactness, the open over $\{X \backslash A_\alpha : \alpha \in \Delta\}$ has a finite subcover $X \backslash A_{\alpha_1}, X \backslash A_{\alpha_2}, \ldots,$ $X \backslash A_{\alpha_n}$, and, consequently $\bigcap_{k=1}^{n} A_{\alpha_k} = \bigcap_{k=1}^{n} X \backslash (X \backslash A_{\alpha_k}) = X \backslash (\bigcup_{k=1}^{n} (X \backslash A_{\alpha_k}))$ $= X \backslash X = \phi$, which contradicts the fact that $\{A_\alpha : \alpha \in \Delta\}$ has the finite intersection property. Therefore $\bigcap_{\alpha \in \Delta} A_\alpha \neq \phi$.

Conversely, take the hypothesis that every family of closed sets in X having the finite intersection property has a nonempty intersection. We are to show X is compact. To this end, let $\{U_\alpha : \alpha \in \Delta\}$ be any open cover of X. Then $\{X \backslash U_\alpha : \alpha \in \Delta\}$ is a family of closed sets such that $\bigcap_{\alpha \in \Delta} (X \backslash U_\alpha) = X \backslash (\bigcup_{\alpha \in \Delta} U_\alpha) = X \backslash X = \phi$. Consequently, our hypothesis implies the collection $\{X \backslash U_\alpha : \alpha \in \Delta\}$ does not have the finite intersection property. Therefore, there is some finite subcollection $X \backslash U_{\alpha_1}, X_{\alpha_2}, \ldots, X \backslash U_{\alpha_n}$, of this collection such that $\bigcap_{k=1}^{n} (X \backslash U_{\alpha_k}) = \phi$ and hence $\bigcup_{k=1}^{n} U_{\alpha_k} = \bigcup_{k=1}^{n} (X \backslash (X \backslash U_{\alpha_k})) = X \backslash \bigcap_{k=1}^{n} (X \backslash U_{\alpha_k}) = X$. Thus, $U_{\alpha_1}, U_{\alpha_2}, \ldots, U_{\alpha_n}$ is a finite subcover of $\{U_\alpha : \alpha \in \Delta\}$ for X, implying X is compact.

Exercises:

1. Let R^1 have the left ray topology.
   (a) Prove that this space is not compact.
   (b) Prove that the subspace $B = \{x : x < 0\}$ is not compact.
   (c) Prove that the subspace $C = \{x : x \leq 0\}$ is compact.

2. Prove Theorem 4.11.

3. Prove Theorem 4.13.

4. Prove that R^2 is not compact.

5. Suppose A and B are compact subspaces of a space X.
   (a) Prove or disprove that $A \cap B$ is compact.
   (b) Prove or disprove that $A \cup B$ is compact.

6. (a) Prove that if $(X, \mathcal{T})$ is compact and $\mathcal{T}_1 \subset \mathcal{T}$, then $(X, \mathcal{T}_1)$ is compact.
   (b) Give an example to show $\mathcal{T} \subset \mathcal{T}_1$ need not imply $(X, \mathcal{T}_1)$ is compact.

7. Let A be a compact set in a regular space X. If U is an open set in X containing A, prove there exists an open set V in X such that $A \subset V \subset \bar{V} \subset U$.

8. Prove that every compact regular space is normal.

9. A space X is called *locally compact at a point* $a \in X$ iff there exists an open set U in X containing a and a compact subset A of X such that $a \in U \subset A$. A space is *locally compact* iff it is locally compact at each point $a \in X$. Prove that every closed subspace of a locally compact space is locally compact.

10. Define a neighborhood of a point x of a space X as a subset M of X containing x such that there exists an open set $U \subset X$ having the property $x \in U \subset M$. Prove that if X is a locally compact space which is either Hausdorff or regular, then the family of all closed compact neighborhoods of a point $x \in X$ is a local base at the point x.

5. More Properties of Compact Spaces

Our first theorem is a stronger result than Theorem 4.13.

5.1. THEOREM: *Let $f: X \to Y$ be a continuous function from the space X to the space Y and A any compact subset of X. Then $f(A) \subset Y$ is compact.*

Proof: Let $\{V_\alpha : \alpha \in \Delta\}$ be an open cover of $f(A)$. Since f is continuous, $\{f^{-1}(V_\alpha) : \alpha \in \Delta\}$ is an open cover of A. The compactness of A implies the existence of a finite subcover $f^{-1}(V_{\alpha_1}), f^{-1}(V_{\alpha_2}), \ldots,$ $f^{-1}(V_{\alpha_n})$ of $\{f^{-1}(V_\alpha) : \alpha \in \Delta\}$ for A. Thus, $V_{\alpha_1}, V_{\alpha_2}, \ldots, V_{\alpha_n}$ is a subcover of $\{V_\alpha : \alpha \in \Delta\}$ for $f(A)$. This shows $f(A)$ is compact.

5.2. THEOREM: *If $f: X \to Y$ is continuous where X is compact and Y is a Hausdorff space, then f is a closed function.*

Proof: For any closed set $A \subset X$, A is compact by Theorem 4.8, and by Theorem 5.1, $f(A)$ is a compact subset of the Hausdorff space Y. Theorem 4.10 gives $f(A)$ closed.

We now give another theorem involving some of the conditions of Theorem 5.2.

5.3. THEOREM: *Let X be a compact Hausdorff space and Y an arbitrary space. If $f: X \to Y$ is a continuous closed surjection, then Y is Hausdorff.*

Proof: Let $f(x)$ and $f(y)$ be any two distinct points in Y. Since X is Hausdorff, $\{x\}$ and $\{y\}$ are closed subsets of X so that $\{f(x)\}$ and $\{f(y)\}$ are both closed subsets of Y, due to the fact that f is a closed function. Now, using the continuity of f, we see that $f^{-1}(f(x))$ and $f^{-1}(f(y))$ are closed subsets of X, which are, of course, disjoint. According to Theorem 4.12, X is normal, and therefore there exist open sets U and V containing $f^{-1}(f(x))$ and $f^{-1}(f(y))$, respectively, such that $U \cap V = \phi$, Using the fact that $X = V \cup X \backslash V$, it follows that $f(X) = Y = f(V) \cup f(X \backslash V)$ and, since f is a closed function, $f(X \backslash V)$ is a closed subset of Y so that $Y \backslash f(X \backslash V)$ is open, contains $f(y)$, and $(Y \backslash f(X \backslash V)) \subset f(V)$. Similarly, $X = U \cup X \backslash U$ so that $Y \backslash f(X \backslash U) \subset f(U)$ is open and contains $f(x)$. By use of DeMorgan's Theorem, we see that $(Y \backslash f(X \backslash U)) \cap (Y \backslash f(X \backslash V)) = Y \backslash (f(X \backslash U) \cup f(X \backslash V))$ and, since U and V are disjoint, $X = (X \backslash U) \cup (X \backslash V)$, which implies $f(X) = Y = f(X \backslash U) \cup f(X \backslash V)$. Therefore, $Y \backslash (f(X \backslash U) \cup f(X \backslash V)) = Y \backslash Y = \phi$. The conclusion that Y is Hausdorff follows.

The condition of compactness gives more information about f as a subset of the product space $X \times Y$. In particular, we have the following two theorems.

5.4. THEOREM: *(Fuller, R. V.) Let X and Y be arbitrary spaces and $f: X \to Y$ a function such that f is a closed subset of $X \times Y$. If $B \subset Y$ is compact, then $f^{-1}(B)$ is a closed subset of X.*

Proof: Let $B \subset Y$ be compact and consider any point $x \in X \backslash f^{-1}(B)$. Then $f(x) \notin B$ and, therefore, for each point $b \in B$, $(x, b) \in X \times Y$ but $(x, b) \notin f$. Since $f \subset X \times Y$ is a closed set, there exist open sets $U_b(x) \subset X$ and $V(b) \subset Y$ containing x and b, respectively, such that $(U_b(x) \times V(b)) \cap f = \phi$. That is, $U_b(x) \times V(b)$ contains no point of the form $(z, f(z))$. If we find such open sets $U_b(x)$ and $V(b)$ for each $b \in B$, then $\{V(b) : b \in B\}$ is an open cover of B, and the compactness of B gives a finite subcover $V(b_1), V(b_2), \ldots, V(b_n)$ of $\{V(b) : b \in B\}$ for B. From this it follows that $\bigcap_{k=1}^{n} U_{b_k}(x) = U(x)$ is an open subset of X containing x such that $U(x) \cap f^{-1}(B) = \phi$. This means $X \backslash f^{-1}(B)$ is open so that $f^{-1}(B)$ is closed.

5.5. THEOREM: *Let X be an arbitrary space, Y a compact space and* $f: X \to Y$ *a function such that* f *is a closed subset of* $X \times Y$. *Then* f *is continuous.*

Proof: Let B be any closed subset of the compace space Y. Then B is compact by Theorem 4.8 and $f^{-1}(B)$ is closed by Theorem 5.4. It follows that f is continuous.

A desirable feature of compact spaces is that the product space $\prod_{\alpha \in \Delta} X_\alpha$ is compact iff each of the spaces X_α is compact. However, we again feel that a proof of this fact is at a level above those of the mainstream of this book. Therefore, after a preliminary theorem, we shall give a proof for finite indexing sets only and then simply state the general theorem. Any of the references in the Bibliography may be consulted for a proof when the indexing set is arbitrary.

5.6. THEOREM: *Let* $\mathscr{B}$ *be a base for the space X. Then X is compact iff each cover* $\{B_\beta \in \mathscr{B} : \beta \in \Delta\}$ *of X by members of* $\mathscr{B}$ *has a finite subcover for X.*

Proof: Let X be compact. If $\{B_\beta \in \mathscr{B} : \beta \in \Delta\}$ is any cover of X by members of $\mathscr{B}$, it is then an open cover of X and hence there is a finite subcover of this cover for X according to the definition of compactness.

Conversely, suppose each cover of X by members of the base $\mathscr{B}$ has a finite subcover for X. It is to be shown X is compact. To this end, let $\{U_\alpha : \alpha \in \Omega\}$ be any open cover of X. Because $\mathscr{B}$ is a base for the topology on X, we have for each $\alpha \in \Omega$, $U_\alpha = \bigcup_{\beta \in \Delta_\alpha} B_\beta$ where the indexing set Δ_α depends upon α. Thus, $X \subset \bigcup_{\alpha \in \Omega} U_\alpha = \bigcup_{\alpha \in \Omega} (\bigcup_{\beta \in \Delta_\alpha} B_\beta)$ so that $\{B_\beta : \beta \in \Delta_\alpha, \alpha \in \Omega\}$ is a cover of X by members of $\mathscr{B}$. By hypothesis, there is a finite subcover $B_{\beta_1}, B_{\beta_2}, \ldots, B_{\beta_n}$ of this cover for X. Now choose one element $U_{\alpha_k} \in \{U_\alpha : \alpha \in \Omega\}$ such that $B_{\beta_k} \subset U_{\alpha_k}$ for each $k = 1, 2, \ldots, n$. This can be done because of the manner in which the cover of X by base elements was constructed. Thus, $X \subset \bigcup_{k=1}^{n} B_{\beta_k} \subset \bigcup_{k=1}^{n} U_{\alpha_k}$ so that $\{U_{\alpha_k} : k = 1, 2, \ldots, n\}$ is a finite subcover of $\{U_\alpha : \alpha \in \Omega\}$ for X, implying X is compact.

5.7. THEOREM: *The product space* $X \times Y$ *is compact iff each of the spaces X and Y is compact.*

Proof: If $X \times Y$ is compact, the projection functions $p_X : X \times Y \to X$ and $p_Y : X \times Y \to Y$ are continuous and surjective so that X and Y are compact by Theorem 5.1.

Now let X and Y be compact spaces, and it will be shown that $X \times Y$ is compact. In view of Theorem 5.6, we need only show that every cover of $X \times Y$ by a collection of basic open sets has a finite subcover. In view of the fact that a basic open set in the space $X \times Y$ has the form $U \times V$ where U is open in X and V is open in Y, we may denote such an open cover by $\{U_\alpha \times V_\alpha : \alpha \in \Delta\}$ where U_α is open in X and V_α is open in Y. For a given point $x \in X$, the subset $\{x\} \times Y$ of $X \times Y$ is homeomorphic to Y by Theorem 2.12 of Chapter 5 and is, therefore, compact by Theorem 4.13. Since $\{x\} \times Y$ is covered by $\{U_\alpha \times V_\alpha : \alpha \in \Delta\}$, a finite subcover $\{U_{\alpha_k} \times V_{\alpha_k} : k \in \Delta_x$ where Δ_x is a finite indexing set depending upon the choice of $x\}$ of this cover for $\{x\} \times Y$ exists where $x \in U_{\alpha_k}$ and $k \in \Delta_x$. Now $\bigcap_{k \in \Delta_x} U_{\alpha_k} = U(x)$ is open, contains x, and $\{U(x) \times V_{\alpha_k} : k \in \Delta_x\}$ is still a finite open cover of $\{x\} \times Y$. Proceeding in this fashion for each $x \in X$, we construct $\{U(x) : x \in X\}$ which is an open cover of X. By compactness of X there is a finite subcover $U(x_1), U(x_2), \ldots, U(x_m)$ of this cover for X. Since each $U(x_j), j = 1, 2, \ldots, m$, is an intersection of open sets in X which were used to form $\{U_\alpha \times V_\alpha : \alpha \in \Delta\}$, we may select an open set $U_{\alpha_{x_j}} \in \{U_\alpha : \alpha \in \Delta\}$ such that $U(x_j) \subset U_{\alpha_{x_j}}$ for $j = 1, 2, \ldots, m$. Therefore, $\{U_{\alpha_{x_j}} : j = 1, 2, \ldots, m\}$ is a finite open

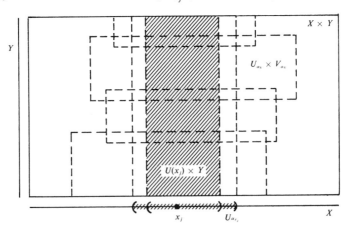

Figure 24

cover of X, and for each j, $1 \leq j \leq m$, $\{U_{\alpha_{x_j}} \times V_{\alpha_k} : k \in \Delta_x\}$ covers the "vertical strip" $U(x_j) \times Y$ in $X \times Y$ as illustrated in Figure 24. By its construction, the collection $\{U_{\alpha_{x_j}} \times V_{\alpha_k} : k \in \Delta_x, j = 1, 2, \ldots, m\}$ is then a finite subcover of $\{U_\alpha \times V_\alpha : \alpha \in \Delta\}$ for $X \times Y$ and, therefore, by Theorem 5.6, $X \times Y$ is compact.

5.8. Corollary: The finite product $\prod\limits_{k=1}^{n} X_k$ is compact iff each space $X_k, k = 1, 2, \ldots, n$, is compact.

Proof: The proof is an application of mathematical induction using Theorem 2.13 of Chapter 5 and Theorem 5.7 above.

We now state, but do not prove, the important Tychonoff Product Theorem.

5.9. THEOREM: (*The Tychonoff Product Theorem.*) *For any family of spaces* $\{X_\alpha : \alpha \in \Delta\}$, $\prod\limits_{\alpha \in \Delta} X_\alpha$ *is compact iff each* $X_\alpha, \alpha \in \Delta$, *is compact.*

Although many important spaces are not compact, it is sometimes useful to construct a compact space containing the original space as a dense subspace. In doing this, a *compactification* of the original space is being formed. The simplest type of compactification is called the *one-point compactification.* It is obtained by taking any noncompact space X and forming the union of X and a single point $a \notin X$ to obtain the set $X^* = X \cup \{a\}$. Then topologize X^* as follows: Open sets consist of all open sets in X together with all subsets U of X^* having the property that $X^* \backslash U$ is a closed compact subset of X. It is an exercise to show X^* is a compact space and that X is dense in the space X^*. We should be made aware that there are other important and useful methods of compactification, but we shall not study them here.

Exercises:

1. Let X and Y be compact Hausdorff spaces. Prove that $f : X \to Y$ is continuous iff for each compact subset $M \subset Y$, $f^{-1}(M)$ is compact in X.

2. Prove that every compact locally connected space has only a finite number of components.

3. Let $p_Y : X \times Y \to Y$ be the projection function where X is a compact space and Y is an arbitrary space. Prove that p_Y is a closed function.

4. Let X be compact and Y Hausdorff. Prove that every continuous bijection $f : X \to Y$ is a homeomorphism.

5. Show that the open sets described for the one-point compactification actually do form a topology for the space X^*.

6. Prove that X is dense in X^* where X^* is the one point compactification of X.

7. Discuss the one point compactification of R^1.

8. Let $f : X \to Y$ be a continuous surjection. If X is compact and Y is Hausdorff, prove that f is an identification.

9. Let $X \neq \phi$ be a set. Prove that a topology making X a compact Hausdorff space is the smallest of all Hausdorff topologies for X.

10. Prove that if $(X, \mathscr{T})$ is a compact Hausdorff space and $\mathscr{T}_1$ is a topology for X such that $\mathscr{T}_1 \not\subset \mathscr{T}$, then $(X, \mathscr{T}_1)$ is not Hausdorff.

6. Compactness in R^n

The major aim of this section is to provide a simple characterization of compact subsets of R^n. The result is that a subset of R^n is compact iff the subset is both closed and bounded. Before proving a series of theorems leading to this conclusion, the definition of a bounded set in R^n is reviewed.

6.1. Definition: The subset $A \subset R^1$ is *bounded* iff there exists a positive real number K such that for each $x \in A$, $|x| \leq K$. If $A \subset R^n$, then A is *bounded* iff there exists a positive real number K such that for each point $(x_1, x_2, \ldots, x_n) \in A$, $|x_j| \leq K$ for each $j = 1, 2, \ldots, n$.

In other words, $A \subset R^1$ is bounded iff every point of A lies within K units of the origin. In R^n, A is bounded iff A lies inside a "box" with center at the origin and having an edge of $2K$ units. Of course, ϕ is bounded by our definition, and, since ϕ is also closed and compact, the ultimate goal of showing a subset of R^n is compact iff it is closed and bounded will certainly hold when that subset is the empty set.

6.2. Example: In R^3, the set $A = \{(x, y, z) : x = 1, y = 2\}$ is not bounded, hence is *unbounded*. The set $B = \{(x, y, z) : x^2 + y^2 = 4, -10 \le z < 4\}$ is bounded and K may be taken as 10 or any real number larger than 10.

6.3. THEOREM: *The closed bounded interval* $[a, b] = \{x : a \le x \le b\} \subset R^1$ *is compact.*

Proof: Let $\mathscr{B} = \{U_\alpha : \alpha \in \Delta\}$ be an open cover of the interval $[a, b]$. The plan is to start at the point a and see how far we can move toward b by using only a finite number of the open sets in $\mathscr{B}$. If we can reach b, the theorem will be proved.

Let A be the set of points $x \in [a, b]$ having the property that the interval $[a, x]$ is contained in a *finite* union of members of $\mathscr{B}$. Certainly $A \ne \phi$ since there is some $U_\beta \in \mathscr{B}$ such that $a \in U_\beta$ and hence $\{a\}$ is covered by the single set U_β. Furthermore, since A is bounded on the right by b, the nonempty bounded set A has a least upper bound which will be called c. Next, it will be shown that $c = b$. Suppose, on the contrary, that $c < b$. Then there is some open set $U_\gamma \in \mathscr{B}$ such that $c \in U_\gamma$, and, by definition of an open set in R^1, there are two points d and e belonging to U_γ such that $d < c < e < b$ and the open interval $(d, e) \subset U_\gamma$. Thus, $d \in A$ and there is some finite number $U_{\alpha_1}, U_{\alpha_2}, \ldots, U_{\alpha_n}$ of members of $\mathscr{B}$ whose union covers $[a, d]$ so that the finite union $U_{\alpha_1}, U_{\alpha_2}, \ldots, U_{\alpha_n}, U_\gamma$ covers the interval $[a, e]$. But this contradicts the fact that c is the least upper bound of A. Consequently, $c = b$.

Only one step now remains, and that is to show $b \in A$. Consider any $U_\delta \in \mathscr{B}$ where $b \in U_\delta$. Then there is an open interval which is a subset of U_δ and contains b and hence contains points less than b in $[a, b]$. Let p be such a point. Then $p \in A$ and hence there exists a finite number $U_{\alpha_1}, U_{\alpha_2}, \ldots, U_{\alpha_n}$ of members of $\mathscr{B}$ whose union covers $[a, p]$. As a consequence, $U_{\alpha_1}, U_{\alpha_2}, \ldots, U_{\alpha_n}, U_\delta$ is a finite collection of members of $\mathscr{B}$ whose union covers $[a, b]$. This shows $b \in A$. Therefore, $\mathscr{B}$ has a finite subcover for $[a, b]$ showing $[a, b]$ is compact.

We are now in a position to prove half of our ultimate goal.

6.4. THEOREM: *If A is a closed bounded subset of R^n, then A is compact.*

Proof: Since A is bounded, the projection functions $p_k : R^n \to R^1$, $k = 1, 2, \ldots, n$, when applied to A, each give a bounded image in R^1.

That is, $p_k(A) = B_k$ where each $B_k, k = 1, 2, \ldots, n$, is bounded in R^1. Let K_k be a real number such that if $x \in B_k$ then $|x| \leq K_k$, and define $K = \max \{K_1, K_2, \ldots, K_n\}$. According to Theorem 6.3, the closed interval $[-K, K]$ is compact, and by Theorem 5.8, the product of $[-K, K]$ with itself n times is a compact subspace of R^n. By construction, A is a subset of this product. Thus, we have the closed set A as a subset of a compact space so that by Theorem 4.8, A is compact.

Finally, our goal is reached with the next theorem.

6.5. THEOREM: *Let $A \subset R^n$ be compact. Then A is closed and bounded.*

Proof: Applying Theorem 4.10 to the Hausdorff space R^n, the compact set A is closed.

We now show that A is bounded. For each $k \in N$, let U_k be the product of the open interval $(-k, k)$ with itself n times. Then each U_k is a basic open set in $R^n, U_1 \subset U_2 \subset U_3 \subset \ldots$, and $\bigcup_{k \in N} U_k = R^n$. Thus, $\mathscr{B} = \{U_k : k \in N\}$ is an open cover of $A \subset R^n$. Since A is compact, some finite number, and hence by their construction some single open set $U_j \in \mathscr{B}$, covers A. Thus, A is a subset of the open "box" U_j so that if $(x_1, x_2, \ldots, x_n) \in A, |x_m| \leq j, m = 1, 2, \ldots, n$. This shows A is bounded.

6.6. THEOREM: *(The Heine-Borel Theorem.) The subset $A \subset R^n$ is compact iff A is closed and bounded.*

Proof: Theorems 6.4 and 6.5.

From studying calculus, we are all aware of the importance of knowing when a continuous function attains a maximum or a minimum. The often stated result is that this happens whenever the function is defined on a closed interval $[a, b]$. We know now that such an interval is compact, and this might suggest the generalization of using any compact domain for the function in place of a closed bounded interval of reals. Indeed, this is the case. Before proving this result, perhaps we should recall the following theorem.

6.7. THEOREM: *Let A be any nonempty compact subset of R^1. Then A has a least upper bound b and a greatest lower bound a and, furthermore, $a \in A$ and $b \in A$.*

Proof: Use Theorem 6.6 to get boundedness, then the completion axiom, and, finally a contradictory argument based on the fact that A is closed. The details are left as an exercise.

6.8. THEOREM: *Let $f: X \to R^1$ be continuous where X is compact and nonempty. Then f attains its maximum and minimum on X. That is, there exists points x_1 and x_2 of X such that $f(x_1) \leqslant f(x) \leqslant f(x_2)$ for all $x \in X$.*

Proof: According to Theorem 5.1, $f(X) \subset R^1$ is compact and hence has a least upper bound b and a greatest lower bound a which both belong to $f(X)$ by Theorem 6.7. Thus, there is some $x_1 \in X$ such that $f(x_1) = a$ and an $x_2 \in X$ such that $f(x_2) = b$. Consequently, for any $x \in X, a = f(x_1) \leqslant f(x) \leqslant f(x_2) = b$.

6.9. THEOREM: *(Another form of the Intermediate Value Theorem.) Let f be a real valued continuous function defined on a closed interval $[a, b] \subset R^1$. Then f takes on every value between its maximum, the least upper bound of $f([a, b])$, and its minimum, the greatest lower bound of $f([a, b])$.*

Proof: The proof is left to the reader.

Exercises:

1. Which of the following subsets of R^1 are compact?
   (a) $\{x : 0 \leqslant x \leqslant 1\}$.
   (b) $\{x : 3 \leqslant x < 5\}$.
   (c) The rational numbers.
   (d) $\{x : 1 \leqslant x \leqslant 2\} \cup \{x : 3 \leqslant x \leqslant 5\}$.
   (e) $\{x : 0 \leqslant x \leqslant 1, x \text{ rational}\}$.
   (f) The natural numbers.

2. Prove Theorem 6.7.

3. Prove Theorem 6.9.

4. Give an example of
   (a) a closed subset of R^3 which is not compact.
   (b) a bounded subset of R^3 which is not compact.

5. Give an example of a family of compact subsets of R^1 whose union is not compact.

6. Let A and B be open subsets of R^n. Prove $A \cap B$ is compact iff $A \cap B = \phi$.

7. (The Cantor Intersection Theorem.) Let A_1 be a nonempty closed bounded subset of R^n and let $A_1 \supset A_2 \supset A_3 \supset \ldots$ be a family of nonempty closed sets. Prove $\bigcap_{n \in N} A_n \neq \phi$.

8. Let $\{A_n : n \in N\}$ be a countable family of closed subsets of R^1 such that $\bigcup_{n \in N} A_n$ contains a nonempty open set. Prove that at least one of the sets A_n must contain a nonempty open set.

9. Let $\{U_\alpha : \alpha \in \Delta\}$ be an open cover of the compact set $A \subset R^1$. Prove that there exists a positive number δ such that if $x, y \in A$ and $|x - y| < \delta$, then there is some U_α containing both x and y.

10. Let X be any compact space and define an equivalence relation on X as follows: For each $x, y \in X$, xRy iff for every continuous $f : X \rightarrow R^1$ such that $f(x) = 0$ and $f(y) = 1$, there is an element $z \in X$ such that $f(z) = 1/2$. Prove that the equivalence classes of X are also the components of X.

7. Other Types of Compactness

There are many types of compactness defined in mathematics. Only two additional kinds will be considered here, with the major aim being to show that they are equivalent to compactness in the R^n spaces. In fact, it will be shown that they are equivalent to compactness in second countable T_1-spaces from which the equivalence in the R^n spaces follows.

7.1. Definition: (a) The space X is *countably compact* iff every countable open cover of X has a finite subcover for X.

(b) The space X has the *Bolzano-Weierstrass property* iff every infinite subset of X has a cluster point belonging to X.

7.2. THEOREM: *Every compact space X is also countably compact.*

Proof: Let $\{U_n : n \in N\}$ be any countable open cover of X. Since X is compact, every open cover has a finite subcover for X and, in particular, $\{U_n : n \in N\}$ has a finite subcover for X, implying X is countably compact.

Our knowledge about compact spaces together with Theorem 7.2 yield several examples of countably compact spaces. The exercises point out that the R^n spaces are not countably compact.

7.3. Example: (A space that is not countably compact but does have the Bolzano-Weierstrass property.) For each $n \in N$, let $B_n = \{2n - 1, 2n\}$. Let $\mathscr{T}(\mathscr{B})$ be the topology on N generated by $\mathscr{B} = \{B_n : n \in N\}$ and having $\mathscr{B}$ as a base. The space $(N, \mathscr{T}(\mathscr{B}))$ is not countably compact. To see this, note that $\mathscr{B}$ is a countable cover of N with no finite subcover for N. The space does, however, have the Bolzano-Weierstrass property. To see this, let A be any infinite subset of N and consider any $a \in A$. If a is odd, let $b = a + 1$, and, if a is even, let $b = a - 1$. Then every open set belonging to $\mathscr{T}(\mathscr{B})$ and containing b also contains a, so that b is a cluster point of A.

One of the important properties of countable compact spaces is that they are characterized by the behavior of sequences, as is shown next.

7.4.   THEOREM: *The space X is countably compact iff each sequence in X accumulates to a point in X.*

Proof: First let X be countably compact. Suppose $f : N \to X$ is a sequence in X that does not accumulate to any point in X. Thus, for each $x \in X$, there exists an open set $U(x)$ containing x and a natural number $n \in N$ such that $U(x) \cap \{f(n + 1), f(n + 2), f(n + 3), \ldots \} = \phi$. Now for each natural number $n \in N$, let $M_n = \{x \in X :$ there exists an open set $U(x)$ containing x such that $U(x) \cap \{f(n + 1), f(n + 2), f(n + 3), \ldots \} = \phi\}$. If $M_n \neq \phi$, let us further define for each $x \in M_n$ the set $W_n(x)$ to be the union of all possible open sets $U(x)$ having the property that $U(x) \cap \{f(n + 1), f(n + 2), f(n + 3), \ldots \} = \phi$. Finally, for $M_n \neq \phi$, let $V_n = \bigcup\limits_{x \in M_n} W_n(x)$. From these definitions, we have $M_1 \subset M_2 \subset \ldots$ and $V_1 \subset V_2 \subset \ldots$. By the well ordering of the natural numbers, there exists a smallest natural number k such that $M_k \neq \phi$ and, therefore, for all $n \geqslant k, V_n \neq \phi$. Therefore $\bigcup\limits_{n \geq k} V_n = X$ so that $\{V_n : n \in N, n \geqslant k\}$ is a countable open cover of X. However, by their construction, if any $V_j, j \geqslant k$ is removed from $\{V_n : n \in N, n \geqslant k\}$, the remaining V_n do not form a cover for X. As a consequence, $\{V_n : n \in N, n \geqslant k\}$ does not contain a finite subcover for X, implying X is not countably compact. This contradiction to the hypothesis means that the supposition that $f : N \to X$ has no accumulation point in X is false, thereby concluding half of the proof.

Conversely, suppose each sequence in X accumulates to some point in X. If X is not countably compact, there exists a countable open cover $\{U_n : n \in N\}$ having no finite subcover for X. Assuming this, we shall now construct a sequence in X having no accumulation point in

X and, in this manner, obtain a contradiction. Let $x_1 \in U_1$. Now let U_{n_2} be the first among $\{U_n : n \in N\}$ which is not a subset of U_1 and choose $x_2 \in U_{n_2} \backslash U_1$. In general, let U_{n_k} be the first among the open cover $\{U_n : n \in N\}$, not a subset of $\bigcup_{m=1}^{k-1} U_{n_m}$, and choose $x_k \in (U_{n_k} \backslash \bigcup_{m=1}^{k-1} U_{n_m})$. The choice of x_k may be made in each case since we are assuming no finite subcover of $\{U_n : n \in N\}$ for X. By this construction, we have defined a sequence $f : N \to X$ given by $f(n) = x_n$ for each $n \in N$. Now if $x \in X$, then x belongs to some U_{n_j} so that $U_{n_j} \cap \{f(j+1), f(j+2), f(j+3), \ldots\} = \phi$, showing that x is not an accumulation point of $f : N \to X$. This contradiction asserts X is countably compact.

It is important to notice in Theorem 7.4 that the accumulation point of the sequence must be in the space X to imply countable compactness, and conversely. The same applies to cluster points of infinite sets in the Bolzano-Weierstrass property. For example, consider the space $X = \{x : 0 < x \leqslant 1\}$ as a subspace of R^1 and the sequence $f : N \to X$ given by $f(n) = 1/n$ for each $n \in N$. The origin is the only accumulation point of this sequence, and it does not belong to X. So by Theorem 7.4, X is not countably compact. The set $f(N)$ is an infinite set in X having no cluster point in X, which implies X does not have the Bolzano-Weierstrass property.

Our next goal will be to show conditions under which countably compact spaces are also compact. To do this, we need the preliminary result of Theorem 7.5.

7.5. THEOREM: *Let X be a second countable space. Then each open cover of X has a countable subcover for X.*

Proof: By definition, the second countable space X has a countable base. Let $\mathscr{B} = \{B_\beta : \beta \in \Delta\}$ be such a countable base. Now let $\{U_\alpha : \alpha \in \Omega\}$ by any open cover of X. For each $x \in X$ and each U_α containing x, there exists a basic open set $B_\beta \in \mathscr{B}$ such that $x \in B_\beta \subset U_\alpha$. The totality of all such B_β chosen in this manner forms a countable family $\{B_\beta : \beta \in \Sigma \subset \Delta\}$. Notice also, that this collection is a cover for X. For each B_β, where β belongs to the countable set Σ, choose a single index $\alpha_\beta \in \Omega$ such that $B_\beta \subset U_{\alpha_\beta}$. Then the totality of open sets $\{U_{\alpha_\beta} : \beta \in \Sigma\}$ is a countable family of open sets and, furthermore, covers X because $\bigcup_{\beta \in \Sigma} B_\beta \subset \bigcup_{\beta \in \Sigma} U_{\alpha_\beta}$ and $\{B_\beta : \beta \in \Sigma\}$ is a cover of X.

7.6. **THEOREM:** *Let X be a countably compact space which is second countable. Then X is compact.*

Proof: To show that X is compact, let $\{U_\alpha : \alpha \in \Delta\}$ be any open cover of X. Since X is second countable, there is a countable subcover $\{U_\alpha : \alpha \in \Sigma \subset \Delta, \Sigma \text{ countable}\}$ according to Theorem 7.5. By definition of countable compactness, this countable subcover has a finite subcover of X. Therefore, the original open cover has a finite subcover of X, which means X is compact.

7.7. Corollary: Compactness and countable compactness are equivalent in second countable spaces.
Proof: Theorems 7.6 and 7.2.

Next, we relate countable compactness to the Bolzano-Weierstrass property.

7.8. THEOREM: *Every countably compact space X has the Bolzano-Weierstrass property.*

Proof: Let A be any infinite subset of X. By Theorem 5.7 of Chapter 2, there exists a sequence $f : N \to X$ of distinct points in A. According to Theorem 7.4, f accumulates to a point $a \in X$, and, since f is injective, a must be a cluster point of $f(N)$ by Exercise 6, Section 3, of Chapter 7. Thus, X has the Bolzano-Weierstrass property.

7.9. THEOREM: *Let X be a T_1-space which has the Bolzano-Weierstrass property. Then X is countably compact.*

Proof: Assume that $\{U_n : n \in N\}$ is a countable open cover of X which has no finite subcover of X. Now construct, as in the proof of Theorem 7.4, a sequence $f : N \to X$ of distinct points in X. As in that proof, this sequence has no accumulation point in X. By the Bolzano-Weierstrass property, the set $f(N)$ has a cluster point $a \in X$. Then use Exercise 8, Section 1, of Chapter 6, to see that in the T_1-space X every open set containing a also contains infinitely many points of $f(N)$. But this means a is an accumulation point of f, which is a contradiction. The details are left to the reader.

7.10. Corollary: In T_1-spaces, countable compactness is equivalent to the Bolzano-Weierstrass property.

Proof: Theorems 7.8 and 7.9.

7.11. THEOREM: *In second countable T_1-spaces, the concepts of compactness, countable compactness, and the Bolzano-Weierstrass property are all equivalent.*

Proof: Corollaries 7.7 and 7.10.

7.12. Corollary: For each $n \in N$, in R^n the concepts of compactness, countable compactness, and the Bolzano-Weierstrass property are all equivalent.

Proof: Corollaries 1.12 and 5.14 of Chapter 6, along with Theorem 7.11.

7.13. THEOREM: (*The Bolzano-Weierstrass Theorem.*) *Every bounded infinite subset of R^n has at least one cluster point.*

Proof: An easy indirect proof is left to the reader.

Exercises:

1. Prove that R^1 and R^2 are not countably compact.

2. Prove Theorem 7.9 in detail.

3. Prove Theorem 7.13.

4. Prove the general Cantor Product Theorem: Let $A_1 \supset A_2 \supset A_3 \supset \dots$ be a family of nonempty closed sets in a T_1-space X where A_1 is countably compact. Prove $\bigcap_{n \in N} A_n \neq \phi$.

5. Prove that every closed subset of a countably compact space is countably compact.

6. Prove that if $f : X \to Y$ is continuous and X is countably compact, then Y is countably compact.

7. Prove that both countable compactness and the Bolzano-Weierstrass property are topological properties.

8. A space X is called a *Lindelöf space* iff each open cover of X has a countable subcover of X.
   (a) Prove that each second countable space is a Lindelöf space.
   (b) Prove that R^1 with the cofinite topology is a Lindelöf space.

9. Prove that every closed subspace of a Lindelöf space is a Lindelöf space.

10. If X is a Lindelöf space and $f: X \to Y$ is a continuous surjection, prove that Y is a Lindelöf space.

8. Non-Countinuous Functions

We have already discussed several conditions of a topological nature for functions other than continuity. For instance, a closed function, an open function, or a function which is a closed subset of $X \times Y$, have been defined and studied in some detail. In defining such functions, we have observed only a few of the many special types of functions that exist from one topological space to another. In this section we are going to introduce several other types of conditions for functions which are of a topological nature and are found in the study of mathematics. As with those already introduced, other than continuity, the conditions we are about to define allow a function to have one or more points of discontinuity. For that reason, a function having one of these restrictions is generally classified as a *non-continuous* function. There functions do have enough structure, however, to enable us to prove many interesting theorems. In fact, such functions have been a source of considerable mathematical research, much of which is, unfortunately, beyond the scope of this book. Our purpose here then is to become acquainted with some of the definitions and to have a sample of some of the elementary theorems of the subject of non-continuous functions.

8.1. Definition: Let X and Y be topological spaces.
(a) A function $f: X \to Y$ is *connected* iff for each connected subset $A \subset X$, $f(A)$ is connected in Y.
(b) A function $f: X \to Y$ is a *connectivity function* iff the function $G: X \to X \times Y$, given by $G(x) = (x, f(x))$, is a connected function.
(c) A function $f: X \to Y$ is *neighborly* iff for each $x \in X$ and every pair of open sets U and V containing x and $f(x)$, respectively, there exists an open set $W \subset U$ (not necessarily containing x) such that $f(W) \subset V$.
(d) A function $f: X \to Y$ is *monotone* iff for each $y \in Y$, $f^{-1}(y)$ is a connected subset of X.

(e) A function $f: X \to Y$ is *compact* iff for each compact subset $B \subset Y, f^{-1}(B)$ is a compact subset of X.

(f) A function $f: X \to Y$ is *compact preserving* iff for each compact subset $A \subset X, f(A)$ is a compact subset of Y.

Our first theorem shows that continuity is a stronger restriction on a function than most of those in Definition 8.1.

8.2. THEOREM: *Let $f: X \to Y$ be continuous. Then f is a connected function, a connectivity function, a neighborly function, and a compact preserving function.*

Proof: That f is a connected and compact preserving function follows from Theorems 2.1 and 5.1. Exercise 9, Section 2, of Chapter 5, gives $G: X \to X \times Y$, a homeomorphism because f is continuous and, therefore, G is a connected function, making f a connectivity function. To see that f is neighborly, let $x \in X$ and U and V be any two open sets containing x and $f(x)$, respectively. Then continuity of f implies that there exists an open set $W \subset U$ containing x such that $f(W) \subset V$, so that f is neighborly by Definition 1.1(c).

The restrictions of the following theorem are but one set which make a continuous function a compact function.

3.3. THEOREM: *Let X be a compact space and Y a Hausdorff space. Then any continuous $f: X \to Y$ is also a compact function.*

Proof: Let $B \subset Y$ be compact. Then Theorem 4.10 gives B closed, and the continuity of f gives $f^{-1}(B)$ closed in the compact space X. Theorem 4.8 implies $f^{-1}(B)$ is compact, which makes f a compact function.

8.4. Example: (A continuous function need not be either monotone or compact.) Let $X = Y = R^1$ and suppose $f: X \to Y$ is given by $f(x) = \sin(x)$. The point $1 \in Y$ is a compact connected subset of Y and $f^{-1}(1) = \{x \in X : x = \pi/2 + 2k\pi, k = 0, \pm 1, \pm 2, \pm 3, \ldots\}$, which is not a compact nor connected subset of $X = R^1$.

Since the functions we are discussing may have points of discontinuity, the reverse implications of Theorem 8.2 need not hold, as the next example shows.

8.5. Example: Let $X = Y = R^1$ and let $f: X \to Y$ be given by $f(x)$ $= \sin(1/x)$ if $x \neq 0$ and $f(0) = 0$, as shown in Figure 25. The function f is continuous at each point of X except $x = 0$. However, f is a connected function, a connectivity function, a neighborly function, and a compact preserving function. This function is not a monotone function or a compact function. To see that these assertions are indeed true, observe that the only difficulty arises when dealing with sets that contain $x = 0$, since f is continuous at all other points of X. Before examining f for these various properties, it would be helpful to note that if $x = 1/(k\pi)$ for all any $k = \pm 1, \pm 2, \pm 3, \ldots$, then $f(x) = 0$.

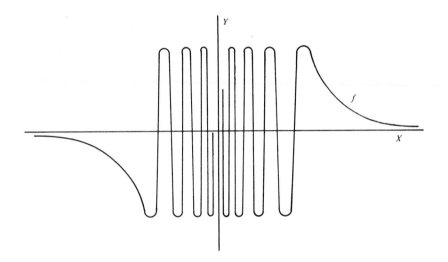

Figure 25

Any connected subset $A \subset X$ containing $x = 0$ is an interval so that $f(A) = \{y \in Y : -1 \leqslant y \leqslant 1\}$, due to the definition of f. This, along with the fact that f is continuous at all other points of X, shows f is a connected function. Also $G(A) = \{(x, f(x)) : x \in A\}$ is a connected subset of $X \times Y$ by use of Theorem 1.11 and, as a consequence, f is a connectivity function.

To see f is a neighborly function, let U and V be any two open sets containing $x = 0$ and $f(0) = 0$, respectively. Then there exists an open interval $W = (a, b)$ such that $0 \in W \subset U$. The definition of f then implies that there exists a $k \in N$ such that $0 < 1/(k\pi) < b$, i.e.,

$1/(k\pi) \in W$, and hence $f(1/(k\pi)) = \sin(k\pi) = 0$. Since f is continuous at $x = 1/(k\pi)$, there exists an open set, and therefore an open interval $Z = \{x : c < x < d\}$, such that $1/(k\pi) \in Z, Z \subset W$ and $f(Z) \subset V$. This shows f is neighborly at $x = 0$, and the continuity of f elsewhere means f is a neighborly function.

A consideration of the various forms that a compact subset of $X = R^1$ may have leads to the conclusion that f is compact preserving.

It is easy to give examples of subsets of Y which will that show f is not a monotone or a compact function.

The connection between connected and connectivity functions is given now.

8.6. Every connectivity function $f : X \to Y$ is a connected function.

Proof: Let $A \subset X$ be connected. Then $G(A) \subset X \times Y$ is connected, and therefore $p_Y(G(A)) = f(A)$ is connected because p_Y is a continuous function.

For a continuous function $f : X \to Y$ and any closed subset $B \subset Y, f^{-1}(B)$ is a closed subset of X. While we cannot prove this result for connected functions, we can establish that each component of $f^{-1}(B)$ is a closed subset of X provided Y is a T_1-space.

8.7. THEOREM: (*Sanderson, D. E.*) *Let $f : X \to Y$ be a connected function and B a closed subset of the T_1-space Y. Then each component of $f^{-1}(B)$ is a closed subset of X.*

Proof: Suppose $A \subset X$ is a component of $f^{-1}(B)$ that is not closed. Then there exists a cluster point $a \in X$ of A such that $a \notin A$. Now $A \cup \{a\}$ is connected and $f(A \cup \{a\}) = f(A) \cup \{f(a)\}$. However, $f(A) \cup \{f(a)\}$ is not connected in the T_1-space Y because $f(A)$ is a subset of the closed set B and $f(a) \notin B$ so that $f(A)$ and $\{f(a)\}$ are separated in the space X. This contradiction to our hypothesis means A must be closed.

In view of Theorem 8.6, the results of Theorem 8.7 also hold for connectivity functions.

8.8. THEOREM: (*Pervin, W. J. and Levine, N.*) *Let Y be a T_1-space and suppose $f : X \to Y$ is a connected and monotone function. Then for each $y \in Y, f^{-1}(y)$ is a closed subset of X.*

Proof: Assume for some $y \in Y$, that $f^{-1}(y) \neq \phi$ and $f^{-1}(y)$ is not closed. Then there exists a cluster point $x \in X$ of $f^{-1}(y)$ such that $x \notin f^{-1}(y)$. Since f is monotone, $f^{-1}(y)$ is connected so that $f^{-1}(y) \cup \{x\}$ is connected. However, $f(f^{-1}(y) \cup \{x\}) = f(f^{-1}(y)) \cup \{f(x)\} = \{y\} \cup \{x\}$ is not connected because $y \neq x$ and Y is a T_1-space. This contradicts the hypothesis that f is a connected function.

Examples easily show that even if $f: X \to Y$ is continuous, the induced function $f^{-1} : \mathscr{P}(Y) \to \mathscr{P}(X)$ need not be a connected function. To overcome this, the following conditions are offered.

8.9. THEOREM: *Let $f: X \to Y$ be a closed, monotone surjection. If $B \subset Y$ is connected, then $f^{-1}(B)$ is connected.*

Proof: If B is a single point, the monotone hypothesis gives the desired result. Otherwise, assume $f^{-1}(B) = H \cup K$ where H and K are separated subsets of X. Then $f(H \cup K) = f(H) \cup f(K) = B$ where $f(H) \cap f(K) = \phi$, and, since B is connected, one of the sets $f(H)$ or $f(K)$ contains a cluster point of the other. Suppose $f(H)$ contains a cluster point p of $f(K)$ and note that the following remarks could be made in case $f(K)$ contains a cluster point of $f(H)$ with a change in notation. Since f is monotone, $f^{-1}(p) \subset H$ and, because $H \cap \bar{K} = \phi$, $p \notin f(\bar{K})$. This means f is not a closed function, contradicting our hypothesis. The conclusion is that $f^{-1}(B)$ is connected.

We have shown earlier that if $A \subset X$, X is Hausdorff and $f: X \to A$ is a continuous retract of X onto A, then A is a closed subset of A. If we replace the continuous f with a connectivity function f, then $f: X \to A$ such that $f(x) = x$ for all $x \in A$, is called a *connectivity retraction* and A a *connectivity retract of X*. Our next theorem shows how this affects the closedness of A.

8.10 THEOREM: (*Hildebrand, S. K., and Sanderson, D. E.*) *Let X be a Hausdorff space, $A \subset X$ and $f: X \to A$ a connectivity retraction of X onto A. Then each component of A is a closed subset of X.*

Proof: Suppose some component M of A is not closed. Then there exists a cluster point $p \in X$ of M such that $p \notin M$, and, furthermore, $p \notin A$ because M is a component of A. Therefore, $f(p) \neq p$, so that there exist open disjoint sets U and V in the Hausdorff space X containing p and $f(p)$, respectively. The disjointness of U and V means the open set $U \times V$ in $X \times X$ contains no point of $G(M) = \{(x, f(x)) : x \in M\} = \{(x, x) : x \in M\}$. It follows that for the connected set $M \cup \{p\}$, $G(M \cup \{p\}) = G(M) \cup \{G(p)\}$ is not connected because $G(M)$ and $\{G(p)\}$ are separated sets in the Hausdorff space $X \times Y$. This contradicts the hypothesis that f is a connectivity function, and, therefore, our assumption is false. Consequently, M is closed.

We now have our following final theorem on connectivity functions.

8.11. THEOREM: *Let X be a connected, locally connected T_1-space, Y a T_1-space, and $f : X \to Y$ a connectivity function. If $V \subset Y$ is open, then each point of $f^{-1}(V) \subset X$ is a cluster point of $f^{-1}(V)$.*

Proof: Assume $a \in f^{-1}(V)$, but a is not a cluster point of $f^{-1}(V)$. Then there exists an open set $U \subset X$ containing a such that $U \cap f^{-1}(V) = \{a\}$. Since X is locally connected at a, there exists a connected open set W such that $a \in W \subset U$. Furthermore, $W \neq \{a\}$. The reason is that $\{a\}$ is closed in the T_1-space X and, if $\{a\}$ is also open, we contradict Theorem 1.7(b). Consequently the open set $U \times V$ of the T_1-space $X \times Y$ contains only the point $(a, f(a))$ of the nondegenerate set $G(W)$, which means, $G(W)$ is not connected. However, this contradicts the hypothesis that f is a connectivity function. It follows that our assumption is false and a is a cluster point of $f^{-1}(V)$.

The final two theorems of this section offer some conditions which imply continuity of functions. The first of these uses a condition on a topological space which we have not previously defined.

8.12. Definition: Let X be a topological space. Then X is called *semi-locally-connected* iff for each $x \in X$ and each open set $U \subset X$ containing x, there exists an open set $V \subset X$ such that $x \in V \subset U$ and $X \backslash V$ consists of a finite number of components.

8.13. THEOREM: *Let Y be a semi-locally-connected space and suppose $f : X \to Y$ is a connected, closed, monotone surjection. Then f is continuous.*

Proof: Let U be any open subset of Y. Then for each point $y \in U$ there exists an open set $V(y) \subset V$ containing y such that $Y \backslash V(y)$ consists of a finite number of components $C_1^y, C_1^y, \ldots, C_n^y$. For each $k = 1, 2, \ldots, n$, C_k^y is closed and connected so that $f^{-1}(C_k^y)$ is closed by

Theorem 8.7 and connected by Theorem 8.9 for each $k = 1, 2, \ldots, n$.
Therefore, $\bigcup_{k=1}^{n} f^{-1}(C_k^y)$ is a closed subset of X and does not contain a
point of $f^{-1}(y)$ so that $X \backslash \bigcup_{k=1}^{n} f(C_k^y) = T_y$ is an open set containing $f^{-1}(y)$
having the property that $T_y = f^{-1}(V(y))$. Repeating this process for
each $y \in U$ yields the open set $\bigcup_{y \in U} T_y = \bigcup_{y \in U} f^{-1}(V(y)) = f^{-1}(\bigcup_{y \in U} V(y))$
$= f^{-1}(U)$, from which the continuity of f follows.

8.14. THEOREM: *Let X be Hausdorff and Y a compact space. If
$f : X \rightarrow Y$ is a compact function, then f is continuous.*

Proof: The proof is an exercise.

Exercises:

1. Let X be an arbitrary space and Y a T_1-space. If $f : X \rightarrow Y$ is a
   connected function and in addition has the property that for each
   closed $B \subset Y, f^{-1}(B)$ has a finite number of components, then
   prove f is continuous.

2. Let X and Y be T_1-spaces and let $f : X \rightarrow Y$ be a connectivity function. Suppose $a \in X$ and U and V are open sets in X and Y containing a and $f(a)$, respectively. Prove that every nondegenerate
   connected subset of U contains a point $x \neq a$ such that $f(x) \in V$.

3. Prove Theorem 8.14.

4. A function $f : X \rightarrow Y$, is called *semiconnected* iff for each closed
   connected $B \subset Y, f^{-1}(B)$ is closed and connected. Prove that if
   $f : X \rightarrow Y$ is a semiconnected surjection and Y is semi-locally-connected, then f is continuous.

5. Let X be arbitrary and Y a T_1-space. If $f : X \rightarrow Y$ is a closed semiconnected function and $B \subset Y$ is connected, then $f^{-1}(B)$ is connected.

6. Let X be an arbitrary space, Y a T_1-space, and $f : X \rightarrow Y$ a connected function. Prove that for every connected subset $C \subset X$,
   $f(\bar{C}) \subset \overline{f(C)}$.

7. Let $f : X \rightarrow Y$ be a function having the property that for every
   closed subset $B \subset Y$ each component of $f^{-1}(B)$ is a closed subset
   of X. Prove that this condition on f is equivalent to the condition
   that for each connected $C \subset X, f(\bar{C}) \subset \overline{f(C)}$.

8. Generalize Theorem 8.11 by replacing connectivity function with connected function.

9. Let X be an arbitrary space and Y a semi-locally-connected T_1-space. If $f: X \to Y$ is bijective where both f and f^{-1} are connected functions, prove that f is continuous.

10. Let X and Y be arbitrary spaces. If $f: X \to Y$ is a bijective connected function where f^{-1} is a connectivity function, prove that f is a connectivity function.

9

Metric Spaces

1. Defining a Metric

Anyone studying the real line R^1, and the R^n spaces in general, is aware of the importance attached to the distance between two points in such spaces. As we shall see, the importance is enhanced by the fact that the standard topologies on these sets have a strong connection with the function defining the distance between two points. In this section we are going to generalize the idea of distance as known in these spaces to the idea of a *metric* for any set and then show how a metric for a set yields a topology for the set in a natural way.

We are all undoubtedly familiar with the standard method of measuring the distance between two points x and y in R^1 by the nonnegative number $|x - y|$. In R^2 the distance between $x = (x_1, x_2)$ and $y = (y_1, y_2)$ is normally given by $\sqrt{(x_1 - y_1)^2 + (x_2 - y_2)^2}$ and, in general, the distance between $x = (x_1, x_2, \ldots, x_n)$ and $y = (y_1, y_2, \ldots, y_n)$ is defined to be $\sqrt{\sum_{k=1}^{n} (x_k - y_k)^2}$. Notice that in each case the distance between a pair of points (they may be the same point) is always a nonnegative real number. Thus, we might view the distance taking process as that of forming a function from $R^n \times R^n$ into the nonnegative reals. That is, to each pair of points in R^n we assign in a particular fashion a real nonnegative number. Of course, to give a meaningful definition of distance which coincides with our knowledge of the R^n spaces, such a function must also obey certain other restrictions to make full use of the idea of a generalized distance, called a *metric*, on a set X. We begin our investigation of these restrictions with a function that is more general than a metric.

1.1. Definition: Let $X \neq \phi$ be a set. Then the function $d : X \times X \to R^1$ is called a *pseudo-metric* for X iff
    (a) For all $x \in X, d(x, x) = 0$.
    (b) For any three points (distinct or not) $x, y, z \in X, d(x, z) + d(y, z) \geq d(x, y)$. (This condition is known as the triangle inequality.)

Recall that, since d is a function into R^1, we may form the sum (as in (b)), difference, etc., of pseudo-metrics. The triangle inequality gets its name from the fact that, if $x, y, z \in R^2$ are vertices of a triangle and d represents the usual distance between two points, then the sum of the lengths of any two sides is greater than or equal to the length of the third side. With Definition 1.1 we can prove the following two important facts about pseudo-metrics.

1.2. THEOREM: *Let $d : X \times X \to R^1$ be a pseudo-metric for the set X. Then*
    (a) For any two points $x, y \in X, d(x, y) \geq 0$.
    (b) For any two points $x, y \in X, d(x, y) = d(y, x)$.

Proof: (a) Consider the three points $x, x, y \in X$. Then, following the pattern of Definition 1.1(b), we have $d(x, y) + d(x, y) \geq d(x, x)$ or that $2d(x, y) \geq 0$ since $d(x, x) = 0$, by Definition 1.1(a). Therefore, $d(x, y) \geq 0$.
    (b) Now, considering the three points $x, y, x \in X$ and again follow-

ing the pattern of Definition 1.1(b), we have $d(x, x) + d(y, x) \geq d(x, y)$ or that $d(y, x) \geq d(x, y)$. In a similar manner the three points $y, x, y \in X$ yield $d(y, y) + d(x, y) \geq d(y, x)$ or $d(x, y) \geq d(y, x)$. The two inequalities shown then give $d(x, y) = d(y, x)$.

This brings us to the definition of a metric, which is our primary concern.

1.3. Definition: Let $X \neq \phi$ be a set and $d : X \times X \to R^1$ a pseudo-metric for X. Then d is a *metric* for X iff for any two points $x, y \in X, d(x, y) = 0$ implies $x = y$.

Perhaps we should summarize the conditions for a metric at this point in the form of a theorem. In so doing, we should observe that these conditions could have been stated at the outset as the definition for a metric. However, the present route was chosen to point out the concept of a pseudo-metric, which has importance in mathematics.

1.4. THEOREM: *Let X be any nonempty set. A function $d : X \times X \to R^1$ is called a metric for X iff d satisfies the following properties:*
  (a) *For any two points $x, y \in X, d(x, y) = 0$ iff $x = y$.*
  (b) *(The triangle inequality.) For all $x, y, z \in X, d(x, y) + d(z, y) \geq d(x, z)$.*

Proof: Definitions 1.1 and 1.3.

Often the conditions $d(x, y) \geq 0$ and $d(x, y) = d(y, x)$ for all $x, y \in X$ are included in the defining conditions for a metric. However, Theorem 1.2 shows that these conditions always hold if d is a metric for X by our present definition. On the other hand, (a) and (b) alone of Theorem 1.4 imply d is a metric for X by Definition 1.3. Therefore, nothing is gained by including in the conditions for a metric the conditions $d(x, y) \geq 0$ for all $x, y \in X$, and the only positive result from including $d(x, y) = d(y, x)$ for all $x, y \in X$ is that the triangle inequality may be written $d(x, y) + d(y, z) \geq d(x, z)$ for all $x, y, z \in X$ at the outset. At any rate, once (a) and (b) of Theorem 1.4 are established for a function $d : X \times X \to R^1$, we may do this anyway if we so choose.

1.5. Example: (The discrete metric for X.) Let X be any nonempty set. For each $x, y \in X$, define $d(x, y) = 0$ if $x = y$ and $d(x, y) = 1$

if $x \neq y$. Condition (a) of Theorem 1.4 is easy to verify for the function d. To establish the triangle inequality, note that if $x = z$, then $d(x, z) = 0$ and hence $d(x, y) + d(z, y) \geq 0$ regardless of the choice of y. Otherwise, $d(x, z) = 1$, which implies $x \neq z$ and, hence, for any $y \in X$, either $y \neq x$ or $y \neq z$. Therefore, $d(x, y) = 1$ or $d(z, y) = 1$ so that $d(x, y) + d(z, y) \geq d(x, z)$. This establishes the triangle inequality and shows d is a metric for X. This metric will henceforth be called the *discrete metric* for X. Notice that for any point $p \in X$, the set of all $x \in X$ such that $d(p, x) < 1/2$ is the single point p. The set of all x such that $d(p, x) < 5$ is all of X.

We now show the usual distance definition in R^n is a metric for R^n. That is, if $x = (x_1, x_2, \ldots, x_n)$ and $y = (y_1, y_2, \ldots, y_n)$ belong to R^n, then $d(x, y) = \sqrt{\sum_{k=1}^{n} (x_k - y_k)^2}$ is a metric for R^n. Examination of Theorem 1.4 for this particular function d shows that the only difficulty that arises is in establishing the triangle inequality. In proving the triangle inequality, we make use of the famous *Schwarz inequality*. This inequality is given now.

1.6. THEOREM: (*The Schwarz inequality*) *Let* $x = (x_1, x_2, \ldots, x_n)$ *and* $y = (y_1, y_2, \ldots, y_n)$ *belong to* R^n. *Then*

$$\sum_{k=1}^{n} x_k y_k \leq \left(\sqrt{\sum_{k=1}^{n} x_k^2} \right) \left(\sqrt{\sum_{k=1}^{n} y_k^2} \right).$$

Proof: We shall actually prove $\left(\sum_{k=1}^{n} x_k y_k \right)^2 \leq \left(\sum_{k=1}^{n} x_k^2 \right) \left(\sum_{k=1}^{n} y_k^2 \right)$ from which the conclusion of the theorem follows. To do this, let λ be any real number and form the expression $\sum_{k=1}^{n} (x_k + \lambda y_k)^2$. Then

$$0 \leq \sum_{k=1}^{n} (x_k + \lambda y_k)^2 = \sum_{k=1}^{n} x_k^2 + 2\lambda \sum_{k=1}^{n} x_k y_k + \lambda^2 \sum_{k=1}^{n} y_k^2.$$

In view of the nonnegativeness of the two expressions involved, the quadratic equation $0 = \sum_{k=1}^{n} x_k^2 + 2\lambda \sum_{k=1}^{n} x_k y_k + \lambda^2 \sum_{k=1}^{n} y_k^2$ in λ can have at most one real root so that the discriminant of the quadratic is nonpositive and hence

$$4 \left(\sum_{k=1}^{n} x_k y_k \right)^2 - 4 \left(\sum_{k=1}^{n} x_k^2 \right) \left(\sum_{k=1}^{n} y_k^2 \right) \leq 0$$

or
$$\left(\sum_{k=1}^{n} x_k y_k\right)^2 \le \left(\sum_{k=1}^{n} x_k^2\right)\left(\sum_{k=1}^{n} y_k^2\right).$$

1.7. THEOREM: *Let* $x = (x_1 x_2, \ldots, x_n)$ *and* $y = (y_1, y_2, \ldots, y_n)$ *be any two points in* R^n *and define* $d(x, y) = \sqrt{\sum_{k=1}^{n} (x_k - y_k)^2}$. *Then* d *is a metric for* R^n.

Proof: The verification of part (a) of Theorem 1.4 for d is left to the reader. To establish the triangle inequality, let $x = (x_1, x_2, \ldots, x_n)$, $(y_1, y_2, \ldots, y_n)$, and $z = (z_1, z_2, \ldots, z_n)$ be any three points in R^n. Then we must show $\sqrt{\sum_{k=1}^{n} (x_k - y_k)^2} + \sqrt{\sum_{k=1}^{n} (z_k - y_k)^2} \ge \sqrt{\sum_{k=1}^{n} (x_k - z_k)^2}$. To simplify the notation, let $(x_k - y_k) = u_k$, $(y_k - z_k) = v_k$ so that $u_k + v_k = x_k - z_k$ for $k = 1, 2, \ldots, n$. The problem now is to show $\sqrt{\sum_{k=1}^{n} u_k^2} + \sqrt{\sum_{k=1}^{n} v_k^2} \ge \sqrt{\sum_{k=1}^{n} (u_k + v_k)^2}$. Both sides of the last inequality are nonnegative so that squaring both sides yields the equivalent inequality $\sum_{k=1}^{n} u_k^2 + 2\left(\sqrt{\sum_{k=1}^{n} u_k^2}\right)\left(\sqrt{\sum_{k=1}^{n} v_k^2}\right) + \sum_{k=1}^{n} v_k^2 \ge \sum_{k=1}^{n} (u_k + v_k)^2$
$= \sum_{k=1}^{n} u_k^2 + 2\sum_{k=1}^{n} u_k v_k + \sum_{k=1}^{n} v_k^2$ or $\left(\sqrt{\sum_{k=1}^{n} u_k^2}\right)\left(\sqrt{\sum_{k=1}^{n} v_k^2}\right) \ge \sum_{k=1}^{n} u_k v_k$.
This last inequality is known to be true by Theorem 1.6 and, since all of the steps involved above are reversible (why?), we see that the triangle inequality holds for the function d.

Henceforth, the metric described in Theorem 1.7 will be known as the *standard metric* for R^n. Of course, this metric reduces to the usual absolute value for the case $n = 1$. That is, $d(x, y) = \sqrt{(x - y)^2} = |x - y|$ for any two points x and y of R^1.

Undoubtedly, we have all shown the triangle inequality $|x| + |y| \ge |x + y|$ for the absolute value of real numbers holds using the definition of absolute value only. Knowing that $d(x, y) = |x - y|$ is a metric will also yield this results. To see this, select any two numbers u and v such that $x = u - v$ and in turn find a number w so that $y = v - w$. Then $|x| + |y| = |u - v| + |v - w| = d(u, v) + d(v, w) \ge d(u, w) = |u - w| = |x + y|$.

1.8. THEOREM: *Let* X *be a set and* d *a metric for* X. *For any two points* $x, y \in X$ *define* $e(x, y) = \text{minimum } \{1, d(x, y)\}$. *Then* e *is a metric for* X.

Proof: Part (a) of Theorem 1.4 is easily established and is left as an exercise. Thus, we shall concentrate on showing $e(x, y) + e(z, y) \geq e(x, z)$ for all $x, y, z \in X$. Note that $e(x, z) \leq 1$, so if $e(x, y) = 1$ or $e(z, y) = 1$, the desired inequality holds. If $e(x, y) \neq 1$ and $e(z, y) \neq 1$, then $e(x, y) = d(x, y)$ and $e(z, y) = d(z, y)$ by definition of e. Therefore, $e(x, y) + e(z, y) = d(x, y) + d(z, y) \geq d(x, z)$ since d is a metric for X. But $d(x, z) \geq$ minimum $\{1, d(x, z)\} = e(x, z)$. Thus, $e(x, y) + e(z, y) \geq e(x, z)$.

1.9. Example: Let d be the standard metric for R^1. As in Theorem 1.8, define the metric e as $e(x, y) =$ minimum $\{1, |x - y|\}$ for all $x, y \in R^1$. Let $x = 2$ and consider all points $y \in R^1$ such that $e(2, y) < 3$. In other words, we want all $y \in R^1$ such that minimum $\{1, |2-y|\} < 3$. This inequality is satisfied by all $y \in R^1$. Contrast this with finding the set of all $y \in R^1$ satisfying $d(2, y) = |2 - y| < 3$. In the latter case, $-1 < y < 5$ must hold. Compare the sets $\{y \in R^1 : d(2, y) < 1/2\}$ and $\{y \in R^1 : e(2, y) < 1/2\}$.

Exercises:

1. Given an example of a pseudo-metric for a set X that is not a metric for X.

2. Prove that (a) of Theorem 1.4 holds for the metrics defined in Theorems 1.7 and 1.8.

3. For any two real numbers x and y, define $d_1(x, y) = x^2 - y^2$ and $d_2(x, y) = |x^2 - y^2|$. Show that neither d_1 nor d_2 are metrics for R^1.

4. Give an example of two metrics for R^1 different from those already discussed.

5. Let X be the set of all real valued continuous functions defined on the interval $\{x \in R^1 : 0 \leq x \leq 1\}$. If $f(x)$ and $g(x)$ belong to X, define $d(f(x), g(x)) = \int_0^1 |f(x) - g(x)| \, dx$. Prove that d is a metric for X.

6. (a) Let d be the standard metric for R^1 and describe $\{x \in R^1 : d(x, 1) < 3\}$.
   (b) Let d_1 be the discrete metric for R^1 and describe $\{x \in R^1 : d_1(x, 1) < 3\}$.

7. Let R^2 be the plane and define $d((x, y), (z, w)) = |x - z| + |y - w|$ for each pair of points (x, y) and (z, w) in R^2. Prove that d is a metric for R^2.

8. For each pair of points (x, y) and (z, w) in R^2, define $d((x, y), (z, w))$ = maximum $\{|x - z|, |y - w|\}$. Prove that d is a metric for R^2.

9. (a) For the standard metric for R^2, describe graphically the set $\{(x, y) : d((x, y), (0, 0)) < 2\}$.
   (b) For the metric defined in Exercise 7, describe graphically the set $\{(x, y) : d((x, y), (0, 0)) < 2\}$.
   (c) For the metric defined in Exercise 8, describe graphically the set $\{(x, y) : d((x, y), (0, 0)) < 2\}$.

10. Let d be a pseudo-metric for X and $A \subset X$. Prove that the restriction of d to $A \times A, (d|A \times A) : (A \times A) \to R^1$ is a pseudo-metric for the set A.

2. Metric Topologies

If d is a metric for X, there is a natural way to define a topology on X by use of the function d. To this end we make the following definition.

2.1. Definition: Let d be a metric for X and $r > 0$ a real number. Then for any point $p \in X$ the set $S_r^d(p) = \{x \in X : d(p, x) < r\}$ is called an *open r-sphere* (or open r-ball), center at p of radius r with respect to the metric d. When there is no confusion about different metrics, we sometimes omit the superscript d in the open r-sphere notation.

The word "sphere" in Definition 2.1 is a natural one if we use the standard metric in R^3. There, an open r-sphere with center at p consists of the points in R^3 which are "inside" a sphere of radius r and center p, as normally thought of in solid geometry. In R^2 with standard metric, an open r-sphere consists of the points in the plane which are inside a circle of radius r and center at p, while in R^1 with standard metric, an open r-sphere is an open interval of length $2r$ and center at p. For other sets and metrics the word "sphere" may have no such natural geometric meaning but may merely denote a certain subset of the set upon which the metric is defined. Figure 26 shows open 1-spheres for two metrics defined for R^2.

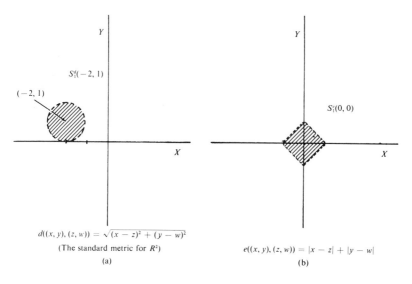

$$d((x, y), (z, w)) = \sqrt{(x - z)^2 + (y - w)^2}$$

(The standard metric for R^2)

(a)

$$e((x, y), (z, w)) = |x - z| + |y - w|$$

(b)

Figure 26

The next theorem shows that the set of all open r-spheres in a set X with metric d may be used as a base for a topology on X. Thus, each set X upon which is defined a metric may be thought of as a topological space.

2.2. THEOREM: *Let X be a nonempty set and d a metric for X. The set of all open r-spheres in X is a base for a topology on X.*

Proof: Let $\mathscr{B}$ be the set of all open r-spheres in X. We now establish conditions (a) and (b) of Theorem 1.7 of Chapter 4. For each $x \in X$ and any $r > 0$, $S_r(x) \in \mathscr{B}$ so that (a) is satisfied. To see (b), let $S_r(x)$ and $S_q(y)$ belong to $\mathscr{B}$ and let $z \in S_r(x) \cap S_q(y)$. Now let $t = \text{mini-}$ mum $\{r - d(x, z), q - d(y, z)\}$ and consider any point $w \in S_t(z)$. It follows immediately that $d(w, x) \leq d(w, z) + d(z, x) < t + d(z, x) < r$. Therefore, $d(w, x) < r$ and $w \in S_r(x)$ so that $S_t(z) \subset S_r(x)$. Similarly, $S_t(z) \subset S_q(y)$ so that $S_t(z) \subset S_r(x) \cap S_q(y)$ and condition (b) is established. Therefore, by Theorem 1.7 of Chapter 4, $\mathscr{B}$ is a base for the topology $\mathscr{T}(\mathscr{B})$ on X, and this topology is unique by Theorem 1.8 of Chapter 4.

The results of Theorem 2.2 now allow us to define a metric space.

2.3. Definition: Let X be a nonempty set X and d a metric for X. The unique topology on X generated by the set of all open r-spheres in X and having these open r-spheres as a base is called the *d-metric topology* for X. The d-metric topology for X will henceforth be denoted by $\mathscr{T}(d)$. The topological space $(X, \mathscr{T})$ is called a *metric space* or *metrizable* iff there exists a metric d for X such that the d-metric topology $\mathscr{T}(d)$ on X is the same as $\mathscr{T}$. The notation (X, d) will be used to denote a metric space.

2.4. Example: Let d be the discrete metric (Example 1.5) on X. The d-metric topology is the discrete topology since open r-spheres are either single points or all of X.

2.5. Example: If d is the standard metric for R^1, then the d-metric topology is the standard topology on R^1. This is because open r-spheres are nonempty open intervals and hence form a base for the topology on R^1.

2.6. Example: Let $X = \{0, 1\}$ with topology $\mathscr{T} = \{\phi, X, \{0\}\}$. Then $(X, \mathscr{T})$ is not metrizable. The reason is that any metric d for X has the property that $d(0, 1) = r > 0$ and hence $S_{r/2}(0) = \{0\}$ and $S_{r/2}(1) = \{1\}$ are open sets which belong to the d-metric topology on X. That is, the d-metric topology for X is always discrete and unequal to $\mathscr{T}$.

We will now discuss the closure of open r-spheres in metric spaces.

2.7. THEOREM: *Let (X, d) be a metric space and let $r > 0$. Then, for each $p \in X$, the set $A = \{x \in X : d(x, p) \leq r\}$ is a closed subset of X.*

Proof: We shall show $X \backslash A$ is open in X. To this end, let $x \in X \backslash A$. Then $x \notin A$ so that $d(x, p) = s > r$. Now let $s - r = t$ and consider $S_t(x)$. If $y \in S_t(x) \cap A$, then $d(y, x) < t$ and $d(y, p) \leq r$ so that $d(x, p) \leq d(x, y) + d(y, p) < t + r = s$. This contradiction to $d(x, p) = s$ means $S_t(x) \cap A = \phi$. Consequently, $X \backslash A$ is open and A is closed in X.

If X is non-degenerate and has the discrete metric, it is easily seen that $\bar{S}_r(p)$ is not necessarily equal to $A = \{x \in X : d(x, p) \leq r\}$. We can discuss this point when considering Exercise 6 at the end of this section.

To help shed more light on the metric topology, the distance between two nonempty subsets of a metric space is introduced.

2.8. Definition: Let $A \neq \phi$ and $B \neq \phi$ be two subsets of the metric space (X, d). Then the *distance between A and B*, denoted by $d(A, B)$, is the greatest lower bound of the set $\{d(x, y) : x \in A, y \in B\}$. If $A = \{a\}$, we write $d(a, B)$ for $d(A, B)$.

2.9. Example: Consider the metric space R^2 Let $A = \{(x, y): 0 \leq x \leq 1, 0 < y < 1\}$, $B = \{(x, y) : 2 < x < 3, 0 < y < 1\}$, $C = \{0, 0)\}$, and $D = \{(1/2, 1/2)\}$. Then $d(A, B) = 1, d(A, C) = 0, d(C, B) = 2, d(D, A) = 0$, and $d(D, B) = 3/2$.

2.10. THEOREM: *Let (X, d) be a metric space and $A \neq \phi$ a subset of X. Then $x \in \bar{A}$ iff $d(x, A) = 0$.*

Proof: Suppose $x \in \bar{A}$. Then, for each $r > 0$, $S_r(x) \cap A \neq \phi$. Therefore, for each $r > 0$ there exists a point $a_r \in A$ such that $d(x, a_r) < r$ and, as a consequence, the greatest lower bound of $\{d(x, a) : a \in A\}$ is zero. Our conclusion is that $d(x, A) = 0$.

For the converse, assume that $x \in X$ and $d(x, A) = 0$. It must be shown that $x \in \bar{A}$. Since $d(x, A) = 0$, for each $r > 0$, there exists a point $a_r \in A$ such that $d(x, a_r) < r$. It follows that for each $r > 0$, $S_r(x) \cap A \neq \phi$ showing $x \in \bar{A}$.

As a consequence of Theorem 2.10, if a subset A of a metric space (X, d) is closed and $x \notin A$, then $d(x, A) > 0$. This fact will be helpful in showing that a metric space is also a T_4-space and thus satisfies all of our separation axioms.

2.11. THEOREM: *Every metric space is a T_4-space.*

Proof: We first show (X, d) is a Hausdorff space and hence is a T_1-space. Let $x \neq y$ be two points of X. Then $d(x, y) = r > 0$ so that $S_{r/2}(x) \cap S_{r/2}(y) = \phi$, implying (X, d) is Hausdorff.

To show X is normal, let A and B be two closed disjoint subsets of X and define

$$U = \{x \in X : d(x, A) < d(x, B)\}$$
$$V = \{x \in X : d(x, A) > d(x, B)\}$$

From the definition of U and V, $U \cap V = \phi$. Furthermore, if $a \in A$, then $d(a, A) = 0$, $a \notin B$ and, since B is closed, $d(a, B)$ is a positive real number according to Theorem 2.10. Therefore, $0 = d(a, A) < d(a, B)$ so that $a \in U$. It follows that $A \subset U$. Similarly, $B \subset V$. Now, if we can show U and V are open, X will be normal.

To show U is open, let $x_0 \in U$ and use the notation $q = d(x_0, A) < d(x_0, B) = r$ where r is a positive real number. Since $r - q$ is a positive real number, we may define $p = 1/2(r - q)$ and consider the basic open set $S_{p/2}(x_0)$. If $x \in S_{p/2}(x_0)$ and we can show $d(x, A) < d(x, B)$, then $S_{p/2}(x_0) \subset U$, showing that U is open. To this end, let $\epsilon > 0$ be a real number. By the definition of $d(x_0, A)$, there exists an $a \in A$ such that $d(x_0, a) < q + \epsilon$. Therefore, $d(x, A) \leq d(x, a) \leq d(x, x_0) + d(x_0, a) < p/2 + (q + \epsilon) = (p/2 + q) + \epsilon$, from which it follows that $d(x, A) \leq p/2 + q = 1/4(r + 3q)$. Now, for every point $b \in B$, we have $d(b, x) + d(x, x_0) \geq d(b, x_0)$ and, since $d(b, x_0) \geq d(x_0, B)$ and $d(x, x_0) < p/2$, we may write $d(b, x) + p/2 > d(x_0, B) = r$. Thus, $d(b, x) > r - p/2 = 1/4(3r + q)$ and, as a consequence, $d(x, B) \geq 1/4(3r + q)$. Since $q < r$, $2q < 2r$ and thus $3q + r < 3r + q$, which means $d(x, A) \leq 1/4(r + 3q) < 1/4(3r + q) \leq d(x, B)$. This completes the proof that U is open.

The same reasoning shows V is also an open subset of X. Therefore, we have found open disjoint sets U and V of X containing A and B, respectively, which shows (X, d) is normal.

2.2. THEOREM: *Every metric space (X, d) is first countable.*

Proof: We must show that there is a countable local base at each point $p \in X$. The collection $\{S_{1/n}(p) : n \in N, p \in X\}$, together with the definition of the d-metric topology, give the desired results. The details are left to the reader.

Examples easily show that a metric space need not be second countable.

We make a final note for this section of a fact that is probably apparent to the reader. If A is a nonempty subset of the metric space (X, d) with metric $d : X \times X \to R^1$, then A is itself a metric space and inherits its metric from X. A check of Theorem 1.4 quickly verifies that the restriction of d to $A \times A$, $(d|A \times A) : (A \times A) \to R^1$ is indeed a metric for A. Thus, the inherited subspace topology on A is the same as that obtained by restricting d to $A \times A$ and then forming the metric topology on A. To see this, we need only show that the basic open sets in the inherited subspace topology on A are the same as those obtained by restricting d to $A \times A$.

2.13. THEOREM: *Let (X, d) be a metric space and A a nonempty X. Then the subspace topology on A is the same as the $(d|(A \times A))$- metric topology on A.*

Proof: The details should be verified by the reader.

Exercises:

1. Let $X \neq \phi$ be any finite set and d a metric for X. What is the d- metric topology for X?

2. Complete the details of the proof to Theorem 2.12.

3. Prove Theorem 2.13.

4. Let (X, d) be a metric space and $x \in X$. If r and q are real numbers such that $0 < r < q$, prove $S_r(x) \subset \bar{S}_r(x) \subset S_q(x)$.

5. Give an example of

   (a) A disconnected metric space.
   (b) A compact metric space.
   (c) A metric space that is not second countable.

6. In a metric space (X, d), prove that if $y \in \bar{S}_r(x)$, then $d(x, y) \leq r$ and give an example to show the converse is not always true.

7. Let (X, d) be a metric space, p a point in X and $r > 0$ a real number. Prove that $\{x \in X : d(x, p) \geq r\}$ is a closed subset of X.

8. Give an example to show that, for an open r-sphere $S_r(p)$ in a metric space (X, d), it is not necessarily true that $\mathrm{Bd}(S_r(p)) = \{x \in X : d(x, p) = r\}$.

9. Prove that R^2 with the standard metric is second countable.

10. (a) Let (X, d) be a metric space and A and B nonempty subsets of X. Prove that if $d(A, B) \neq 0$, then $A \cap B = \phi$.
    (b) Give an example of two closed, disjoint, nonempty subsets A and B of a metric space where $A \cap B = \phi$ and $d(A, B) = 0$.

11. Let (X, d) be a metric space and A a nonempty subset of X. If $x, y \in X$, prove that $d(x, A) \leq d(x, y) + d(y, A)$.

3. Equivalent Metric Topologies

Example 2.5 has already pointed out that the topology generated by the standard metric for R^1 is the same as the standard topology for R^1. We now show that the topology generated by the standard metric for R^2 is the standard topology for R^2 and then generalize this result to the R^n spaces. We know the standard topology on R^2 has open rectangles for a base and the standard-metric topology on R^2 has a base consisting of open r-spheres which are the interiors of circles. Thus, to use Theorem 1.11 of Chapter 4, the following result must be established.

3.1. THEOREM: *Let R^1 and R^2 each have their standard metrics. (a) Let $p = (x_0, y_0) \in R^2$ and $S_r(q)$ an open r-sphere in R^2 with center at q such that $p \in S_r(q)$. Then there exists an open rectangle containing p and lying entirely within $S_r(q)$. (b) Let $U \times V$ be an open rectangle in R^2 containing $p = (x_0, y_0)$. That is, $x_0 \in U$ and $y_0 \in V$ where U and V are both open intervals in R^1. Then there exists a real number $r > 0$ such that $S_r(p) \subset U \times V$.*

Proof: (a) Let $s = r - d(p, q)$. Then $S_s(p) \subset S_r(q)$ and $(x_0 - s/\sqrt{2}, x_0 + s/\sqrt{2}) \times (y_0 - s/\sqrt{2}, y_0 + s/\sqrt{2}) \subset S_s(p)$. See Figure 26(a).

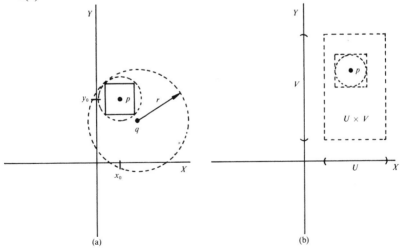

(a)   (b)

Figure 27

(b) First, find an open square having p as a center and lying inside $U \times V$. Then choose r so that $S_r(p) \subset U \times V$. See Figure 27(b). The details are left as an exercise.

3.2. THEOREM: *The standard-metric topology on R^2 is the same as the standard topology on R^2.*

Proof: Let $\mathscr{B}_1$ be the set of all open rectangles in R^2 and $\mathscr{B}_2$ the set of all open r-spheres in R^2. Then $\mathscr{T}(\mathscr{B}_1)$, the topology generated by $\mathscr{B}_1$ and having $\mathscr{B}_1$ as a base, is the standard topology on R^2. From the results of Theorem 3.1, $\mathscr{B}_1$ and $\mathscr{B}_2$ are equivalent bases according to Theorem 1.11 of Chapter 4. Therefore, $\mathscr{T}(\mathscr{B}_1) = \mathscr{T}(\mathscr{B}_2)$ by Theorem 1.12 of Chapter 4. Since $\mathscr{T}(\mathscr{B}_1)$ is the standard metric-topology on R^2, our theorem is proved.

3.3. THEOREM: *Let d be the standard metric for R^n. Then the standard topology on R^n is the same as the d-metric topology on R^n.*

Proof: First, note that a base for the standard topology on R^n is the collection of sets of the form $U_1 \times U_2 \times \cdots \times U_n$ where U_k, $k = 1, 2, \ldots, n$, may be taken as an open interval in R^1. If $p = (x_1, x_2, \ldots, x_n)$ and $S_r(q)$ is an open r-sphere containing the point p and we let $s = r - d(p, q) > 0$, then we have $S_s(p) \subset S_r(q)$. To see this, observe that for any $x \in S_s(p)$, it follows that $d(x, q) \leq d(x, p) + d(p, q) < s + r - s = r$ so that $x \in S_r(q)$. Now consider the product $U = \prod\limits_{k=1}^{n} (x_k - s/\sqrt{n}, x_k + s/\sqrt{n})$ of open intervals in R^1. We shall show $U \subset S_s(p)$. If $y = (y_1, y_2, \ldots, y_n) \in U$, then $d(y, p) = \sqrt{\sum\limits_{k=1}^{n} (y_k - x_k)^2}$ and, since $x_k - s/\sqrt{n} < y_k < x_k + s/\sqrt{n}$ or $-s/\sqrt{n} < y_k - x_k < s/\sqrt{n}$, it follows that $(y_k - x_k)^2 < s^2/n$ for $k = 1, 2, \ldots, n$. Thus, $d(y, p) < \sqrt{(ns^2)/n} = s$ so that $y \in S_s(p)$.

Now let $U_1 \times U_2 \times \ldots \times U_n$ be a basic open set in R^n containing the point p where $U_k = (a_k, b_k)$ is an open interval in R^1 for each $k = 1, 2, \ldots, n$. Define t_1 to be the smallest number in the set $\{x_k - a_k : k = 1, 2, \ldots, n\}$ and t_2 to be the smallest number in the set $\{b_k - x_k : k = 1, 2, \ldots, n\}$. Now, if $t = \text{minimum } \{t_1, t_2\}$, then $S_t(p) \subset U_1 \times U_2 \times \ldots \times U_n$. To see this, let $y = (y_1, y_2, \ldots, y_n) \in S_t(p)$ and note that $d(x, p) < t \leq t_i, i = 1, 2$. Thus, $a_k < y_k < b_k$ for $k = 1, 2, \ldots, n$ and therefore $y \in U_1 \times U_2 \times \ldots \times U_n$.

If we now apply Theorems 1.11 and 1.12 of Chapter 4 to what we

have shown thus far, it follows that the two topologies are the same.

For a given metric space (X, d), there may be other metrics for X which will yield the same metric topology as d does. Consequently, the topology of a metric space is not uniquely determined by the metric. The remainder of this section investigates when two metrics for a set generate the same metric topology for that set.

3.4. Definition: Let d and e be two metrics for the set X. Then d and e are *equivalent* iff $\mathscr{T}(d) = \mathscr{T}(e)$.

After a preliminary theorem we will give a characterization of when two metrics are equivalent.

3.5. THEOREM: Let (X, d) be a metric space, $p \in X$ and $r > 0$ a real number. If $a \in S_r(p)$, there exists a real number $s > 0$ such that $S_s(a) \subset S_r(p)$.

Proof: Let $s = r - d(p, a)$. Then if $x \in S_s(a)$, $d(x, a) < s$ and $d(x, p) \leq d(x, a) + d(a, p) < s + d(a, p) = r$. Therefore $x \in S_r(p)$ and $S_s(a) \subset S_r(p)$.

3.6. THEOREM: Let d and e be two metrics for the set X. Then d and e are equivalent iff for each point $p \in X$ and real number $r > 0$, there exist two positive real numbers s and t such that
   (a) $S_s^d(p) \subset S_r^e(p)$ and
   (b) $S_t^e(p) \subset S_r^d(p)$ both hold.

Proof: Use Theorem 3.5 along with Theorems 1.11 and 1.12 of Chapter 4. The details are a straightforward exercise.

3.7. Definition: A subset $A \subset X$ of a metric space (X, d) is *bounded* with respect to d iff $A = \phi$, or there exists a positive real number M such that $d(x, y) \leq M$ for every choice of $x, y \in A$. A metric d for the set X is a *bounded metric* iff X is bounded with respect to the metric d.

The discrete metric is a bounded metric for R^1. The standard metric for R^1 is not a bounded metric.

3.8. THEOREM: Every metric has an equivalent bounded metric. That is, if (X, d) is a metric space, there exists a bounded metric e

for X such that the d-metric topology is the same as the e-metric topology for X.

Proof: Let (X, d) be a metric space and consider the metric e defined in Theorem 1.8 by $e(x, y) = $ minimum $\{1, d(x, y)\}$ for all $x, y \in X$. Certainly e is a bounded metric and we shall show that d and e are equivalent by showing that the two conditions of Theorem 3.6 hold.

Let $r > 0$ be a real number which is fixed throughout our discussion. Let $s = r$ and $x \in S_s^d(p)$. Then $d(x, p) < s$ and by definition of the e-metric, $e(x, p) \leq d(x, p)$. Therefore, $e(x, p) < s = r$ and $x \in S_r^e(p)$. Hence $S_s^d(p) \subset S_r^e(p)$.

Now let $t = $ minimum $\{1, r\}$ and suppose $x \in S_t^e(p)$. We consider the two cases $r < 1$ and $r \geq 1$. First, if $r < 1$, then $t = r$ and, therefore, $e(x, p) < t$ implies $e(x, p) = d(x, p)$. Hence $d(x, p) < t = r$ so that $x \in S_r^d(p)$. Next, if $r \geq 1$, then $t = 1$. For $x \in S_t^e(p)$, minimum $\{1, d(x, p)\} = e(x, p) < t = 1$ implies again that $e(x, p) = d(x, p)$ so that $d(x, p) < t = 1 < r$. Thus $x \in S_r^d(p)$. Therefore, both cases lead to the fact that $S_t^e(p) \subset S_r^d(p)$. We now have (a) and (b) of Theorem 3.6 satisfied, which means d and e are equivalent metrics for X.

3.9. Corollary: *Every metric space is homeomorphic to a bounded metric space.*

Proof: The proof is a consequence of Theorem 3.8.

3.10. THEOREM: *If (X, d) is a compact metric space, then d is a bounded metric.*

Proof: The proof is an application of definitions after considering any point $p \in X$ and the open cover $\{S_n(p) : n \in N\}$ of open r-spheres for X. The details are left as an exercise.

Exercises:

1. Give an example of a metric d for R^1 such that d is not equivalent to the standard metric for R^1.

2. Complete the details of the proof to Theorem 3.1.

3. Prove Theorem 3.6.

4. Prove Theorem 3.9.

5. Prove Theorem 3.10.

6. Let (X,d) be a metric space and define $e(x,y)=d(x,y)/(1+d(x,y))$. Prove that e is a bounded metric for X. Then prove that d and e are equivalent metrics.

7. Give an example to show that the converse of Theorem 3.10 is not true.

8. Prove the metric $e((x, y), (z, w)) = |x - z| + |y - w|$ defined for R^2 (see Exercise 7, Section 1) is equivalent to the standard metric for R^2.

9. Is the metric defined for R^2 in Exercise 8 of Section 1 equivalent to the standard metric for R^2?

10. If d is a pseudo-metric for X, then d defines a topology for X in exactly the same manner as Definition 2.3 for a metric. Similarly two pseudo-metrics d and e for X are equivalent if the pseudo-metric topologies are equal.
    (a) If d and e are equivalent pseudo-metric topologies for X and d is also a metric for X, prove e is a metric for X.
    (b) Prove that a pseudo-metric space is a T_1-space iff it is a metric space.

4. Continuity of Functions Between Metric Spaces

We have noted before that if $f: R^1 \rightarrow R^1$, a standard way of defining continuity of f at $a \in R^1$ is as follows: For each real number $\epsilon > 0$ there exists a real number $\delta > 0$ (depending upon the choice of ϵ and the point a) such that if $|x - a| < \delta$, then $|f(x) - f(a)| < \epsilon$. Since the absolute value defines the standard metric on R^1, we see that $|x - a| < \delta$ and $|f(x) - f(a)| < \epsilon$ define open spheres in R^1 with centers at a and $f(a)$ having radii δ and ϵ, respectively. Therefore, f is continuous at the point a iff for each real number $\epsilon > 0$ there exists an open δ-sphere with center at a such that the image of this sphere under f is a subset of an open ϵ-sphere with center at $f(a)$. We now generalize this notion to arbitrary metric spaces. This is easily done since each metric space (X, d) has a d-metric topology and we have studied continuity for general topological spaces.

4.1. THEOREM: *Let (X, d) and (Y, e) be metric spaces. Then $f : X \rightarrow Y$ is continuous at $a \in X$ iff for each real number $\epsilon > 0$ there* *exists a real number* $\delta > 0$ *(depending upon the choice of* ϵ *and the point a) such that* $f(S_\delta^d(a)) \subset S_\epsilon^e(f(a))$.

Proof: It is easy to prove the theorem by use of the definition of continuity. The details are left as an exercise.

There has been a great deal of research about when an arbitrary topological space $(X, \mathscr{T})$ is metrizable. That is, when is there a metric d for X whose d-metric topology is the same as $\mathscr{T}$? We know any finite space X with nondiscrete topology is not metrizable since any metric d on X has a discrete d-metric topology. A few moments reflection on when a space is metrizable will probably convince us that the problem is not a trivial one. Although we shall not go into details, there are several different conditions on $(X, \mathscr{T})$ which make the space metrizable. For example, one such condition is that all second countable, regular, Hausdorff spaces are metrizable. The important fact that metrizability is a topological property is given next.

4.2. THEOREM: *Metrizability is a topological property.*

Proof: Let (X, d) be a metric space which is homeomorphic to the space $(Y, \mathscr{T})$ and let $h: X \to Y$ be a homeomorphism. It is to be shown that $(Y, \mathscr{T})$ is metrizable. We first define $e(y, z) = d(h^{-1}(y),$ $h^{-1}(z))$ for each $y, z \in Y$ and show that e is a metric for Y. By definition of e, $e(y, z) = 0$ iff $d(h^{-1}(y), h^{-1}(z)) = 0$, which is true iff $h^{-1}(y) = h^{-1}(z)$ and this, in turn, is true iff $y = z$, since h is a homeomorphism. For the triangle inequality, $e(y, z) + d(w, z) = d(h^{-1}(y),$ $h^{-1}(z)) + d(h^{-1}(w), h^{-1}(z)) \geq d(h^{-1}(y), h^{-1}(w)) = e(y, w)$. Thus, e is a metric for Y by Theorem 1.4.

Next, we shall show that h is a homeomorphism from (X, d) onto (Y, e) by use of Theorem 4.1. First, the function h is a bijection from X onto Y. To show h is continuous at $a \in X$, let $\epsilon > 0$ and $x \in S_\delta^d(a)$ where $\delta = \epsilon$. Then $e(h(x), h(a)) = d(x, a) < \delta = \epsilon$ so that $h(x)$ $\in S_\epsilon^e(h(a))$. Hence $h(S_\delta^d(a)) \subset S_\epsilon^e(h(a))$ so that, by Theorem 4.1, h is continuous at $a \in X$. A similar argument shows h^{-1} continuous at each point of Y. Thus, h is a homeomorphism from (X, d) onto (Y, e).

Since (X, d) and $(Y, \mathscr{T})$ are homeomorphic by hypothesis and we have just shown (X, d) and (Y, e) are homeomorphic, $(Y, \mathscr{T})$ and (Y, e) are homeomorphic by the symmetric and transitive properties of the homeomorphism relation. Therefore, the e-metric topology is precisely $\mathscr{T}$ so that $(Y, \mathscr{T})$ is metrizable by Definition 2.3.

We will now show that the distance function itself is continuous.

4.3. THEOREM: *Let (X, d) be a metric space and $X \times X$ have the product topology. Then $d : X \times X \to R^1$ is continuous.*

Proof: Let $(a, b) \in X \times X$ and $\epsilon > 0$. New let (x, y) be any point in the basic open subset $S_{\epsilon/2}(a) \times S_{\epsilon/2}(b)$ of $X \times X$. By two uses of the triangle inequality, we have $d(x, y) \leq d(x, a) + d(a, b) + d(b, y) < \epsilon/2 + d(a, b) + \epsilon/2 = d(a, b) + \epsilon$. Similarly, $d(a, b) \leq d(a, x) + d(x, y) + d(y, b) < \epsilon/2 + d(x, y) + \epsilon/2$ or that $d(a, b) - \epsilon < d(x, y)$. Thus, for any $(x, y) \in S_{\epsilon/2}(a) \times S_{\epsilon/2}(b)$, it follows that $d(a, b) - \epsilon < d(x, y) < d(a, b) + \epsilon$ or that $|d(x, y) - d(a, b)| < \epsilon$. Consequently, d is continuous at (a, b) and therefore at each point of $X \times X$.

If (X, d) is a metric space and A is a nonempty subset of X, we may define the function $f : X \to R^1$ given by $f(x) = d(x, A)$. This function is also continuous.

4.4. THEOREM: *Let A be any nonempty subset of the metric space (X, d). Then the function $f : X \to R^1$ given by $f(x) = d(x, A)$ is continuous.*

Proof: Our proof uses the fact that f is continuous iff for each subbasic open set W in R^1, $f^{-1}(W)$ is open in X. Since W is either an open left ray or an open right ray, we must show for each $r \in R^1$ the sets $B = \{x \in X : f(x) > r\}$ and $C = \{x \in X : f(x) < r\}$ are open in the space (X, d). The proof that B and C are open is essentially the same as the proof of (X, d) being normal in Theorem 2.11. Hopefully, the repetition in the present setting will strengthen the understanding of both Theorem 2.11 and the present theorem.

To show that $B \subset X$ is open, we first let $b \in B$ and observe that $d(b, A) = f(b) = r_1 > r$. Then we take $p = (r_1 - r)/2$ and consider the basic open set $S_p(b)$ of (X, d). Now if $x \in S_p(b)$ and $a \in A$, then $d(x, a) + d(x, b) \geq d(a, b), d(b, a) \geq d(b, A)$ and $d(b, x) < p$ all hold, which implies that $d(x, a) + p > d(b, A)$ or $d(x, a) > d(b, A) - p = r_1 - p$. The validity of this set of inequalities for all $x \in S_p(b)$ means we have $f(x) = d(x, A) \geq r_1 - p = r_1 - (r_1 - r)/2 = r_1/2 + r/2 > r$ because $r_1 > r$. Therefore, $x \in S_p(b) \subset B$, that showing B is open in (X, d).

To show C open, let $c \in C$ and note that $d(c, A) = f(c) = r_2 < r$. Now let $p = (r - r_2)/2$. From the definition of $d(c, A)$ and the fact that $d(c, A) = r_2 < r$, it follows that there is at least one point $a \in A$ such that $d(c, a) < (r_2 + r)/2 = r_2 + p$. If we now consider the basic open set $S_p(c)$, then for each $x \in S_p(c)$ we have $f(x) = d(x, A) \leq d(x, a) \leq d(x, c) + d(a, c) < p + r_2 + p = r$ or that $x \in C$. This means $x \in S_p(c) \subset C$ and C is open.

In the discussion that follows, suppose (X, d) and (Y, e) are metric spaces, $f: X \to Y$ is continuous, and $\epsilon > 0$ is a fixed real number. For $a \in X$, we know that there is a $\delta > 0$ such that $f(S_\delta^d(a)) \subset S_\epsilon^e(a))$. If we now choose a point $b \neq a$ in X, there is a $\delta_1 > 0$ such that $f(S_{\delta_1}^p(b)) \subset S_\epsilon^e(f(b))$, where δ_1 in all likelihood will be of neccessity different from δ. In other words, if we fix ϵ first and consider different points of X, we will usually have to pick different values of δ to satisfy the continuity definition. Therefore, δ depends not only upon ϵ but also upon the point of X in question. There are functions, however, in which the choice of δ is independent of the point under consideration. In such cases, the function is called *uniformly countinuous*. A formal definition is given next.

4.5. Definition: Let (X, d) and (Y, e) be metric spaces and $f: X \to Y$. Then f is *uniformly continuous* on X iff for each real number $\epsilon > 0$ there exists a real number $\delta > 0$ (depending upon ϵ only) such that for every choice of $x, w \in X$, where $d(x, w) < \delta$, then $e(f(x), f(w)) < \epsilon$.

4.6. Example: Let R^1 have the standard metric and suppose $f: R^1 \to R^1$ is given by $f(x) = x$ for all $x \in R^1$. For each $\epsilon > 0$ choose $\delta = \epsilon$ and note that, if $d(x, w) = |x - w| > \delta$, then $d(f(x), f(w)) = d(x, w) < \epsilon$. Thus, f is uniformly continuous on R^1.

4.7. THEOREM: *Let (X, d) and (Y, e) be metric spaces. If $f: X \to Y$ is uniformly continuous on X, then f is continuous.*

Proof: The proof is an easy application of Definitions 4.1 and 4.5. The details are an exercise.

An extremely useful characterization of a function not being uniformly continuous is proved at this point. It will be helpful in showing that the converse of Theorem 4.7 is false.

4.8. THEOREM: *Let (X, d) and (Y, e) be metric spaces. Then the function $f: X \to Y$ is not uniformly continuous on X iff there exists a*

positive real number ϵ_0 and two sequences $g : N \to X$ and $h : N \to X$ in
X such that for each $n \in N$, $d(g(n), h(n)) < 1/n$ and $d(f(g(n)), f(h(n)))$
$\geq \epsilon_0$.

Proof: First, suppose that f is not uniformly continuous on X. Then
by Definition 4.5, there exists an $\epsilon_0 > 0$ such that for any choice
whatever of $\delta > 0$, there are at least two points $x, w \in X$ such that
$d(x, w) < \delta$ and $e(f(x), f(w)) \geq \epsilon_0$. Thus, for each $n \in N$, let $\delta_n = 1/n$
and find two points $x_n, w_n \in X$ such that $d(x_n, w_n) < 1/n$ and $e(f(x_n),$
$f(w_n)) \geq \epsilon_0$. The sequences $g : N \to X$ and $h : N \to X$ defined by $g(n)$
$= x_n$ and $h(n) = w_n$, respectively, satisfy the required conditions.

Conversely, suppose there exists an $\epsilon_0 > 0$ and two sequences g and
h in X such that for each $n \in N$, $d(g(n), h(n)) < 1/n$ and $e(f(g(n)),$
$f(h(n))) \geq \epsilon_0$. It follows that, for any choice of $\delta > 0$, we may find
an $n \in N$ such that $1/n < \delta$ and hence two points $g(n), h(n) \in X$ so
that $d(g(n), h(n)) < 1/n < \delta$ but $e(f(g(n)), f(h(n))) \geq \epsilon_0$. Therefore, f
cannot be uniformly continuous on X.

We are now ready to give an example which shows that the converse
of Theorem 4.7 is not true.

4.9. Example: (A function that is continuous but not uniformly
continuous.) Let $X = \{x \in R^1 : 0 < x \leq 1\}$ with the standard metric
and consider $f : X \to R^1$ given by $f(x) = 1/x$ for each $x \in X$. The
function f is continuous at each point of X but is not uniformly continuous on X. To see that f is not uniformly continuous on X, note
that the sequences $g(n) = 1/n$ and $h(n) = 1/(n + 1)$ have the property
that $d(g(n), h(n)) = |g(n) - h(n)| = |1/n - 1/(n+1)| = |1/(n(n+1))|$
$< 1/n$ while $d(f(g(n)), f(h(n))) = |n - (n + 1)| = 1$. If we take
$\epsilon_0 = 1$, the conditions in Theorem 4.8 are satisfied, which means that
f is not uniformly continuous on X.

Even though the continuous function $f : X \to Y$ may not be uniformly continuous on X, restriction to a certain type of subset will
ensure that f is uniformly continuous on that subset. This is established
by the next theorem.

4.10. THEOREM: (*The Uniform Continuity Theorem.*) *Let (X, d)*
and (Y, e) be metric spaces. If $f : X \to Y$ is continuous and (X, d) is
compact, then f is uniformly continuous on X.

Proof: Let $\epsilon > 0$ and consider $\epsilon/2 > 0$ as fixed throughout our discussion. Since f is continuous at each point $a \in X$, there exists a $\delta(a) > 0$, depending upon a and $\epsilon/2$, such that $f(S^d_{\delta(a)}(a)) \subset S^e_{\epsilon/2}(f(a))$. In other words, for each $a \in X$, if $d(x, a) < \delta(a)$, then $e(f(x), f(a)) < \epsilon/2$. The set of open spheres $\{S^d_{\delta(a)/2}(a) : a \in X\}$ forms an open cover for the compact space X, and therefore a finite number of them $\{S^d_{\delta(a_k)/2}(a_k) : k = 1, 2, \ldots, n\}$ cover X. Now let $\delta = \text{minimum}\{\delta(a_1)/2, \delta(a_2)/2, \ldots, \delta(a_n)/2\}$ and suppose $x, w \in X$ are such that $d(x, w) < \delta$. Then $x \in S^d_{\delta(ak)/2}(a_k)$ for some $1 \leq k \leq n$, which implies $d(x, a_k) < (\delta(a_k))/2 < \delta(a_k)$, which in turn implies $d(w, a_k) \leq d(w, x) + d(x, a_k) < \delta + (\delta(a_k))/2 < 2(\delta(a_k))/2 = \delta(a_k)$. It now follows from our above notation on continuity that we have both of the inequalities $e(f(x), f(a_k)) < \epsilon/2$ and $e(f(a_k), f(w)) < \epsilon/2$ holding. Finally, $e(f(x), f(w)) < e(f(x), f(a_k)) + e(f(a_k), f(w)) < \epsilon/2 + \epsilon/2 = \epsilon$, and the definition of uniform continuity is satisfied.

Exercises:

1. Let (X, d) be a metric space and $f : X \to R^1$ a continuous function. If $f(a) > 0$ for some $a \in X$, prove there exists an open set $U(a)$ containing a such that $f(x) > 0$ for all $x \in U(a)$.

2. Prove Theorem 4.1.

3. Prove the statement in the proof to Theorem 4.2 that $h^{-1} : Y \to X$ is continuous at each point of Y.

4. Prove Theorem 4.7.

5. Prove that the function defined in Theorem 4.9 is continuous at each point of X.

6. Let d be any pseudo-metric for X and define an equivalence relation on X as follows: For each $x, y \in X$, xRy iff $d(x, y) = 0$. Prove the quotient space X/R is a metric space.

7. Let $f : R^1 \to R^1$ be given by $f(x) = x^2$ for all $x \in R^1$.
   (a) Prove that f is not uniformly continuous on R^1.
   (b) If $A = \{x \in R^1 : -2 \leq x \leq 2\}$, is f uniformly continuous on A? Why?

8. Let A be a bounded subset of R^1 and let $f : A \to R^1$ be uniformly continuous, on A. Prove or disprove that $\{f(x) : x \in A\}$ is a bounded subset of R^1.

9. Let $f: R^1 \to R^1$ be a continuous periodic function. That is, f is continuous and there exists a positive real number p such that $f(x + p) = f(x)$ for all $x \in R^1$. Prove that $\{f(x): x \in R^1\}$ is bounded and f is uniformly continuous on R^1.

10. Let $A = \{x \in R^1 : 0 < x < 1\}$ and $f: A \to R^1$ be uniformly continuous. Prove or disprove that f can be defined at the points 0 and 1 in such a way that f is continuous on the closed interval $[0, 1]$.

5. Compactness and Products of Metric Spaces

One of our objectives for this section will be to show that if a metric space has any one of the properties of compactness, countable compactness, or the Bolzano-Weierstrass property, then it has the other two. We first consider the closed and bounded properties for a metric space.

5.1. THEOREM: *Let (X, d) be a metric space. If $A \subset X$ is compact, then A is closed and bounded.*

Proof: We know that all metric spaces are Hausdorff, and, therefore, the compact set A is closed by Theorem 4.10 of Chapter 8. If we assume that A were unbounded, the open cover $\{S_n(p): n \in N,\ p$ a fixed point in $X\}$ of X also covers A but has no finite subcover which covers A. This would contradict the hypothesis that A is compact. Therefore, A is bounded.

The converse of Theorem 5.1 is not true in general, however. That is, there are closed bounded metric spaces which are not compact, as the following example shows.

5.2. Example: Consider R^1 with the discrete metric. This metric is bounded by $M = 1$ and is also closed since any space is closed. But (X, d) is not compact because the open cover $\{S_{1/2}(x) = x : x \in R^1\}$ does not have a finite subcover.

Thus, the property of being closed and bounded will not characterize compact metric spaces as it did in the case of the R^n spaces. Metric spaces do have other familiar characterizations, as we shall see after some preliminary work concerning ϵ-nets.

5.3. Definition: Let (X, d) be a metric space and ϵ a positive real number. The finite set $A \subset X$ is called an ϵ-*net* for X iff for each point $x \in X$ there exists a point $a \in A$ such that $d(x, a) < \epsilon$.

5.4. THEOREM: *Every countably compact metric space (X, d) has an ϵ-net for each $\epsilon > 0$.*

Proof: Suppose, on the contrary, that there is some $\epsilon > 0$ for which X has no ϵ-net. Let $x_1 \in X$ be an arbitrary point and form $S_\epsilon(x_1)$. Since $\{x_1\}$ is not an ϵ-net for X by assumption, there exists a point $x_2 \in X$ such that $d(x_1, x_2) \geq \epsilon$. Thus $x_1 \notin S_\epsilon(x_2)$. Since $\{x_1, x_2\}$ is not an ϵ-net for X, there exists a point x_3 such that $d(x_1, x_3) \geq \epsilon$ and $d(x_2, x_3) \geq \epsilon$ and hence $x_3 \notin S_\epsilon(x_1) \cup S_\epsilon(x_2)$. Continuing by induction, if $\{x_1, x_2, \ldots, x_n\}$ is not an ϵ-net for X, there exists a point x_{n+1} such that $d(x_k, x_{n+1}) \geq \epsilon$ for $k = 1, 2, \ldots, n$, so that $x_{n+1} \notin \bigcup_{k=1}^{n} S_\epsilon(x_k)$. Thus, we have defined a sequence $f : N \to X$ given by $f(n) = x_n$ in X such that $d(f(i), f(j)) \geq \epsilon$ for $i \neq j, i, j \in N$. Since X is countably compact, f accumulates to a point x_0 in X by Theorem 7.4 of Chapter 8. Thus, $S_{\epsilon/2}(x_0)$ contains infinitely many points of $f(N) = \bigcup_{n \in N} x_n$ and thus if $f(i), f(j) \in S_{\epsilon/2}(x_0)$, then $d(f(i), f(j)) < \epsilon$. However, this contradicts the construction of f, i.e., $d(f(i), f(j)) \geq \epsilon, i \neq j$, for $i, j \in N$. The conclusion is that X has an ϵ-net for each $\epsilon > 0$.

5.5. THEOREM: *Every countably compact metric space (X, d) is separable.*

Proof: We need to exhibit a countable dense subset of X. By use of Theorem 5.4, for each $n \in N$ there is a finite set A_n which is an $1/n$-net for X. The set $A = \bigcup_{n \in N} A_n$ is a countable subset of X, and we also shall show that it is dense in X. To this end, let $z \in X$ and U be any open set containing z. Then there is a $\delta > 0$ such that the open sphere $S_\delta(z) \subset U$. We now choose a natural number $n(\delta) \in N$ such that $1/n(\delta) < \delta$. Since $A_{n(\delta)}$ is a $1/n(\delta)$-net for X, we can find a point $w \in A_{n(\delta)}$ such that $d(z, w) < 1/n(\delta) < \delta$. Therefore, $S_\delta(z)$ contains a point of A so that U contains a point of A and thus $z \in \bar{A}$. Consequently, $\bar{A} = X$ and (X, d) is separable.

5.6. THEOREM: *A metric space (X, d) is second countable iff it is separable.*

Proof: If (X, d) is second countable, then (X, d) is separable according to Theorem 5.7 of Chapter 6.

Now let (X, d) be separable and $A \subset X$ be a countable dense subset. We must show that (X, d) is second countable. To do this, let $\mathscr{B}$ be the set of all open r-spheres with centers at points of A and rational radii. Certainly $\mathscr{B}$ is countable, and it remains only to show that $\mathscr{B}$ is a base for X. For any $z \in X$ and U open in X containing z, there exists a positive real number δ such that the open sphere $S_\delta(z) \subset U$. Now let r be a rational number such that $0 < r < \delta/2$. Since A is dense in X, there exists a point $a \in A$ such that $d(z, a) < r$. Thus, $z \in S_r(a) \subset S_\delta(z) \subset U$, which implies $\mathscr{B}$ is a base for the d-metric topology on X.

We are now ready to state a theorem concerning equivalent forms of compactness for metric spaces.

5.7. THEOREM: *In a metric space, compactness, countable compactness, and the Bolzano-Weierstrass property are all equivalent.*

Proof: We know metric spaces are T_1 so that countable compactness and the Bolzano-Weierstrass property are equivalent according to Theorem 7.10 of Chapter 8. By Theorem 7.2 of Chapter 8, compactness always implies countable compactness. If the metric space is countably compact, it is also separable by Theorem 5.5, hence second countable by Theorem 5.6, and therefore compact by Theorem 7.6 of Chapter 8. It follows that in metric spaces, compactness and countable compactness are equivalent. This proves the theorem.

We next turn our attention to products of metric spaces. It is not true that $\prod_{\alpha \in \Delta} X_\alpha$ is a metric space iff each X_α is a metric space. For information regarding arbitrary indexing sets, see reference 2 in the Bibliography, for example. For finite products we have the following result.

5.8. THEOREM: *The space $X \times Y$ is metric iff both X and Y are metric spaces.*

Proof: If $X \times Y$ is metric space, we may easily show that both X and Y are metric spaces by the technique we have employed before.

Now let X and Y be metric spaces with metrics d and e, respectively. Of course, $X \times Y$ is a topological space with the product topology.

Thus, we need to prove this space is metrizable. To do this, define $f((x_1, y_1), (x_2, y_2)) = d(x_1, x_2) + e(y_1, y_2)$ for each (x_1, y_1) and (x_2, y_2) in $X \times Y$. It is an exercise to show f is a metric for $X \times Y$. We now need to show that the product topology on $X \times Y$ is the same as the f-metric topology on $X \times Y$. Let $\mathscr{B}_1$ and $\mathscr{B}_2$ represent bases of the product topology and the f-metric topology, respectively. Then a typical element in $\mathscr{B}_1$ has the appearance of $B_1 = S_r^d(x_0) \times S_q^e(y_0) = \{(x, y) : x \in X, y \in Y, d(x_0, x) < r, d(y_0, y) < q\}$ where $x_0 \in X$, $y_0 \in Y, r > 0$ and $q > 0$. An element $B_2 \in \mathscr{B}_2$ has the form $B_2 = S_r^f(x_0, y_0) = \{(x, y) : f(x_0, x_0) (x, y)) < r\} = \{(x, y) : d(x_0, x) + e(y_0, y) < r\}$ where $(x_0, y_0) \in X \times Y$ and $r > 0$. We now show conditions (a) and (b) of Theorem 1.11 of Chapter 4 are satisfied. First, suppose $(x_1, y_1) \in S_r^d(x_0) \times S_q^e(y_0) \in \mathscr{B}_1$. Then there exist real positive numbers r_1 and q_1 such that $S_{r_1}^d(x_1) \subset S_r^d(x_0)$ and $S_{q_1}^e(y_1) \subset S_q^e(y_0)$ by Theorem 3.5 so that $S_{r_1}^d(x_1) \times S_{q_1}^e(y_1) \subset S_r^d(x_0) \times S_q^e(y_0)$. Define $t =$ minimum $\{r_1, q_1\}$ and consider any point $(x, y) \in S_t^f(x_1, y_1)$. It follows that $d(x_1, x) \leq d(x_1, x) + e(y_1, y) = f((x_1, y_1), (x, y)) < t \leq r_1$. Similarly, $e(y_1, y) < q_1$. Therefore $(x, y) \in S_{r_1}^d(x_0) \times S_{q_1}^e(y_0)$ and $(x, y) \in S_t^f(x_1, y_1) \subset S_r^d(x_0) \times S_q^e(y_0)$. We have now satisfied part (a) of Theorem 1.11 of Chapter 4.

Now let $(x_1, y_1) \in S_r^f(x_0, y_0) \in \mathscr{B}_2$. Then there exists a real number $q > 0$ such that $(x_1, y_1) \in S_q^f(x_1, y_1) \subset S_r^f(x_0, y_0)$ by Theorem 3.5. Let $r_1 = q_1 = q/2$ and consider the basic open set $S_{r_1}^d(x_1) \times S_{q_1}^e(y_1) \in \mathscr{B}_1$. If $(x, y) \in S_{r_1}^d(x_1) \times S_{q_1}^e(y_1)$, then $f((x_1, y_1), (x, y)) = d(x_1, x) + e(y_1, y) < r_1 + q_1 = q$. Therefore $(x, y) \in S_q^f(x_1, y_1)$ and hence $(x, y) \in S_{r_1}^d(x_1) \times S_{q_1}^e(y_1) \subset S_q^f(x_1, y_1) \subset S_r^f(x_0, y_0)$. Condition (b) of Theorem 1.11 of Chapter 4 is now satisfied.

Now $\mathscr{T}(\mathscr{B}_1)$ is the product topology on $X \times Y$, and, from what we have just shown, $\mathscr{B}_1$ and $\mathscr{B}_2$ are equivalent bases according to Theorem 1.11 of Chapter 4. Therefore, $\mathscr{T}(\mathscr{B}_1) = \mathscr{T}(\mathscr{B}_2)$ by Theorem 1.12 of Chapter 4, and it follows that $X \times Y$ is metrizable.

5.9. Corollary: The space $\prod_{k=1}^{n} X_k$ is a metric space iff each space X_k, $k = 1, 2, \ldots, n$, is a metric space.

Proof: The proof follows the familiar pattern.

We close this section with an interesting result concerning the distance between two sets in a metric space.

5.10. THEOREM: *Let (X, d) be a metric space. If $A \subset X$ is compact, $F \subset X$ is closed and $A \cap F = \phi$, then $d(A, F) > 0$.*

Proof: By Theorem 4.4, the function $f : X \to R^1$ defined by $f(x) = d(x, F)$ is continuous and, therefore, the restriction $f|A$ is continuous. Since A is compact, $f|A$ attains its minimum according to Theorem 6.8 of Chapter 8. In other words, there is some point $a \in A$ such that $d(a, F)$ equals greatest lower bound $\{d(x, F) : x \in A\} = d(A, F)$. Since $a \notin F = \bar{F}$, it follows from Theorem 2.11 that $d(A, F) = d(a, F) > 0$.

Exercises:

1. Prove, as asserted at the first of the proof to Theorem 5.8, that if $X \times Y$ is a metric space, then X and Y are both metric spaces.

2. Prove that f, as defined in the proof of Theorem 5.8, is a metric for $X \times Y$.

3. Give an example of two closed subsets A and B of a metric space such that there exist no points $a \in A$, $b \in B$ for which $d(a, b) = d(A, B)$.

4. Prove that if A and B are nonempty compact subsets of a metric space, there exist points $a \in A$ and $b \in B$ for which $d(a, b) = d(A, B)$.

5. Let (X, d) be a compact metric space and $\{A_\alpha : \alpha \in \Delta\}$ be a family of closed subsets of X such that $\bigcap_{\alpha \in \Delta} A_\alpha = \phi$. Prove that there exists a real number $\epsilon > 0$ such that for each $x \in X, d(x, A_\alpha) \geq \epsilon$ for at least one $\alpha \in \Delta$.

6. Let A be a bounded subset of the metric space (X, d). Define the *diameter* of A to be the least upper bound of $\{d(x, y) : x, y \in A\}$ if $A \neq \phi$ and define the diameter of A to be zero if $A = \phi$. Denote the diameter of A by $\delta(A)$. If $A \neq \phi$ is compact and bounded, prove that there exist points $a, b \in A$ such that $d(a, b) = \delta(A)$.

7. Let A and B be bounded subsets of a metric space (X, d). Prove

   (a) $\delta(\bar{A}) = \delta(A)$.
   (b) If $A \cap B \neq \phi$, then $\delta(A \cup B) \leq \delta(A) \cup \delta(B)$.

8. Prove that a metric space is second countable iff it is a Lindelöf space. (See Exercise 8, Section 7, of Chapter 8.)

9. Let $\{(X_\alpha, d_\alpha) : \alpha \in \Delta\}$ be a family of metric spaces. Prove that if $\prod_{\alpha \in \Delta} X_\alpha$ is a metric space then X_α is a metric space for each $\alpha \in \Delta$.

10. Let $\{U_\alpha : \alpha \in \Delta\}$ be an open cover for a compact subset A of a metric space (X, d). Prove that there exists a positive real number δ such that for every $x \in A$ there is some $\alpha \in \Delta$ for which $S_\delta(x) \subset U_\alpha$.

6. Sequences in Metric Spaces

Since metric spaces are topological spaces, much of our earlier discussion of sequences applies here. We shall add to our previous knowledge some results which apply in metric spaces and in particular to the R^n spaces. For metric spaces, the condition for a sequence to converge may be stated as in the next theorem.

6.1. THEOREM: *The sequence $f : N \to X$ in the metric space (X, d) converges to the point $a \in X$ iff for each real number $\epsilon > 0$ there exists an $n_0 \in N$ such that if $n > n_0$ then $d(f(n), a) < \epsilon$. Stated differently, $f \to a$ iff for each $\epsilon > 0$ there exists $n_0 \in N$ such that if $n > n_0$ then $f(n) \in S_\epsilon(a)$.*

Proof: The proof is an evident application of Definition 1.3 of Chapter 7.

The next definition describes the concept of a bounded sequence in a metric space (X, d).

6.2. Definition: Let (X, d) be a metric space. A sequence $f : N \to X$ in X is bounded iff $f(N)$ is a bounded subset of X.

6.3. THEOREM: *Every convergent sequence $f : N \to X$ in the metric space (X, d) is bounded.*

Proof: Let $f \to a \in X$. Then the open set $S_1(a)$ contains all points of $f(N)$ except possibly a finite number, $f(n_1), f(n_2), \ldots, f(n_j)$. If we use the notation $d(f(n_k), a) = z_k$ for $k = 1, 2, \ldots, j$, and let $M = $ maximum $\{1, z_1, z_2, \ldots, z_j\}$, we see $f(N) \subset S_M(a)$ and, therefore, $d(f(n), f(m)) < 2M$ for all $n \in N$. Thus, $2M$ is a bound for $f(N)$, showing that f is a bounded sequence.

6.4. Example: (A bounded sequence in a metric space may not converge.) Let (X, d) be an infinite metric space with the discrete metric and therefore the discrete topology. Every sequence is bounded in this space due to the definition of the discrete metric. However, the sequence $f: N \rightarrow X$, where f is injective, does not converge to any point in X. Furthermore, this sequence has no convergent subsequence.

In contrast to the sequence in Example 6.4, the following result holds for the R^n spaces.

6.5. THEOREM: *Every bounded sequence $f: N \rightarrow R^n$ in R^n has a convergent subsequence.*

Proof: We distinguish between the cases where $f(N)$ is finite and infinite. If $f(N)$ consists of only a finite number of distinct points $\{a_1, a_2, \ldots, a_n\}$, then at least one of these points a_{n_0}, $1 \le n_0 \le n$ is such that $f(n) = a_{n_0}$ for infinitely many $n \in N$. Letting $\{n \in N : f(n) = a_{n_0}\} = A \subset N$ and using the well ordering of N, there is a smallest element $n_1 \in A$, a smallest element $n_2 \in A \backslash \{n_1\}$, and, in general, a smallest element $n_k \in A \backslash \bigcup_{j=1}^{k-1} n_j$ for each $k \in N$. This construction may be done since A is infinite and hence $A \backslash \bigcup_{j=1}^{k-1} n_j$ is infinite. Now define $g: N \rightarrow N$ by $g(k) = n_k$ for each $k \in N$ and form the subsequence $fg: N \rightarrow X$ of f for which $f(g(k)) = f(n_k) = a_{n_0}$. The subsequence fg clearly converges to a_{n_0}.

If the bounded set $f(N)$ contains infinitely many distinct points, then $f(N)$ has at least one cluster point $a \in R^n$ according to the Bolzano-Weierstrass Theorem 7.13 of Chapter 8. Therefore, we may select a point $f(n_1)$ from $f(N)$ such that $f(n_1) \in S_1(a)$. Now consider the set $Q_1 = \{f(n) : n > n_1\}$. An indirect argument quickly shows a to be a cluster point of Q_1 and, consequently, we may select a point $f(n_2) \in S_{1/2}(a) \cap Q_1$. Again, a is a cluster point of $Q_2 = \{f(n) : n > n_2\}$, and we may select a point $f(n_3) \in S_{1/3}(a) \cap Q_2$. Proceeding inductively, we construct a subsequence fg of f converging to the point a. The details should be verified by the reader.

6.6. Definition: Let $f: N \rightarrow X$ be a sequence in the metric space (X, d). Then f is called a *Cauchy sequence* iff for each real number

$\epsilon > 0$ there is an $n_0 \in N$ such that if $n, m > n_0$, then $d(f(n), f(m)) < \epsilon$.

Thus, a Cauchy sequence is one in which the images under f of elements in N eventually become close together. Convergent sequences in metric spaces have this property, as shown in the following theorem.

6.7. THEOERM: *Every convergent sequence $f : N \rightarrow X$ in the metric space (X, d) is a Cauchy sequence.*

Proof: Let $f \rightarrow p \in X$ and $\epsilon > 0$. Then there exists an $n_0 \in N$ such that for all $n > n_0$, $d(f(n), p) < \epsilon/2$. Thus, if $n, m > n_0$ $d(f(n), f(m)) \leq d(f(n), p) + d(p, f(m)) < \epsilon/2 + \epsilon/2 = \epsilon$. Consequently, f is a Cauchy sequence.

The converse of Theorem 6.7 is false in an arbitrary metric space, as the next example will show. However, we shall show the converse is true in the R^n spaces, which will then give a characterization of convergent sequences in those spaces.

6.8. Example: Let R^1 have the standard metric d and consider the subspace $R^1 \backslash \{0\}$. Define $f(n) = 1/n$, $n \in N$. Then f is a sequence in $R^1 \backslash \{0\}$ but does not converge to a point in that subspace. (The sequence tries to converge to 0, but 0 is not in the subspace under consideration.) The sequence is Cauchy, however. To see this, let $\epsilon > 0$ be any positive real number and then let $k \in N$ be such that $1/k < \epsilon/2$. Now if $n, m > k$, then $1/n$ and $1/m$ are both less than $\epsilon/2$ so that $d(f(n), f(m)) = |1/n - 1/m| = |1/n + (-1/m)| \leq |1/n| + |1/m| < \epsilon/2 + \epsilon/2 = \epsilon$ by use of the triangle inequality.

6.9. THEOREM: *Every Cauchy sequence in a metric space (X, d) is bounded.*

Proof: The proof is left as an exercise.

6.10. THEOREM: *Let $f : N \rightarrow X$ be a Cauchy sequence in the metric space (X, d). If the subsequence $fg : N \rightarrow X$ of f in X converges to a point $a \in X$, the f converges to the point a.*

Proof: Let $\epsilon > 0$ be a real number. Since f is Cauchy, there is an $n_0 \in N$ such that if $n, m > n_0$, then $d(f(n), f(m)) < \epsilon/2$. By hypothesis, the subsequence fg converges to a and, therefore, for the same $\epsilon > 0$ there is a natural number $t \in N$ for which $g(t) > n_0$ and $d(f(g(t)), a)$

$< \epsilon/2$. Thus, if $n > n_0$, $d(f(n), a) \leq d(f(n), f(g(t))) + d(f(g(t)), a)$
$< \epsilon/2 + \epsilon/2 = \epsilon$, which implies f converges to a.

Finally, we can characterize convergent sequences in R^n by use of Cauchy sequences.

6.11. THEOREM: *(The Cauchy convergence criterion.)* *A sequence in R^n is convergent iff it is a Cauchy sequence.*

Proof: If a sequence $f: N \to R^n$ is convergent, it is a Cauchy sequence by Theorem 6.7. If f is a Cauchy sequence, it is bounded by Theorem 6.9. According to Theorem 6.5, the bounded sequence f has a convergent subsequence and Theorem 6.10 then implies the entire sequence f converges.

Exercises:

1. Prove Theorem 6.1.

2. Give the details of the proof to Theorem 6.5.

3. Prove that the sequence in Example 6.8 does not converge to a point in the subspace $R^1 \backslash \{0\}$.

4. Prove Theorem 6.9.

5. Use the definition to prove or disprove each of the following sequences in R^1 is a Cauchy sequence:
   (a) $f(n) = (-1)^n + 1/n$.
   (b) $f(n) = 1 + (-1)^n/n$.
   (c) $f(n) = 1 - n(-1)^n$.
   (d) $f(n) = 1/(n(n+1))$.
   (e) $f(n) = (-1)^n$.

6. If A is an infinite subset of a compact metric space, prove that there is a Cauchy sequence $f: N \to A$ in A such that if $n \neq m$, then $f(n) \neq f(m)$.

7. If $f: N \to R^n$ is not a Cauchy sequence, does there exist an unbounded subsequence of f?

8. Prove Theorem 4.10 indirectly by use of sequences.

9. (The monotone convergence theorem.) Let $f: N \to R^n$ be a monotone increasing sequence. That is, if $n \geq m$, then $f(n) \geq f(m)$.

(a) Prove that f converges iff f is bounded.

(b) If f does converge, prove f converges to the least upper bound of $f(N)$.

10. Let $f: N \to R^1$ be a sequence. Prove that f must have either a monotone increasing or a monotone decreasing subsequence.

7. A Glimpse of Function Spaces

Let X and Y be topological spaces and consider the set of all possible functions from X to Y. For such a set, we shall use the standard notation of the mathematical literature, $Y^X = \{f : f : X \to Y\}$. With the set Y^X at our disposal, it is natural to ask whether topologies for this set exist which have significance within the realm of general mathematics. This is indeed the case. Once a topology $\mathscr{T}$ for Y^X is defined, we call $(Y^X, \mathscr{T})$ a *function space*. After becoming acquainted with some of the usual topologies for Y^X, our interest will turn to sequences in Y^X which will bring forth the main reason for our introduction to function spaces. We should bear in mind, however, that our discussion here is only a brief glimpse at the rich and rewarding subject of function spaces.

7.1. Definition;   Let $A \subset X$ and $B \subset Y$. Then the notation $(A, B) = \{f \in Y^X : f(A) \subset B\}$ is used to describe a subset of Y^X determined by the sets A and B.

With the aid of the notation in Definition 7.1, three of the several well-known topologies for Y^X now will be defined.

7.2. Definition: (a) (The closed-open topology) The topology $\mathscr{T}(\mathscr{B}[\mathscr{S}])$ on Y^X having the subbase $\mathscr{S} = \{(F, V) : F$ is a closed subset of X and V is open in $Y\}$ is called the *closed-open* topology.

(b) (The point-open topology) The topology $\mathscr{T}(\mathscr{B}[\mathscr{S}])$ on Y^X having the subbase $\mathscr{S} = \{(p, V) : p$ is a point in X and V is open in $Y\}$ is called the *point open* topology.

(c) (The compact-open topology) The topology $\mathscr{T}(\mathscr{B}[\mathscr{S}])$ on Y^X having the subbase $\mathscr{S} = \{(A, V) : A$ is a compact subset of X and V is open in $Y\}$ is called the *compact-open* topology.

A simple example will illustrate the topologies of Definition 7.2.

7.3. Example: Let $X = \{a, b\}$ with $\mathscr{T}_X = \{\phi, X, \{a\}\}$ and take $Y = X$ and $\mathscr{T}_Y = \mathscr{T}_X$. Then $Y^X = \{f_1, f_2, f_3, f_4\}$ where $f_1(X) = a$, $f_2(X) = b$, $f_3(a) = a$ and $f_3(b) = b$, and $f_4(a) = b$ and $f_4(b) = a$. For the closed-open topology on Y^X, the subbasic open sets are $\mathscr{S} = \{(\{b\}, \{a\}), (\{b\}, Y), (X, \{a\}), (X, Y)\}$ where $(\{b\}, \{a\}) = \{f_1, f_4\}$, $(\{b\}, Y) = Y^X$, $(X, \{a\}) = \{f_1\}$, and $(X, Y) = Y^X$. The closed-open topology $\mathscr{T}(\mathscr{B}[\mathscr{S}])$ is now easily constructed.

For the point-open topology, $\mathscr{S} = \{(a, \{a\}), (a, Y), (b, \{a\}), (b, Y)\}$ where, for example, $(a, \{a\}) = \{f_1, f_3\}$. Here again, it is easy to construct the point-open topology $\mathscr{T}(\mathscr{B}[\mathscr{S}])$.

For the compact-open topology, observe first that every subset of X is compact. Thus $\mathscr{S}$, and hence $\mathscr{T}(\mathscr{B}[\mathscr{S}])$, is easily constructed.

Definition 7.2 suggests that any type of subset may be used to replace closed subsets, points, or compact subsets, respectively, of X. In this manner we could describe an entire class of "set-open" topologies for Y^X. Other then these, we might mention one other topology for Y^X that is called the *graph topology*. It is defined by letting $F_u = \{f \in Y^X : f \subset U$ where U is open in $X \times Y\}$ and then using $\mathscr{B} = \{F_U : U$ is open in $X \times Y\}$ to generate the topology $\mathscr{T}(\mathscr{B})$ having $\mathscr{B}$ as a base.

With reference to the topologies defined in Definition 7.3, the point-open and the compact-open are the ones of major interest to us. For these topologies, the point-open is smaller than the compact-open because each point in X is a compact subset of X. Two results of general interest concerning these topologies will be given next. Theorem 7.4 is given for the compact-open topology, although it is a straightforward exercise to show that it can be generalized to any set-open topology for Y^X.

7.4. THEOREM: *Let Y^X have the compact-open topology. For each $y \in Y$, let $f_y : X \to Y$ be defined as the constant function $f_y(x) = y$ for all $x \in X$. Then the subspace $M = \{f_n : y \in Y\}$ of Y^X is homeomorphic to the space Y.*

Proof: Define $F : Y \to M$ as $F(y) = f_y$ for each $y \in Y$. Certainly F is bijective. Now let A be any compact subset of X and V an open set in Y. Then $f_y \in (A, V)$ iff $f_y(A) \subset V y \in V$. Thus, for each open $V \subset Y$, $F(V) = (A, V) \cap M$ is a subbasic open set in the subspace $M \subset Y^X$. This implies that both F and F^{-1} are continuous so that $F : Y \to M$ is a homeomorphism.

7.5. THEOREM: *Let Y^X have the point-open topology. Then Y^X is Hausdorff iff Y is Hausdorff.*

Proof: We have remarked that Theorem 7.4 is true for any set-open topology and hence the point-open topology. Therefore, if Y^X is Hausdorff, then the subspace $M = \{f_y \in Y^X : f_y(x) = y \text{ for all } x \in X\}$ is Hausdorff. By Theorem 7.4, $Y \cong M$ gives Y Hausdorff.

Conversely, suppose Y is Hausdosff and let $f \neq g$ be any two points in Y^X. Then $f(x_0) \neq g(x_0)$ for at least one $x_0 \in X$. Since Y is Hausdorff, there exist open disjoint subsets U and V of Y containing $f(x_0)$ and $g(x_0)$, respectively. Therefore, the subbasic open set $(x_0, U) \subset Y^X$ and $(x_0, V) \subset Y^X$ are disjoint and contain f and g, respectively, making Y^X Hausdorff.

7.6. THEOREM: *Let Y^X have the compact-open topology. Then Y^X is Hausdorff iff Y is Hausdorff.*

Proof: Half of this result follows because the compact-open topology is larger than the point-open topology. The other half is an exercise.

As was indicated earlier, our reason for introducing function spaces was to study sequences of functions and, in particular, sequences of functions in Y^X where Y^X has either the point-open or the compact-open topology. In these spaces, as in any function space, convergence of a sequence of functions offers nothing essentially new since we have already discussed convergence of sequences in topological spaces in Chapter 7. We will hereafter refer to this type of convergence as convergence in the topological sense to distinguish from two other types of convergence we shall introduce in a moment. For the sequence $g : N \rightarrow Y^X$, if we use the notation $g(n) = f_n \in Y^X$ for each $n \in N$, then g may be described by the symbolism (f_n). We discussed this briefly in Chapter 7 but continued to use the function notation. Now, however, there are so many functions involved that we are going to henceforth use (f_n) to denote a sequence in Y^X.

7.7. Definition: Let X and Y be topological spaces and let (f_n) be a sequence of functions in Y^X. Then (f_n) *converges pointwise* to $f_0 \in Y^X$ iff for each fixed $x_0 \in X$ the sequence $G : N \rightarrow Y$ given by $G(n) = f_n(x_0)$ for each $n \in N$ converges in the topological sense to the point $f_0(x_0) \in Y$.

7.8. Example: Let $X = Y = R^1$ and consider the sequence (f_n) in Y^X where for each $n \in N$, $f_n(x) = x/n$ for all $x \in X$. Figure 28 describes some of the functions in question.

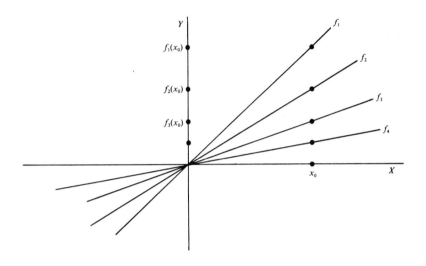

Figure 28

For each fixed $x_0 \in X$, the sequence in Y given by $G(n) = f_n(x_0) = x_0/n$ for all $n \in N$ converges in the topological sense to zero. Therefore, the sequence (f_n) converges pointwise to the function $f_0 : X \rightarrow Y$ given by $f_0(x) = 0$ for all $x \in X$.

The significance of the point-open topology is given in the following theorem.

7.9. THEOREM: *Let Y^X have the point-open topology. Then the sequence (f_n) in Y^X converges in the topological sense to $f_0 \in Y^X$ iff the sequence (f_n) converges pointwise to the function $f_0 \in Y^X$.*

Proof: First, let (f_n) be a sequence of functions in Y^X which converges in the topological sense to the function $f_0 \in Y^X$. Let $x_0 \in X$ and let V be any open set in X containing x_0. Then (x_0, V) is a subbasic open set in Y^X containing f_0, and the convergence in the topological sense of (f_n) to f_0 implies there exists an $n_0 \in N$ such that for all $n > n_0$, $f_n \in (x_0, V)$. In other words, for all $n > n_0$, $f_n(x_0) \in V$, which is precisely the statement we need to prove (f_n) converges pointwise to $f_0 \in Y^X$.

Conversely, suppose the sequence (f_n) converges pointwise to the function $f_0 \in Y^X$ and let (x_0, V) be any subbasic open set in the space Y^X which contains f_0. Then $f_0(x_0) \in V$ by definition of the point-open topology and, due to the fact (f_n) converges pointwise to f_0, there exists an $n_0 \in N$ such that if $n > n_0$, then $f_n(x_0) \in V$. Now, according to Exercise 7, Section 1, of Chapter 7, the sequence (f_n) converges in the topological sense to the function f_0.

The second kind of special convergence in function spaces is introduced next.

7.10. Definition: Let X be an arbitrary topological space and (Y, d) a metric space. Then the sequence (f_n) in Y^X *converges uniformly on the subset $A \subset X$* to the function $f_0 \in Y^X$ iff for each real number $\epsilon > 0$ there is an $n_0 \in N$ (depending only upon ϵ and not on $x \in A$) such that if $n > n_0$, then $d(f_n(x), f_0(x)) < \epsilon$ for all $x \in A$.

Figure 29 gives visual help with Definition 7.10.

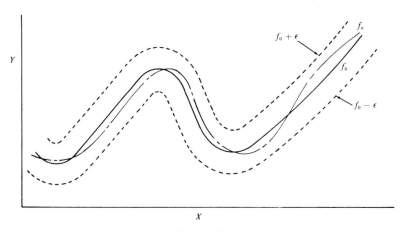

Figure 29

7.11. Example: The sequence of functions given by $f_n(x) = x/n$ of Example 7.8 converges pointwise to $f_0(x) = 0$ but does not converge uniformly on $A = R^1$ to $f_0(x) = 0$. To see this, let $\epsilon > 0$ and n_0 be any natural number. Then there exists an $x \in R^1$ snch that $x > (n_0 + 1)\epsilon$ so that $x/(n_0 + 1) > \epsilon$. That is, $x > 0$ and $d(f_{n_0+1})(x), f_0(x)) = d(x/(n_0 + 1), 0) > \epsilon$, which means that for every natural number n_0 there exists an $m = n_0 + 1 > n_0$ and an $x \in R^1$ such that $d(f_m(x), f_0(x)) > \epsilon$, showing that (f_n) does not converge uniformly to f_0.

The significance of the compact-open topology is given in the next theorem.

7.12. THEOREM: *Let X be a Hausdorff space, let (Y, d) be a metric space, and suppose that Y^X has the compact-open topology. Then a sequence (f_n) in Y^X converges uniformly on every compact subset $A \subset X$ to a continuous function $f_0 \in Y^X$ iff (f_n) converges to f_0 in the topological sense.*

Proof: Take the hypothesis that (f_n) converges uniformly on every compact subset $A \subset X$ to the continuous f_0 and let (A, V) be a sub-basic open set in Y^X containing f_0. Then $f_0(A) \subset V$ and, since f_0 is continuous, $f_0(A)$ is compact. According to Theorem 5.10, $d(f_0(A), Y \backslash V)$ is a positive number which we shall call ϵ. Now we use Definition 7.10 to obtain the existence of an $n_0 \in N$ such that if $n > n_0$ then $d(f_n(x), f_0(x)) < \epsilon$ for all $x \in A$. This means $f_n(A) \subset V$ so that $f_n \in (A, V)$ for all $n > n_0$. Therefore, (f_n) converges in the topological sense to the function $f_0 \in Y^X$ with the compact-open topology.

Conversely, suppose (f_n) converges in the topological sense to the continuous function $f_0 \in Y^X$ with the compact-open topology. Let $A \subset X$ be any compact set, $\epsilon > 0$ a real number, and observe that for each $x \in A$, $\bar{S}_{\epsilon/2}(f_0(x)) \subset S_\epsilon(f_0(x))$ by Exercise 4 of Section 2 in this chapter. Since f_0 is continuous at each point $x \in A$, there exists an open set $U(x) \subset X$ containing x such that $f_0(U(x)) \subset S_{\epsilon/2}(f_0(x))$. Thus, $U(x)$ is a subset of the closed set $f_0^{-1}(\bar{S}_{\epsilon/2}(f_0(x)))$ so that $\bar{U}(x) \subset f_0^{-1}(\bar{S}_{\epsilon/2}(f_0(x)))$ and $f_0(\bar{U}(x)) \subset \bar{S}_{\epsilon/2}(f_0(x)) \subset S_\epsilon(f_0(x))$ for each $x \in A$. Now $U(x) \cap A$ is open in the subspace A and $\bar{U}(x) \cap A$ is closed because A is compact, hence closed, in the Hausdorff space X, according to Theorem 4.10 of Chapter 8. Therefore, by Theorem 4.8 of Chapter 8, $\bar{U}(x) \cap A$ is compact in A and $U(x) \cap A \subset \bar{U}(x) \cap A$. The family $\{U(x) : x \in A\}$ is an open cover of A, and compactness of A implies that there exists a finite subcover $U(x_1), U(x_2), \ldots, U(x_m)$ of this family for A so that $\bar{U}(x_1), \bar{U}(x_2), \ldots, \bar{U}(x_m)$ also covers A. Now, due to the fact that (f_n) converges to f_0 for the compact open topology on Y^X, for each $1 \leq k \leq m$ there exists an $n_k \in N$ such that if $n > n_k$, then $f_n \in (\bar{U}(x_k) \cap A, S_\epsilon(x_k))$. In other words, for each $k = 1, 2, \ldots, m$, there exists an $n_k \in N$ such that if $n > n_k$, then $f_n(\bar{U}(x_k) \cap A \subset S_\epsilon(x_k)$. Therefore, letting $n_0 = $ maximum $\{n_1, n_2, \ldots, n_m\}$, we have for $n > n_0$ and each $x \in A$, $d(f_n(x), f_0(x)) < 2\epsilon$, showing that (f_n) converges uniformly on the compact set A to the function f_0.

We will now give an example to show that a sequence of continuous functions may not converge pointwise to a continuous function. Stated differently, the point-open topology on Y^X allows a sequence of continuous functions to converge to a non-continuous function. After the example, our final theorem shows how uniform convergence remedies this situation.

7.13. Example: Let $X = \{x \in R^1 : 0 \leq x \leq 1\}$ and for each $n \in N$ define $f_n : X \to R^1$ by $f_n(x) = x^n$ for all $x \in X$. With the aid of Figure 30, we can see that (f_n) converges pointwise to the function $f_0(x) = 0$ if $0 \leq x < 1$ and $f_0(1) = 1$. The function $f_0 : X \to R^1$ is not continuous at the point $x = 1$.

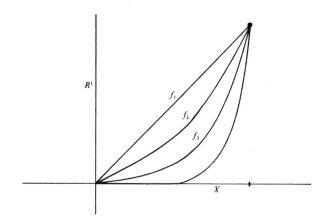

Figure 30

7.14. THEOREM: *Let (X, d) and (Y, e) be metric spaces. If (f_n) is a sequence of continuous functions in Y^X, which converges uniformly on X to the function $f_0 \in Y^X$, then f_0 is a continuous function.*

Proof: Since (f_n) converges uniformly on X to f_0, for each $\epsilon > 0$, there exists an $n_1 \in N$ such that if $n > n_1$, then $e(f_n(x), f_0(x)) < \epsilon/3$ for all $x \in X$, by Definition 7.10. Now let $a \in X$, and we shall show that f_0 is continuous at the point a. For any $n > n_1$ the triangle inequality gives $e(f_0(x), f_0(a)) \leq e(f_0(x), f_n(x)) + e(f_n(x), f_n(a)) + e(f_n(a), f_0(a)) < \epsilon/3 + e(f_n(x), f_n(a)) + \epsilon/3$. By the continuity of f_n at $a \in X$, there exists a positive number δ (depending upon n and $\epsilon/3$) such that if $d(x, a) > \delta$, then $e(f_n(x), f_n(a)) < \epsilon/3$. As a consequence, $e(f_0(x), f_0(a)) < \epsilon$ for all $x \in X$ such that $d(x, a) < \delta$,

An example will show the converse of Theorem 7.14 does not hold.

Exercises:

1. Generalize Theorem 7.4 to any set-open topology on Y^X.

2. State a characterization of when a sequence of functions (f_n) in Y^X does not converge to f_0 uniformly on a subset $A \subset X$.

3. If X is a Hausdorff space, how do the three topologies of Definition 7.2 compare?

4. Prove Theorem 7.6.

5. Let X and Y be spaces and let $C(X, Y)$ denote the set of all continuous functions from X to Y. Give an example of two spaces X and Y, neither of which are finite, such that the subspace $C(X, Y) \subset Y^X$ consists of only constant functions.

6. Let $X = \{x \in R^1 : x \geq 0\}$ and $Y = R^1$. Which of the following sequences of functions in Y^X converges pointwise to a function in Y^X? Which of the sequences converge uniformly on X to a function in Y^X?
   (a) $f_n(x) = 1/n$.
   (b) $f_n(x) = x^n/n$.
   (c) $f_n(x) = x^n/(1 + x^n)$.
   (d) $f_n(x) = \begin{cases} nx & \text{if } 0 \leq x \leq 1/n \\ 1/(nx) & \text{if } 1/n < x \end{cases}$.

7. Can a sequence of non-continuous functions converge uniformly on their domain to a continuous function?

8. Give an example to show that the converse of Theorem 7.14 does not hold.

9. For the compact-open topology on Y^X, prove that $(\overline{A}, V) \subset (A, \overline{V})$.

10. Suppose X has the discrete topology. Prove that Y^X is homeomorphic to $\prod_{x \in X} Y_x$ where $Y_x = Y$ for each $x \in X$.

11. Let Y^X have the point-open topology and consider the product $\prod_{x \in X} Y_x$ where $Y_x = Y$ for each $x \in X$. That is, each point $\prod_{x \in X} Y_x$ is a function $g : X \to Y$. Define $F : Y^X \to \prod_{x \in X} Y_x$ as $F(f) = g \in \prod_{x \in X} Y_x$ such that $g(x) = f(x)$ for all $x \in X$. Prove that F is a homeomorphism from Y^X to $F(Y^X) \subset \prod_{x \in X} Y_x$.

Bibliography

Listed below are four books on topology of a more advanced nature which offer additional information on the subject matter contained in this book.

Cullen, Helen F., *Introduction to General Topology*. Boston, Massachusetts: D. C. Heath and Company, 1968.

Dugundji, James, *Topology*. Boston, Masachusetts: Allyn and Bacon, Inc., 1966.

Kelley, John L., *General Topology*. Princeton, New Jersey: D. Van Nostrand Company, Inc., 1955.

Thron, Wolfgang J., *Topological Structures*. New York, New York: Holt, Rinehart and Winston, Inc., 1966.

Index